MODELING FOR
ALL SCALES

MODELING FOR ALL SCALES

An Introduction to System Simulation

Howard T. Odum

Environmental Engineering Sciences
University of Florida
Gainesville, Florida

Elisabeth C. Odum

Santa Fe Community College
Gainesville, Florida

ACADEMIC PRESS

San Diego   London   Boston   New York   Sydney   Tokyo   Toronto

Cover photograph: Eyewire Images.

This book is printed on acid-free paper.

Copyright © 2000 by ACADEMIC PRESS

All Rights Reserved.
No part of this publication may be reproduced or transmitted in any form or by any
means, electronic or mechanical, including photocopy, recording, or any information
storage and retrieval system, without permission in writing from the publisher.

Requests for permission to make copies of any part of the work should be mailed to:
Permissions Department, Harcourt Inc., 6277 Sea Harbor Drive,
Orlando, Florida, 32887-6777

Academic Press
A Harcourt Science and Technology Company
525 B Street, Suite 1900, San Diego, California 92101-4495, USA
http://www.apnet.com

Academic Press
24-28 Oval Road, London NW1 7DX, UK
http://www.hbuk.co.uk/ap/

Library of Congress Catalog Card Number: 99-64627

International Standard Book Number: 0-12-524170-4

PRINTED IN THE UNITED STATES OF AMERICA
99 00 01 02 03 04 QW 9 8 7 6 5 4 3 2 1

Contents

$\blacklozenge$

PART ONE
MODELING

Chapter One
MODELING AND SIMULATION

Chapter Two
ENERGY SYSTEMS DIAGRAMMING

v

Chapter Three

NUMBERS ON NETWORKS

PART TWO

SIMULATION

Chapter Four

SIMULATION WITH PICTURE BLOCKS
FOR EXTEND

Chapter Five

SIMULATION WITH GENERAL SYSTEMS
BLOCKS FOR EXTEND

Chapter Six

EQUATIONS FROM DIAGRAMS

Chapter Seven

CALIBRATING MODELS

Chapter Eight

SIMULATING WITH SPREADSHEET

Chapter Nine

PROGRAMMING IN BASIC

Chapter Ten

SIMULATING WITH STELLA

Chapter Eleven

SIMULATING EMERGY
AND TRANSFORMITY

Chapter Fifteen

MODELS OF SERIES AND OSCILLATION

Chapter Sixteen

MINIMODELS OF SUCCESSION
AND EVOLUTION

Chapter Seventeen

MODELS OF MICROECONOMICS

PART FOUR

APPLICATION

Chapter Twenty-One

MODELING PROJECTS AND COMPLEXITY

Chapter Twenty-Two

SIMULATION APPROACHES

Appendix A

PROGRAMMING ENERGY SYSTEMS BLOCKS
FOR SIMULATING WITH EXTEND

Appendix B

NOTES ON COMPUTER USE AND
BASIC PROGRAMMING

Appendix C

ANSWERS TO "WHAT IF"
EXPERIMENTAL PROBLEMS

Appendix D
USE OF ENERGY SYSTEMS SYMBOLS

Appendix E
CONTENTS OF THE CD SUPPLIED WITH THIS BOOK

Locations of Models

(*continues*)

(*continued*)

Model	Simulation program	Chapter	Page
FIRE	BASIC	Chapter 15	234
FISH	EXCEL	Chapter 7, 8	92, 105
FREEMARK	BASIC	Chapter 19	308
INFOBEN	BASIC	Chapter 19	318
INTERACT	BASIC	Chapter 14	221
INTLIMIT	BASIC	Chapter 12	177
LAGOON	BASIC	Chapter 21	347
LOGISTIC	BASIC	Chapter 13	206
MACROEC	BASIC	Chapter 18	297
MONEYGRO	BASIC	Chapter 18	292
NETPROD	BASIC	Chapter 12	170
NONRENEW	BASIC	Chapter 13	201
NUTRSPEC	BASIC	Chapter 16	258
OPENAQ	BASIC	Chapter 12	184
OSCILLAT	BASIC	Chapter 15	230
PC&CYCLE	EXTEND, BASIC, STELLA	Chapter 5, 7, 9	69, 128, 145
PEOPLE	BASIC	Chapter 20	328
PEXPO	EXTEND	Chapter 4	56
PONDWATR	EXTEND	Chapter 4	53
PREYPRED	BASIC	Chapter 15	228
PULSE	BASIC	Chapter 15, 22	237
RAMP	BASIC	Chapter 6	82
RENEMGY	BASIC	Chapter 11	160
RENEW	BASIC	Chapter 13	194
RESERVE	BASIC	Chapter 17	286
ROTATION	BASIC	Chapter 18	300
SALES	BASIC	Chapter 17	272
SLOWREN	BASIC	Chapter 13	198
SPECAREA	BASIC	Chapter 16	262
SPECIES	BASIC	Chapter 16	265
STATECON	BASIC	Chapter 20	337
TANK	EXTEND, BASIC, STELLA	Chapter 5, 6, 9	10, 68, 134
TANKSALE	BASIC	Chapter 17	274
TWOPOP	BASIC	Chapter 14	215
WAR	BASIC	Chapter 19	316
WETLAND	BASIC	Chapter 16	250
WORLDCO2	BASIC	Chapter 20	324

Preface

————————————◆————————————

Modeling and simulation are intellectually creative and quantitatively rigorous, mainstream ways of connecting ideas with reality. The time has come when everyone engaged in intellectual inquiry needs to model and simulate the phenomena of his or her interest. Models help us understand how things are organized and function. Simulating models is a way to learn how systems and their components grow and change.

A powerful and rigorous understanding of systems results when normal verbal thinking is connected to quantitative simulation using the systems diagrams and simulation programs. The process of representing the fundamentals of science also shows how similar all the branches of knowledge are, often using similar functions under different names. Thus, this book helps simplify knowledge by showing that many of the great variety of processes and events in our world are special cases of a relatively few principles commonly found on all scales of size and time.

Since 1989, versions of this workbook were used at the University of Florida and Santa Fe Community College to introduce modeling and simulation to students of varied backgrounds. A diagrammatic "energy systems" language is used to translate word models into the quantitative models to be used with numerical data and computers. The diagrams organize any system and its parts from left to right in order of the natural hierarchy of energy.

By studying systems models and then running the models on the computer, a student gains understanding of how things work. This process provides insight different from that provided by most kinds of study. The dynamic medium of simulation provides time perspective. The computer simulation programs make systems "come alive."

As shown in Figure P.1, the diagramming methodology in this book leads students to make models and write equations with less mathematics. Several ways of simulation are given with and without the need to program.

◆

LEARNING TO PROGRAM

A computer simulation program is a set of sequential, logical steps that represent a system process. Simulation programs are often compared to cookbook recipes. Setting out a sequence of statements represents a system in a stepwise language. Learning some computer programming is necessary to understand the time dimension of systems. When you run a program, you can check whether it is correct and does what you think it should.

FIGURE P.1.
Approaches to modeling and simulation in this book, starting with word models on the left. Location by chapter: making models as energy diagrams; Chapters 1–3 and 21; writing equations and calibrating, Chapters 6 and 7; writing programs, Chapter 9 and Appendix A; simulating with EXTEND, Chapters 4 and 5; simulating with BASIC, Chapters 9, 11, and 12–20 and Appendixes B and C; simulating with spreadsheets, Chapter 8; simulating with STELLA, Chapter 10; and mathematics approaches, Chapter 22.

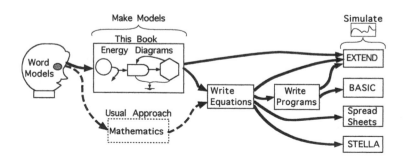

In the early days of computing, most students learned to program as part of their introduction to computers. Unfortunately, in recent years, the majority of students have learned to manipulate menus of complex software packages, without learning a programming language. In the opinion of many, everyone needs to be able to program to be well educated.

This book teaches BASIC, the simplest programming language. Some form of BASIC is available for most computers so that the model programs can be simulated in BASIC, QUICKBASIC, VISUALBASIC, etc. The programs for each model in the book are listed so the reader can study the lines in relation to the diagrams and equations.

USE IN TEACHING

This book is intended to introduce modeling and simulation without requiring a mathematical background. It can be used either as a stand-alone introduction to systems modeling and simulation or as a companion book for courses in other fields. Figure P.1 shows several pathways that can be taken to use this book to learn simulation. Select Chapters 1–5 to introduce models and elementary simulation. Use Chapters 6–11 to write equations and programs from the diagrams. Teach general systems concepts by running minimodels from Part III (Chapters 12–20). Develop a project that applies methods with Chapter 21. Survey some of the excitement in the varied field of modeling and simulation with Chapter 22. Refer to the appendixes for details on programming BASIC and EXTEND, the answers to the "What If" questions, and a summary of rules for use of energy systems symbols.

This book on simulation is not a general introduction to systems, which was the subject of an earlier volume (Odum, 1995, *Ecological and General Systems,* University Press of Colorado, Niwot). That book relates models, their diagrams, and their mathematics and may be a helpful reference for more advanced study.

Some other introductions to simulation are listed in the references section at the end of the book. The approaches, model selection, and programming languages vary.

◆

BOOK DISK

The CD supplied with this book contains versions of the programs Chipmunk BASIC and EXTEND, all the simulation programs used in this book including BASIC minimodels, the object-oriented programs for EXTEND, and spreadsheets used for calibration and simulation. See the listing Appendix E.

◆

ACKNOWLEDGMENTS

We thank Mark T. Brown and the many students who used models and gave their responses. T. E. Bullock helped us start examining this approach in the 1970s. The PEOPLE model was developed with Graeme Scott in New Zealand. Developing object-oriented blocks for EXTEND started with the BioQUEST workshops at Beloit College with John Jungck, Ethel Stanley, and Patti Soderberg. Nils Peterson collaborated on the programming. The application section, Part IV, was added at the suggestion of the publisher's referee. Ms. Joan Breeze was editorial assistant.

Howard T. Odum and Elisabeth C. Odum
Gainesville, Florida

PART ONE

---◆---

MODELING

Part I introduces modeling concepts, starting with a simple storage tank model and its simulation in Chapter 1. Chapter 2 introduces the energy systems symbols to represent parts and relationships in systems diagrams. The symbols and their use are also summarized in Appendix D. Chapter 3 explains ways of making models quantitative by writing numerical values of materials, money, and/or energy on flow pathways and in storage tanks.

Chapter One

◆

MODELING AND SIMULATION

$\mathbf{A}$lthough the universe in which we live is far too complex for the human mind to visualize in detail all at once, we can understand simplifications. The simpler concepts by which we think are often called *models*. As children early in life we develop models about the way the world works around us, usually expressing them in words. For example, in thinking about a water tank we could say "the more water we store in a tank the faster it flows out." In this chapter we explain how to give our word models a system view, and make them quantitative so that they can be computer simulated.

Models represent systems. *A system is a set of parts and their connected relationships.* Typical parts of our planet are the lakes, rivers, oceans, mountains, organisms, people and cities, some large and some small. Processes connect everything directly and indirectly to everything else. Our world is really one huge, complex system. But in order for humans to understand it, we have to simplify it by creating models. To do that, we first put an imaginary box in our minds around the subjects of our interest, thus defining a system. Next we draw symbols representing the outside influences, the inside parts, and the connecting lines that represent relationships and flows. Then we add numerical values to make the model quantitative. Finally, we use one of several methods for simulating the model with a computer. Simulation usually means letting the computer calculations show what the model does over time. Let's consider a simple example of the process of modeling and simulation.

◆

1.1 A SIMPLE SYSTEM OF STORAGE AND THE MODEL TANK

Let's start by modeling a system containing the storage process. Although we use water as an example of the medium being stored, the model applies to any type of storage. First, we draw a shaded frame with rounded corners as the boundary to define a simple system containing a storage tank, an inflow, and an outflow (Figure 1.1).

Examples of the storage system are sketched in Figure 1.1a, pictures showing storage containers with inflows and outflow. Expressed in words, the verbal model states:

Flows of something from outside go into a storage reservoir. Some of this quantity flows out through another route in proportion to the quantity stored.

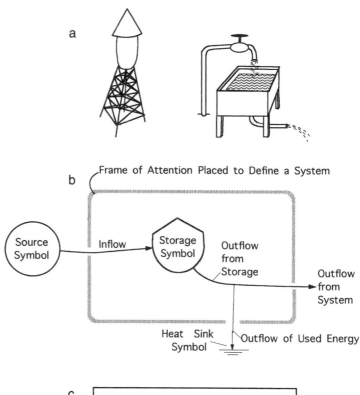

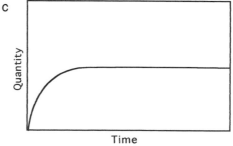

FIGURE 1.1.

Example of an energy systems diagram—a system containing storage, inflows, and outflows. (a) Examples of water tanks storing water; (b) energy systems diagram; (c) typical graph of growth of storage after starting with an initially small value. Also see Figure 6.1.

The available energy driving the flows comes in with
the inflow, but some is used in the process and disperses from
the system, no longer able to do work.

The model is made more precise by drawing an energy systems
diagram that considers the materials and energy (Figure 1.1b).
The energy systems diagram uses three of the symbols of the
energy systems language listed in Figure 1.2. The system includes

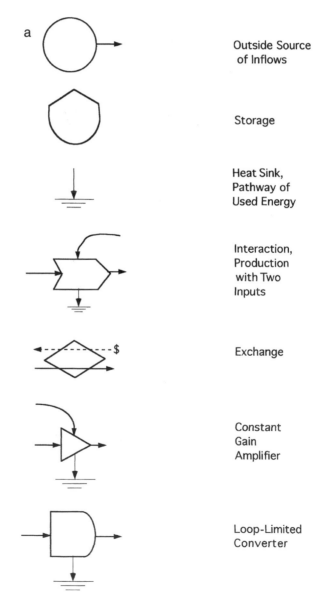

Outside Source
of Inflows

Storage

Heat Sink,
Pathway of
Used Energy

Interaction,
Production
with Two
Inputs

Exchange

Constant
Gain
Amplifier

Loop-Limited
Converter

FIGURE 1.2.
Symbols of the energy systems
language. (a) Sources, storages,
and pathway interactions;
(b) composite symbols for
representing units and
processes in the aggregate. To
show exact functions, we must
diagram within these symbols.

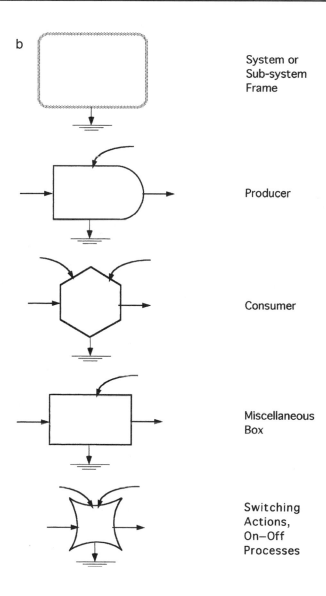

FIGURE 1.2.
(Continued)

connecting flow lines. In Figure 1.1b the inflow from outside the chosen boundary is shown coming from the circular *source symbol*. The *storage symbol* resembles water tanks that covered America earlier in the century (Figure 1.1a). Like the other general systems symbols, the symbol for storage can represent anything on any scale: water, money, fuels, people, books, whatever. The symbol can be used for stocks of chemicals at the scale of microbes or matter on the scale of the planetary processes. We give the model a name: TANK.

The network picture (Figure 1.1b) is supposed to be read

visually from left to right by imagining flows coming into the tank, as in the water tank example, and then flowing out in proportion to the pressure built up in the tank. The water flowing out of the tank continues to the right, flowing past the system boundary. The pathway branch that goes down represents the used energy (but no water). More water coming represents more storage and causes a greater outflow. If the inflow is constant, the water in the tank will increase until the outflows equal inflows (Figure 1.1c). After that the water level is constant.

REPRESENTING ENERGY LAWS

Because energy accompanies all storages and flows, energy systems diagrams are drawn to include the behavior of energy. The energy systems diagrams represent energy laws. The *first energy law (conservation of energy)* states that energy flowing into a system must either be accounted for in outflows or in storage within the system. In the tank model (Figure 1.1b), available energy at the source pumps in the water. Energy is stored with the water according to its height above the ground. The more water, the more available energy is stored. As the water flows out, some of the energy is dispersed by friction. According to the *second energy law* dispersed energy cannot do any more work and leaves the defined system degraded. In the diagrams, dispersed energy (used energy, used-up availability) is always shown leaving the system through the *heat sink symbol,* usually drawn with thinner lines. The heat sink has an arrowhead (barb) indicating that it cannot run backwards.

SIMULATION OF THE TANK MODEL

Next, run a simulation of the TANK model to get the idea of simulation. Use one or both of the following procedures.

Simulating the TANK Model with the EXTEND Program

Put the disk that comes with this book in your computer and load the program EXTEND. Next, use that program's FILE menu to open the file TANK.mox. The program loads a library of symbols and uses them to draw a systems diagram of the tank

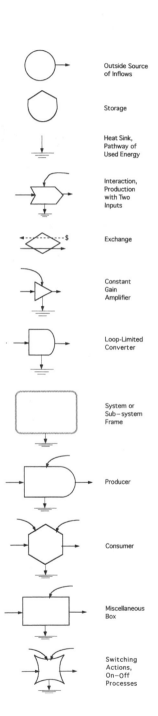

Outside Source of Inflows

Storage

Heat Sink, Pathway of Used Energy

Interaction, Production with Two Inputs

Exchange

Constant Gain Amplifier

Loop-Limited Converter

System or Sub–system Frame

Producer

Consumer

Miscellaneous Box

Switching Actions, On–Off Processes

model on screen (Figure 1.3a). Then run the simulation with the RUN menu. A graph similar to Figure 1.1c appears (Figure 1.3b), confirming the expectation from the verbal thinking.

Bring the diagram window to the front by clicking on it with the mouse. Double click on the circular source symbol. A dialog box appears that has a place for the input value. Type in a higher number. Before you run it, predict what will happen. Then go to the RUN menu and repeat the simulation. The new graph levels off at a higher value. Does this agree with your expectation? With more inflow, the storage and its pressure builds up to a higher value before the outflow equals the inflow.

Simulating the TANK Model with a BASIC Program

Load QBASIC from PC Windows (Macintosh users can load CHIPBASIC from the CD-Rom). Use the FILE menu on the screen

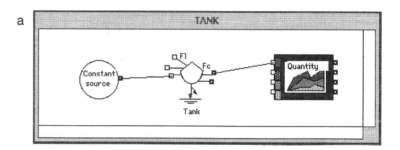

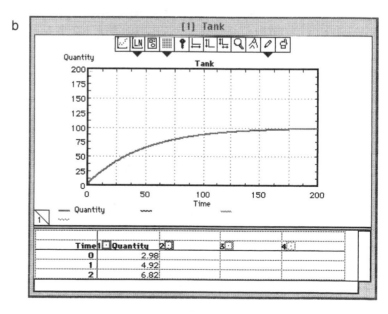

FIGURE 1.3.
Views of the computer screen when the storage model TANK is simulated by EXTEND.
(a) Screen view of the systems window with connected icons;
(b) screen display of the simulated growth.

to load the program TANK (file: TANK.bas; CHIPTANK.bas for Mac). Use the mouse to click on the RUN command on the menu. A graph of storage over time appears on the screen similar to that shown in Figure 1.1c. Next, use the LIST command in the menu to show the statements of the program. The list of numbered statements in Table 1.1 appears on the screen. We will explain how to write such programs in Chapter 9. For now, you can make a change in the program by retyping a line in the program and running the program again to see how the model changes its behavior. Increase the inflow J by retyping line 50 to read: 50 J = 4. The curve levels off at a higher level, as you might expect.

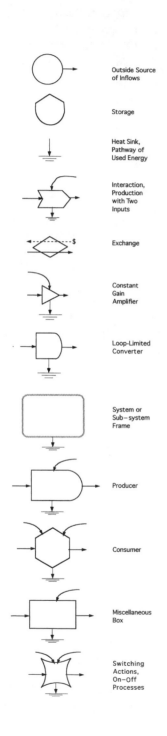

Outside Source of Inflows

Storage

Heat Sink, Pathway of Used Energy

Interaction, Production with Two Inputs

Exchange

Constant Gain Amplifier

Loop-Limited Converter

System or Sub–system Frame

Producer

Consumer

Miscellaneous Box

Switching Actions, On–Off Processes

TABLE 1.1
Program in BASIC for the Storage Model TANK[a]

```
  5  REM PC
 10  REM TANK.bas (Growth of storage)
 15  CLS
 20  SCREEN 1,0
 30  COLOR 0,0
 40  LINE (0,0) - (240,180), 3, B
 50  J = 2
 60  Q = 1
 70  K = .02
 80  Dt = 1
100  REM Remark: Start of Iteration Loop
110  PSET(T,180-Q),1
120  DQ = J - K*Q
130  Q = Q + DQ*DT
140  T = T + DT
150  IF T<240 GOTO 100
```

---

[a] Without scaling factors.

◆

1.2 PARTS AND WHOLES

When we think about a model, we define an arbitrary boundary, usually without considering it consciously, including within it the phenomenon of interest. With the energy systems language we place the boundary consciously (in this book a shaded, rounded rectangle). This defines a whole and the symbols that we put inside are the parts. Those connected from outside of the frame are sources and outflows. See the example shown in Figure 1.1b.

Because of the way science has usually been taught, we most often look from the whole inward to the parts in order to find mechanisms of the way the whole works. By doing this, we move to a smaller scale of time and space, one that may be too small for the questions raised. For example, in many environmental affairs, study of the parts of an ecosystem is not as important to understanding and prediction as studying the ecosystem's interactions as a part of the economy around it.

Starting with a unit of interest, we can either call it the whole, put a frame around it, and diagram its parts, or regard it as a part to be fitted into a model whose frame is part of a larger system. For example, in Figure 1.4 the hatched unit (Figure 1.4a) is shown again in Figure 1.4b with its parts diagrammed. In Figure 1.4c the hatched unit is included in a diagram of the next larger scale where its role at that level shows in the network relationships.

Study of models is both analytic and synthetic. Analysis often means taking something apart to see how things are structured and processed. Synthesis often means putting parts together to identify and understand the whole. Modeling can do both at the same time.

◆

1.3 REPRESENTING SCALE AND ENERGY HIERARCHY

Energy systems diagrams are drawn so as to represent the scales of phenomena and the principles of matter and energy.

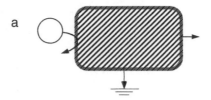

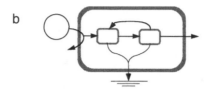

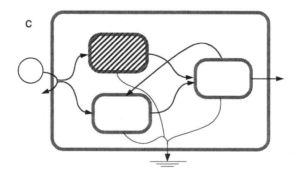

FIGURE 1.4.
Systems and subsystems.
(a) A selected system (shaded);
(b) model of the components
within the selected system;
(c) the selected system as a
subsystem modeled within a
larger network.

As illustrated with the energy systems diagram of Figure 1.5b, symbols representing many small-scale units are on the left and symbols representing larger scale units are on the right. For example, in Figure 1.5c, if the system is a pasture, the source on the left is the inflow of tiny photons of sunlight.

The producer symbol represents the many small blades of grass, the consumer symbol the medium-sized sheep, and the block on the right the human users drawing inputs from the economic sources on a larger scale. The system looks complex when all the units are viewed together (Figure 1.5a). Better understanding results when the components are separated and arranged in order of scale (Figure 1.5b).

In all systems energy accompanies everything, including flows and storage of information. Thus, all pathways on an energy

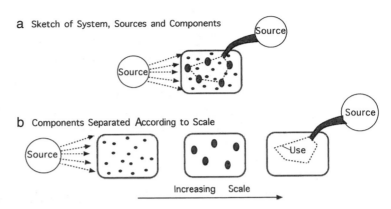

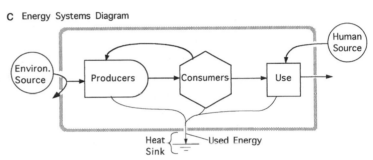

FIGURE 1.5.
Modeling scale and energy transformations. (a) Sketch showing the way items of different size and scale are often depicted together; (b) items of similar scale separated from left to right in order of increasing scale; (c) energy systems diagram of these sources and components.

systems diagram include energy. According to one of the universal energy laws (second energy law), when usable energy is transformed into a new form, most of it is degraded and dispersed in order to create a small amount of the new type. Dispersed energy cannot do any more work. The dispersal of energy is represented with the heat sink symbol (Figure 1.2). Notice the flows out the bottom of the diagram in Figure 1.5c.

In a chain of energy transformations like that shown in Figure 1.5, many joules of energy flowing from the left support fewer joules of energy in the middle and even fewer joules on the right. In other words, there is an energy hierarchy that is recognized by the arrangement of symbols in an energy systems diagram from left to right.

Energy flows to the right are said to be of higher quality because more energy of one kind is used to make them. The ratio of energy of one kind required for the higher quality type is called the *transformity*, a concept explained further in Chapter 11. Transformities increase from left to right, a measure of position in the energy hierarchy.

Figure 1.6 shows a diagram commonly used to relate the scale of events in time to the scale of phenomena in space. On the lower left are things that occupy small territories and have fast turnover (short replacement time). The upper right has items of large territory and a slow rate of turnover. For example, the scale of chemical reactions is on the lower left, the scale of environmental and geological processes is in the middle, and the scale of astronomical events is at the upper right.

SCALES OF MODELING AND THE MACROSCOPE

Because the real world operates simultaneously on many scales, models should include all scales pertinent to the phenomena of interest. There is a tendency, through long habit and the desire to simplify, to concentrate on models of one scale. No scale is more basic than another, but people concentrating their work think of their scale as special. However, limiting the scale of view limits understanding, because every scale is part of the scale above and composed of the smaller scaled items below. We cannot understand one scale without studying its relation to that above and below. Control on the longer time scale comes from the larger scale to the right, but understanding of parts comes from the organization of their relationships. One safe way to model is to always include two or three scales in modeling a phenomenon of interest.

When emphasis on looking smaller is carried to the extreme, people on one scale deny there is any science at the next larger scale. Many people are taught that looking for *purpose* is not scientific, that teleology is bad. However, what is viewed as purpose of a unit at one scale is its part in mechanisms of the next larger scale. Avoiding purpose is really hiding from the larger scale, which controls the longer scale of events.

Modeling that starts by identifying and studying parts first and then connecting them into a system is sometimes called *bottom-up modeling*. Because of the analytic (take apart) emphasis in science education, many people tend to approach models from the bottom up. Sometimes this approach includes too many parts, not aggregated enough, so that the system becomes too complex for easy simulation or understanding.

Top-down modeling first defines the scale of space and time that is of interest because of the problems, questions, and pur-

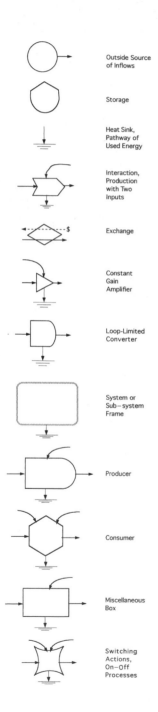

Outside Source
of Inflows

Storage

Heat Sink,
Pathway of
Used Energy

Interaction,
Production
with Two
Inputs

Exchange

Constant
Gain
Amplifier

Loop-Limited
Converter

System or
Sub–system
Frame

Producer

Consumer

Miscellaneous
Box

Switching
Actions,
On–Off
Processes

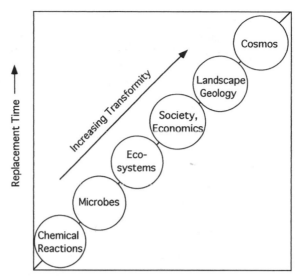

FIGURE 1.6.
Realms of different scale from chemical reactions to the stars referred to a diagram where territory, replacement time, and transformity increase together.

poses of the study. After the systems attention frame is defined, a limited number of symbols is added, each representing aggregation of many smaller parts. When this approach is used, the models do not get lost in detail. A model is finished and running sooner. On the other hand, the aggregation may not bring out everything that is important in that system.

We sometimes refer to top-down modeling as looking through the *macroscope* (Odum, 1971). The human mind is good at changing its scale of mental attention, much like a zoom microscope. Whereas the authors of this book tend to teach top-down thinking, perhaps good modeling requires zoom perspectives, some top down and some bottom up. Only with simulation testing, comparing, and revising with real-world observations can oversimplifications be corrected to make a model appropriate (Chapter 21).

1.4 SUMMARY OF PROCEDURE FOR MAKING A CONCISE MODEL

Let's review the process of translating a word model into a more precise form, whether you are working alone or conducting a group session and using a blackboard for this purpose. What this procedure does is take the knowledge of individuals or the group, which is in verbal form, and collect it into lists. Then the words are translated into symbols to form a network diagram in energy systems language.

1. Draw the frame of attention that selects the boundary.
2. Make a list of all the important input pathways that cross the boundary. Put each in a source symbol and place them around the frame from left to right roughly in order of transformity (sun, wind, rain, rivers, tide, fuels, high-quality foods, impact chemicals, labor, technical services, information, etc.). Label the symbols with the words.
3. Make a list of components believed to be important. Place them within the boundary from left to right in order of transformity. (For example, ecosystem components in food chain order: producers, consumers, etc.) Label the symbols with the words.
4. Make a list of the processes believed important within the system defined. Then use these processes to indicate the pathways connecting the symbols. For example, for photosynthesis draw pathways to connect the ingredients to a production symbol. Label processes of special interest.
5. Remember that matter is conserved. (Something is conserved if the matter flowing in is either accounted for in storages within or in outflows.) Therefore, storages and flows of each material you put into the model should be examined to see if anything is missing. For example, is the water incoming from rain and rivers fully accounted for in lake storage, evapotranspiration pathways, and outflows?
6. Check to see that money flows form a closed loop within the frame and that money inflows across the boundary lead to money outflows. Use dashed lines to distinguish money flows from other flows.
7. Check all pathways to see that energy flows are appropriate. In general, all symbols should have a pathway of degraded, used energy going out to the heat sink at the bottom of the frame.
8. If color is used, the following color scheme is suggested:
    Yellow—sunlight, heat flows including used energy flows
    Blue—circulating materials of the biosphere such as water, air, nutrients

Outside Source of Inflows

Storage

Heat Sink, Pathway of Used Energy

Interaction, Production with Two Inputs

Exchange

Constant Gain Amplifier

Loop-Limited Converter

System or Sub–system Frame

Producer

Consumer

Miscellaneous Box

Switching Actions, On–Off Processes

Brown—geological components, fuels, mining

Green—environmental areas, producers, production

Red—consumers (animal and economic), population, industry, cities

Purple—money

9. If a complex diagram has resulted (more than 25 symbols), redraw it to make it neat and save it as a useful inventory and summary of the input knowledge. Because for most purposes a simpler model is desirable for an overview for policy discussion or for simulation, redraw the diagram with the same boundary definition, aggregating symbols, and flows to obtain a model of the desired complexity (perhaps 3 to 10 symbols). You can combine them, but do not leave out any of the inflows and outflows.

Often when we show someone the complex diagram, they may be "turned off," saying that it is too complex for the time they have to study it or for a human's ability to absorb its meaning. But if you show someone an aggregated diagram they may dismiss it as simplistic. If you show them both, they can see what has gone into the model and how simplification may be necessary to hold human credibility and interest. Then they may join you in modifications, calibration, testing, and revising.

In Chapter 2 we explain the symbols and their use in more detail, and in Chapter 3 we explain how to assign numerical values to the diagrams.

Chapter Two

ENERGY SYSTEMS DIAGRAMMING

Chapter 2 contains details for making energy systems diagrams, the first step in this book's procedure for modeling and simulation. Rules are given for pathways and the dozen symbols of the language to represent causal relationships, materials, energy, money, and information in a system.

◆

2.1 TRANSLATING WORDS INTO SYMBOLS

Two groups of energy systems symbols are given in Figure 1.2 and Appendix D (*group symbols* and *precise symbols*). Each of the group symbols—producer, consumer, switching unit, and miscellaneous box—represents categories in the same way as their word descriptions. For example, there are many different models for the structure and function of producers.

The precise symbols of the energy language are source, storage, heat sink, interaction, exchange, constant gain amplifier, and loop-limited converter. These are used for representing a particular part of a model and its quantitative relationships. They have specific energy and mathematical meanings necessary for writing equations and simulating.

Influences from outside the system are sources, which are drawn with a circular symbol placed outside the defined boundary, with a pathway crossing into the system. A source symbol can be made precise by defining the nature and timing of its input.

Diagrams are drawn to illustrate the first and second energy laws and concepts as already explained in Chapter 1. Every system has to have at least one source symbol for energy input (and the force to input the resource). Where there is available energy (energy capable of doing work), ordinary width lines are used for flow pathways. Outflows that still have energy available to do work are shown going out the side or the top of the frame. Only degraded energy goes out the bottom.

First an overview diagram is made for a word model using

group symbols. Then an exact model can be made for it by diagramming details inside with precise symbols. In the process you make decisions about the structure and functions you want in the model. For example, let's draw an energy systems diagram for the following word model:

Environmental energy sources support producers, and these support consumers symbiotically. Some of the consumers are used by humans.

Figure 2.1a represents the model with group symbols. Symbols for producer, consumer, and miscellaneous box are shown from left to right. This is the same diagram already used to discuss the scale and energy hierarchy in Figure 1.5.

Conversely, if the diagram in Figure 2.1a were supplied to you, it could be verbalized as follows:

Resources on the left stimulate the producer, and its products go to the consumer from which something is shown going back to the left to contribute to producers. Human energy enters on the right, interacting with the consumers to produce some kind of yield.

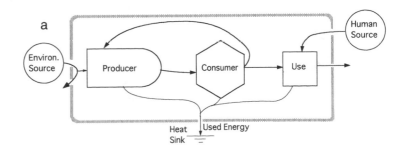

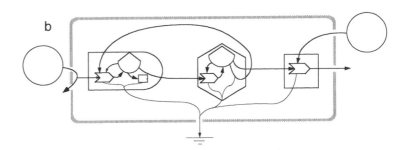

FIGURE 2.1.
Highly aggregated diagram of an ecosystem and its use. (a) Using group symbols; (b) details of structure and functions shown by including precise symbols from Section 2.5 within group symbols.

Outside Source
of Inflows

Storage

Heat Sink,
Pathway of
Used Energy

Interaction,
Production
with Two
Inputs

Exchange

Constant
Gain
Amplifier

Loop-Limited
Converter

System or
Sub–system
Frame

Producer

Consumer

Miscellaneous
Box

Switching
Actions,
On–Off
Processes

In Figure 2.1a, the details are unspecified, just as they were unspecified by the word descriptions. The exact functions of the group symbols are not indicated. Diagrams using groups symbols are useful before details are known, or when general kinds of relationships are being discussed.

The diagram of general concepts in Figure 2.1a is made into a particular model by using precise symbols to draw details within the group symbols (Figure 2.1b). Each group symbol there has an inside network showing flows, storages, and interactions. The explicit model in Figure 2.1b is only one of the many that could fit the model of general categories shown in Figure 2.1a.

As explained in Chapter 1, symbols on diagrams are arranged from left to right according to their position in the energy hierarchy. This makes one person's diagrams similar to that of another. If a flow is becoming more concentrated, as in rains converging into streams, the pathway should be drawn from left to right. However, if the flow is decreasing its concentration, as with a river dispersing into the ocean, or nutrient wastes recycling into a forest, then the pathway should be drawn from right to left (see Figure 2.3 later in this chapter).

2.2   PATHWAYS, FORCE AND FLOW

Representing the way most systems form structure and process, energy systems diagrams are a network of units (symbols) connected with pathway lines (Figure 2.1). A pathway line is used for the flow of anything: materials, information, organisms, people, or whatever. All the pathways of an energy systems diagrams flow at the same time. Because everything, including information, has some available energy associated with it, there is energy flow on all pathways. Figure 2.2 shows types of pathways and branches.

Flows are pushed by "forces," where the word *force* has its more general meaning that includes physical force, chemical concentration, or any other property that has the energy content capable of causing a flow. For example, electrical flow measured in amperes is driven by electrical force measured in volts. Diffusion of chemicals in fluids is driven by the differences of concentrations. A force either comes from outside the system marked with a source symbol or from a storage tank within a system.

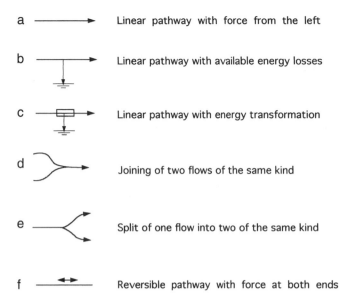

FIGURE 2.2.
Pathways in energy systems diagrams.

Where the flow depends only on the force behind it, an arrowhead (barb) is used where the pathway comes to an end, usually at the next symbol (Figure 2.2a). A simple pathway with arrowhead means that the flow is proportional to one driving force (Figures 2.2a and 2.2b). This kind of pathway is said to be linear, whereas a pathway that is proportional to other inputs as well is said to be nonlinear.

Where the flow depends on the difference between the force at one end and the back force from the other end, a line is used without a barb, and this pathway may flow in either direction (Figure 2.2f). Sometimes we put a two-headed arrow next to, but not on, a pathway to indicate that flow reversals occur.

According to the rules of the language, two pathways can merge (Figure 2.2d) if the flows on these pathways carry the same kind of material. Similarly, a pathway that divides as a simple branch represents a splitting of a flow of one kind into two flows of the same kind (Figure 2.2e).

2.3 THE CIRCULATION OF MATERIAL

Materials flow on many pathways but not all. For example, flows of sunlight, sound, and heat are pure energy, not accompanied by matter. Materials usually circulate in the same direction as the accompanying energy. For example, phosphorus is used in plant production, passed in organic matter to consumers, and then released and recycled. In Figure 2.3a material flows are highlighted with heavy lines; the thin lines are energy flows without materials. Another way to represent this difference is to use color for the materials.

Many materials are conserved, neither created nor destroyed, while in a system. Diagramming the pathways of material of one kind is a useful way to represent data on that material (Figure 2.3b). In Figure 2.3b the material pathways are isolated without changing the shape of the network so that materials can be readily related to the rest of the system. If that network is drawn on transparent film, it becomes an overlay.

However, a material diagram should not be made without the

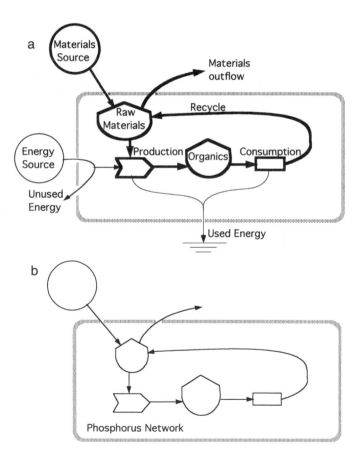

FIGURE 2.3.
Representing material flows in energy systems diagrams. (a) Systems diagram for a whole ecosystem with phosphorus pathways emphasized; (b) overlay of the model in Part a to show the pathways of phosphorus only.

rest of the system because the interactions with the rest of the system are responsible for the flows. Hence, the best procedure is to diagram the whole system first and then use color or an overlay for the materials. For example, phosphorus in an ecosystem is driven by photosynthetic production and it by a number of other inputs and various other interactions with animals, circulation, and the economy. To understand the phosphorus cycle, the network of other units affecting the phosphorus cannot be omitted.

2.4 THE CIRCULATION OF MONEY

Money usually circulates in the opposite direction from the flows of materials and energy. In energy systems diagrams it is shown with dashed lines (Figure 2.4) connected to its material and energy flow by a diamond symbol. Sometimes network diagrams in economics show only one pathway for the flow of commodities, while evaluating the flow with money. In energy systems language flows of money are kept separate from the flows of real commodities and services which they buy. Both are drawn. For many short-term models, money is conserved.

2.5 PRECISE SYMBOLS

The symbols with precisely defined meanings are explained in more detail using Figures 2.5 through 2.8.

SOURCES

The circular symbol for outside sources can be made precise by indicating if it is supplying a force or a flow and the pattern of its influence over time. For example, the source of energy from sunlight seasonally may be defined as an inflow of energy that varies seasonally with the pattern of a sine wave (Appendix B).

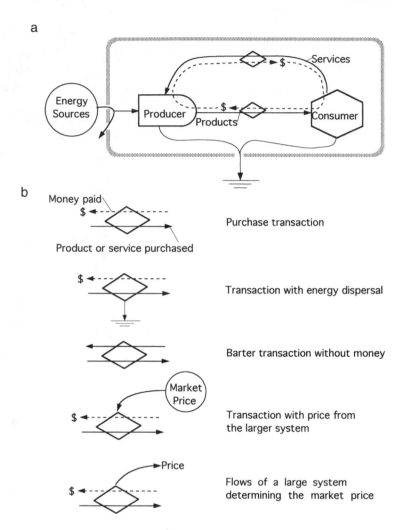

FIGURE 2.4.
Representing money and
the exchange symbol.
(a) Circulation of money in an
economy with exchange
symbols used for purchase of
products and purchase of
services; (b) various uses of the
exchange symbol.

STORAGE SYMBOL

The storage tank symbol (Figure 2.5) was used to explain the
simulation of the TANK model (Figure 1.1). The rate of change
of the quantity in the tank is the balance between rates of inflow
and outflow. Every flow into or out of storage must be of the
same type. For example, if water flows into the stored quantity
of water, the outflow pathways have to be water also. The storage
represented by this symbol does not have a limit, and can grow
to any size.

Because a storage tank represents a concentration of something (energy, materials, structure, information), there has to be

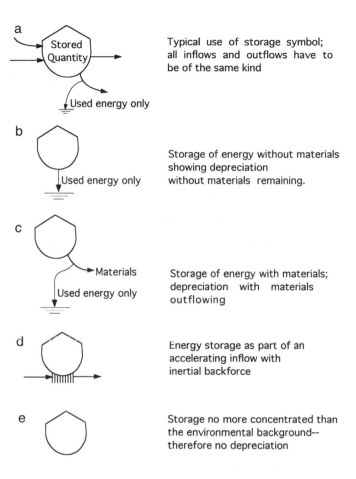

FIGURE 2.5.
Uses of the storage symbol.
(a) Typical use; (b) energy
storage; (c) materials storage;
(d) storage for inertial
acceleration; (e) storage without
depreciation.

some spontaneous dispersal (second energy law). Therefore, a model of a storage tank should have at least one linear outflow pathway (flow out proportional to storage) that connects with the heat sink. The necessary linear pathway is sometimes called *diffusion, dispersal,* or *depreciation.*

When the contents of the storage drain through a linear pathway, the pathway of dispersal of degraded energy is drawn as a branch with a thin line to the heat sink. Materials and any remaining available energy are shown with pathways passing elsewhere within the system or out a lateral boundary (Figure 2.5c).

INERTIAL STORAGE

A special kind of storage and pathway is inertial storage (Figure 2.5d), which exerts a backforce to inputs in proportion to the acceleration. (Acceleration is the change in velocity with time.) As long as the input is accelerating, there is an increase in backforce, and during that period, flow is resisted and storage of energy increases. Later, if the acceleration stops, the storage can drive the flow in the same direction of the original acceleration. An example is the surge of electrical current through a coil of wire, which builds up a magnetic field as it resists the surge. Another example is the backforce created when a pitcher accelerates a baseball throw: The backforce is from the energy being stored in the baseball's forward momentum during the short time when the arm is accelerating the ball. Another example is the resistance to change by people who store information against the change while someone is trying to persuade them.

INTERACTION SYMBOL
(PRODUCTION PROCESS)

The interaction symbol is for transformations that use two or more pathways, each carrying something different, and each necessary to a production process (Figure 2.6). The symbol was also called a *work gate*. There is an output of new product. The action represented is called a *production function*. Where there is an energy transformation, the symbol normally points from left to right. Whatever the function, all the flows into and out of the interaction symbol are in proportion to the production function.

The most common kind of interaction is a product of the forces on the input pathways. However, other kinds of interactions may occur such as dilution (division), shown in Figure 2.6d. That arrangement means that the production is proportional to the left input divided by the force from the right. For example, the concentration of plankton in a pool (left input) is changed to a lower concentration by the dilution of water (right input). In other situations the left input is decreased in proportion to the right input, an action that subtracts in producing the output (Figure 2.6e).

Inflows have different transformities and the inputs are arranged on the symbol from left to right in order of transformity (Figure 2.6b). The highest transformity is sometimes regarded as a *control* feedback, usually coming from the right. The lowest transformity is the one with the most energy coming in from the

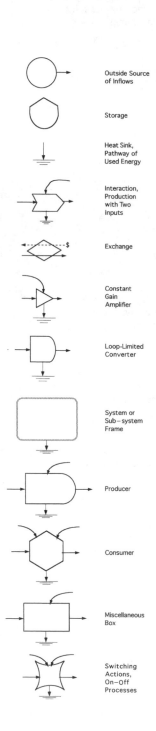

Outside Source of Inflows

Storage

Heat Sink, Pathway of Used Energy

Interaction, Production with Two Inputs

Exchange

Constant Gain Amplifier

Loop-Limited Converter

System or Sub-system Frame

Producer

Consumer

Miscellaneous Box

Switching Actions, On-Off Processes

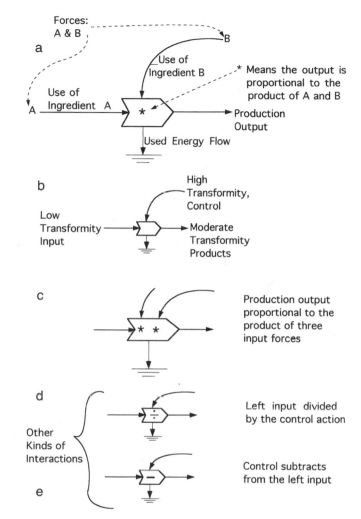

FIGURE 2.6.
Uses of the interaction symbol.
(a) Product interaction;
(b) transformity position;
(c) three product inputs;
(d) divisor action;
(e) subtracting action.

left, whereas the production output is intermediate in quantity of energy and transformity. For more on transformity see Chapter 11.

CONSTANT GAIN AMPLIFIER

With the symbol for the constant gain amplifier (Figure 2.7) control comes from the force on one pathway, but most of the energy for the productive output comes by a pathway from a different source (Figure 2.7a). For example, the sound of a sing-

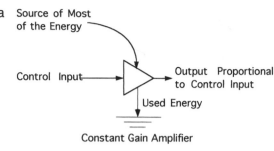

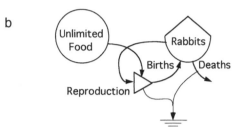

FIGURE 2.7.
Use of constant gain amplifier
symbol including the example
of reproduction at a constant
specific rate. (a) Typical use;
(b) reproduction example.

Example of Constant Gain Amplifier
where the gain is the number
of offspring in the litter

er's voice into a microphone provides the control with small
energy, but the booming amplified output is made possible by
an electrical power supply. Another example is the reproduction
of organisms in which a small amount of energy in high-quality
information is supplied as seeds or eggs, but the energy for a
larger population comes from a food supply (Figure 2.7b).

The constant gain amplifier symbol is defined with the output
a constant factor of the input (constant gain), which assumes
that the main energy supply is unlimited. Of course, no energy
supply is unlimited if the demand on it is too great. Thus, the
symbol is only useful over the range in which energy is not
limiting. It is the appropriate way to diagram population models
that treat reproduction as intrinsic (not limited by energy) (Figure 2.7b).

EXCHANGE SYMBOL

Figure 2.4 shows the symbol for exchange of one flow for
another. Usually money is exchanged for a commodity or service.
The diamond-shaped symbol indicates they are coupled according to a price. Where the transaction is a small one, the price

may come from a market outside the system. Where the flows represent most of the market, the price is determined from the ratio of the flows. Included in Figure 2.4 is barter, in which two commodities are exchanged. If the business transaction is substantial, you can show a heat sink below the exchange symbol. Usually, however, the work of the transaction is done by the units at either end of the transaction, and their heat sinks include the energy dispersed by the process

HEAT SINK

The heat sink (Figure 1.2), showing energy dispersion, was explained in Chapter 1. The second law requires that there be a heat sink symbol at the bottom of each systems diagram. It represents the dispersal of the energy and its export as used energy.

Where the availability (ability to do work) is lost because the energy has been degraded to about the same concentration as that outside the frame, we call it "used energy" and use thinner lines to show the dispersal of this energy down to the heat sink. Many pathways have frictional processes and other losses of available energy. To draw these pathways correctly, heat dispersal branching to the heat sink is required.

Some people confuse the heat sink with the electrical symbol for ground used in electrical engineering. (The electrical ground symbol for flow of electrons in and out of the environment has no arrowhead, and the flow can go either up or down.) However, the heat sink has an arrowhead pointing down and out of the frame to represent the second energy law. Because the heat sink represents second law depreciation, flow through a heat sink is irreversible and can only go in one direction: from concentrated energy with availability to dispersed, unusable energy.

LOOP-LIMITED CONVERTER

The short loop-limited symbol in Figure 2.8a represents the loop subsystem in Figure 2.8b in which the amount of material being recycled is constant. Because the material can only recycle so fast in the closed loop, the output diminishes as the input is increased (Figure 2.8c). In 1913 this performance was discovered

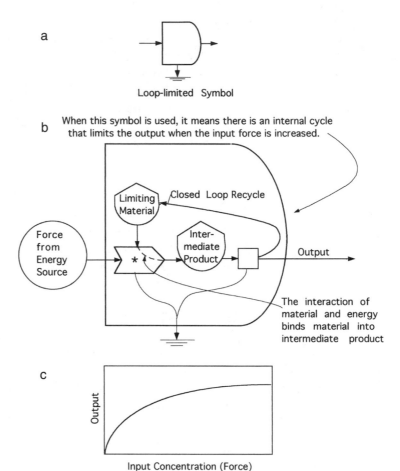

a

Loop-limited Symbol

b When this symbol is used, it means there is an internal cycle
that limits the output when the input force is increased.

Limiting Material

Closed Loop Recycle

Force from Energy Source

Inter-mediate Product

Output

The interaction of material and energy binds material into intermediate product

FIGURE 2.8.
Use of the loop-limited symbol.
(a) Symbol; (b) inside limiting
material cycle, which is implied
when the symbol is used;
(c) diminishing returns on
output as input force is
increased.

c

Output

Input Concentration (Force)

for an energy inflow interacting with a recycling enzyme. We sometimes call this symbol the Michaelis–Menten module, named after the authors of the enzyme study. For simulation of this unit, see the model INTLIMIT in Chapter 12.

The loop-limited symbol is appropriate where pure energy flows are used by units with an internal cycle. For example, in the first step of photosynthesis light interacts with chlorophyll, which is a type of photovoltaic cell, to make plus and minus charges. These are then connected to reactions that make oxygen and organic matter and in the process the chlorophyll is reset to receive more energy.

◆
2.6 GROUP SYMBOLS

The group symbols introduced with Figure 1.2b are described further in Figures 2.9 through 2.12, where several typical inside designs are given for each.

PRODUCER

The bullet-shaped producer symbol (Figure 2.9) usually implies a unit with a production process, often with a storage symbol. It can be used as an outline symbol (Figure 2.9a) with

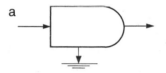

Producer Group Symbol

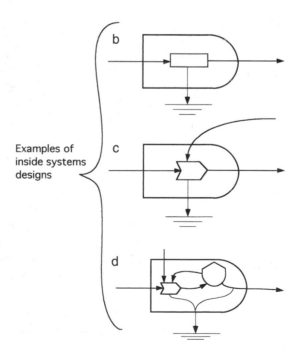

Examples of inside systems designs

FIGURE 2.9.
Use of the producer symbol.
(a) Unspecified details;
(b) production as a transformation of one input such as linear production;
(c) production a product of two input forces; (d) production with autocatalytic use of two inputs.

details unspecified or with precise symbols inside to represent structure and function.

In Figure 2.9b flow is in proportion to the driving force, producing organic matter in proportion to concentration of light. It is appropriate to use this simplification for photosynthesis if the other factors usually considered are constant. In Figure 2.9c use is in proportion to the interactive product of two input forces (example: photosynthesis in proportion to the product of nutrient concentration and light intensity). In Figure 2.9d use is proportional to input forces and feedback from its own storage (example: phytoplankton producing in proportion to nutrient concentration, light intensity and stored biomass). The unit stores the products and exports some.

Outside Source of Inflows

Storage

Heat Sink, Pathway of Used Energy

Interaction, Production with Two Inputs

Exchange

Constant Gain Amplifier

Loop-Limited Converter

System or Sub–system Frame

Producer

Consumer

Miscellaneous Box

Switching Actions, On–Off Processes

CONSUMER

The hexagon-shaped consumer symbol (Figure 2.10), like that for the producer, usually includes a production process and often a storage. Where a consumer is using input from a producer, the consumer is drawn to the right. The production process in the producer symbol is often called *primary production* and that in the consumer *secondary production.*

Figure 2.10a does not have any details specified. Figure 2.10b shows a flow in proportion to driving force (example: microbes consuming in proportion to available sugar). Figure 2.10c has consumption in proportion to the interactive product of two input forces (example: decomposition of organics in proportion to the product of organic concentration and oxygen concentration). In Figure 2.10d use is proportional to input forces and feedback from its own storage (example: zooplankton growing in proportion to food and oxygen concentration).

The names *producer* and *consumer* can be applied on many scales. For example, plants in a forest ecosystem produce food for wildlife consumers. But if the frame of attention is placed over a larger scale of landscape, the production of forest wood is purchased and used by human consumers. The production symbol is usually used for the process using the primary energy (on the left in the diagrams). For example, in a model of manufacturing, industrial production is the first process.

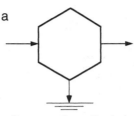

Consumer Group Symbol

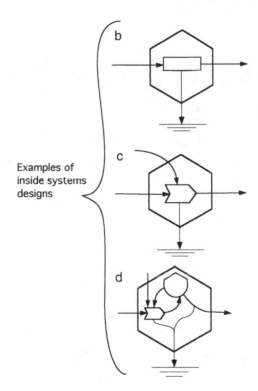

Examples of
inside systems
designs

FIGURE 2.10.
Use of the consumer symbol.
(a) Unspecified details;
(b) consumption as a
transformation of one input
such as a linear process;
(c) consumption a product
of two input forces;
(d) consumption with
autocatalytic use of two inputs.

BOX

It has long been customary to regard the rectangular box as a miscellaneous symbol (Figure 2.11). In cases where the performance and structure within the box are unknown, it is called a black box. If the performance and structure within the box are known, but for purposes of diagramming this complexity is not shown, it is called a white box. We often use the box for a unit that does not really fit the idea of producer or consumer. For example, commerce in the human economy is intermediate between production units and consumers. Details that might be

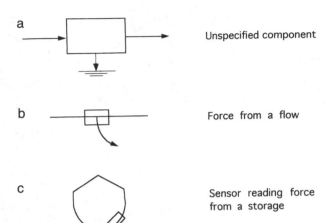

FIGURE 2.11.
Use of the miscellaneous box symbol. (a) As an unspecified component; (b) to indicate a force derived from a flow; (c) to indicate a sensor deriving force from a storage (without appreciable drain from the storage).

used within the box are those also used for producer and consumer symbols (Figures 2.9 and 2.10). The box in Figure 2.9 is used to represent production simplified as a linear process. The one in Figure 2.10 represents consumption simplified to a linear process.

Use a small box on a pathway to send out a force proportional to the flow (Figure 2.11b). Also use a very small box on a storage (Figure 2.11c) to deliver a force to a pathway without drawing much from the storage. These arrangements to deliver a force without much flow are *force sensors.*

SWITCH, DIGITAL BOX

The switching box is used where there are on/off processes (Figure 2.12). Whereas most of the energy language symbols have outputs that increase or decrease with a continuous range of values dependent on their inputs, the switching box has discontinuous actions. Sometimes switching actions are said to be *digital,* whereas the word *analog* is used for those control units with a continuous range of values. The switching box, like other group symbols, is not clear about what switching actions occur until the detail is supplied.

Details can be diagrammed either with standard logic symbols (not shown here) or written as logic statements in words (Figure 2.12). For example, the overflow of a river occurs (switch turns on) when the water level becomes higher than the bank (Figure

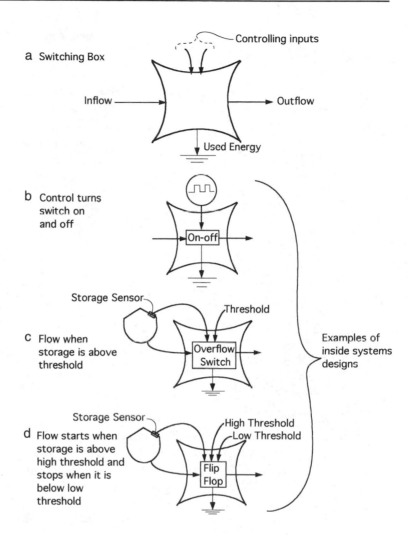

FIGURE 2.12.
Use of the digital switching box. (a) Unspecified details; (b) containing an outside controlled on/off switch; (c) flow when threshold exceeded; (d) flip-flop action.

2.12c). Another example is fire that turns on when biomass reaches a threshold and there is also a pulse of fire-starting heat. In this case fire does not turn off until a low biomass threshold is reached. This switching mechanism is called a *flip-flop* (Figure 2.12d).

2.7 THE MAXIMUM POWER PRINCIPLE

Regarded by many as a fourth energy law, the maximum power principle can be stated as follows:

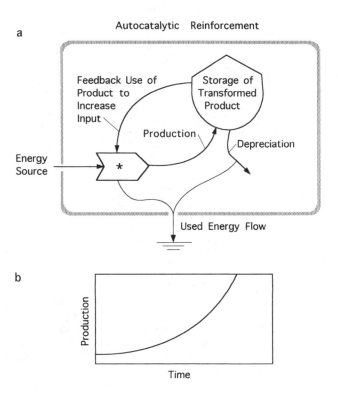

FIGURE 2.13.
Autocatalytic production.
(a) Configuration in energy
systems language;
(b) accelerating growth.

Because designs with greater performance prevail, self-organization selects network connections that feed back transformed energy to increase inflow of resources or use them more efficiently.

One example is the *autocatalytic production* process, in which a stored quantity feeds back to interact as a multiplier with the input (Figure 2.13). For example, the more leaves a plant produces (stored as structure), the more sunlight is caught (by means of feedback interaction) to make more leaves. This is an example of a configuration that *reinforces* performance and sustainability.

If the availability of the input energy is constant and the production is greater than the depreciation of the storage, the production and the storage grow faster and faster (Figure 2.13b).

Another reinforcing design is the feedback of materials recycling for reuse (Figure 2.3). Recognizing and diagramming maximum power designs is an important part of modeling.

Chapter Three

◆

NUMBERS
ON NETWORKS

Providing numbers to the flows and storages makes models quantitative. Assigning numerical values to network diagrams represents concepts, summarizes data, demonstrates turnover rates, aids comparisons, and provides the values for calibrating simulation equations. Chapter 3 explains different ways to use numbers on energy systems diagrams consistent with principles for energy and matter processing.

A systems diagram can be used to show the flows and storages of materials, energy, money, population, or almost anything else. The numbers can be those based on your ideas or those observed in research. With numbers on the diagram you can see at a glance where storages and flows are large or small and where units are turning over rapidly or slowly. Numerical systems diagrams are a good way to summarize systems quantitatively.

3.1 DIAGRAMS WITH NUMBERS OF ONE KIND

To represent the flows and storages of a material substance, numbers on a diagram are of the same kind. For example, a diagram can be drawn in which all the numbers are flows and storages of phosphorus in grams per square meter (Figure 3.1). Biogeochemical studies often use quantitative systems diagrams to show the average flows of one type of chemical substance. Other examples are flows of people, money, and information.

To keep all the causal interactions in view (as explained in Chapter 2), all the pathways of the systems should be retained on a diagram even when it is being used to represent the numerical values for one kind of material. Values are only on the pathways and in storages containing the one kind of material being considered. On an evaluated materials diagram, there are no numbers on the heat sink pathways at the bottom since these are only for the flow of degraded energy.

40

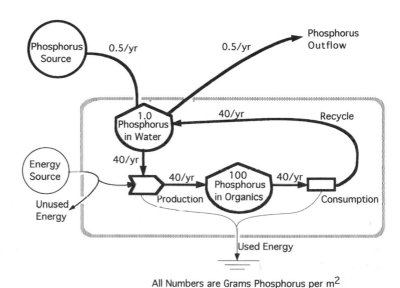

FIGURE 3.1.
Systems diagram with numbers
of one kind, the flows of
phosphorus.

All Numbers are Grams Phosphorus per m^2

ADDING AND SUBTRACTING FLOWS

To add or subtract, numerical values have to be of the same kind (i.e., energy, phosphorus, water, population, etc.). As explained with Figure 2.2, flows of the energy systems language add when pathways join (Figure 3.2a). Flows subtract in a diverging branch as shown in Figure 3.2b, where a flow of 100 per hour is shown split into branching flows of 80 per hour and 20 per hour.

When two flows of the same kind converge into the same storage tank, their flows are additive, contributing to what is already stored there (Figure 3.2c). Outflows from a storage subtract from the quantity stored there (Figure 3.2d).

STEADY STATE

In Figure 3.3a the flow per unit time into the storage (100 per hour) is greater than that in the outflow (30 per hour). There is a net increase in the storage per unit of time. In other words, the stored quantity grows. In contrast, in Figure 3.3b the numerical value of the inflows (100 per hour) is equal to that of the outflows. Therefore, the quantity in storage does not increase

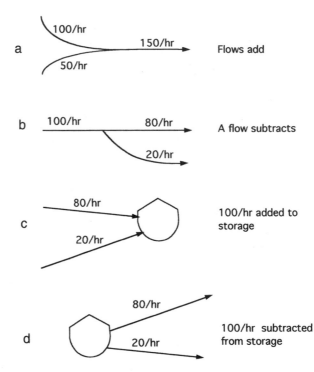

FIGURE 3.2.
Examples of numbers
representing one kind of flow
and the configurations in which
numbers add or subtract.
(a) Joining pathways where the
numbers add; (b) split which
subtracts flow; (c) storage where
converging pathways add;
(d) storage where outflows
subtract from storage.

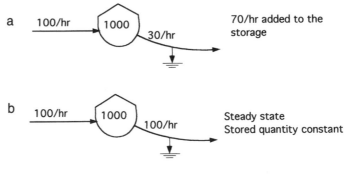

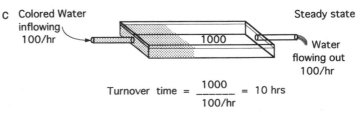

FIGURE 3.3.
Placing numbers on pathways
and within the storage symbol
for the single storage model.
(a) Inflows do not equal
outflows; (b) steady state with
inflows equal to outflows;
(c) inflow of colored water
displacing steady-state storage
in one turnover time.

or decrease even though there are inflows and outflows. This condition is called a *steady state* (also sometimes called a *dynamic equilibrium*).

TURNOVER TIME (REPLACEMENT TIME)

Turnover time is the time required for a flow (quantity flowing per unit time) to replace the stored quantity. For example, visualize a tray of water (storage) in a steady state with inflow equal to outflow (Figure 3.3c). Start with clear water in the tray. Then as new, colored water flows in from the left, the clear water gradually moves out to the right. The time required for the colored water to replace the clear water is the *turnover time*. Therefore, turnover time equals the stored quantity divided by a flow rate:

Turnover time = (stored quantity)/(flow per time).

In the example (Figure 3.3c) the turnover time is equal to:

(1000 liters)/(100 liters per hour) = 10 hours.

As suggested with the scale diagram (Figure 1.6), small units have fast turnover times, and large units have slow turnover times. Estimates of turnover times help you see which parts of a numerical systems diagram system are small, operating rapidly, and which parts are large and slow.

For example, examining the numbers in Figure 3.1 shows the recycle of water replacing the phosphorus in water 40 times in a year, whereas the flow of phosphorus (40/yr) requires 2.5 years to replace the storage of phosphorus in organics (100). Number of turnovers per time is the reciprocal of the time to turnover (replacement time). For the phosphorus flow through the water storage with 40 turnovers per year, the replacement time is 1/40 = 0.025 years.

Because one storage may have several flows in and out at different rates (Figure 3.1), there are as many turnover times for a storage as there are pathways in or out. You can calculate an overall turnover time by dividing the total storage by the sum of flow rates.

Turnover times are sometimes useful for calibrating models where data are not available because people have commonsense knowledge of how long it takes to replace something (example: dog, house, grocery stock).

REPRESENTING CONSERVATION OF MATTER

Because matter is neither created nor destroyed in ordinary processes, a diagram may be used to represent all the matter inflowing, being stored, and outflowing. We call the diagram with numbers a *budget diagram*. For example, Figure 3.1 is a budget diagram for phosphorus. According to the principle of conservation of matter, the numbers inflowing have to equal those going into storage plus those outflowing. For chemical elements like phosphorus we may easily keep track of the element whether it is free in the environment or combined in molecules. Phosphorus in Figure 3.1 is free on the left and bound as organic molecules in central storage.

Representing more complex matter (molecules and more complex substances) is not so easy because the matter under study may be chemically changed, losing identity in chemical recombinations, becoming part of something with a different name. For example, numbers for the flows of water into and out of an ecosystem include rain, runoff, transpiration, and percolation to groundwater, but some water is decomposed in photosynthesis, changing the total water quantity slightly.

3.2 DIAGRAMS WITH ENERGY VALUES

Because all pathways are accompanied by some energy flow, and all storages have some stored energy, numerical values of energy flows and storages can be placed on all the pathways of an energy systems diagram (Figure 3.4).

Writing steady-state values on an energy systems diagram makes a useful energy budget diagram. According to the principle of conservation of energy (first energy law), energy is neither created nor destroyed. Energy inflows equal that stored within the system plus that outflowing. If the numbers are intended to

Outside Source of Inflows

Storage

Heat Sink, Pathway of Used Energy

Interaction, Production with Two Inputs

Exchange

Constant Gain Amplifier

Loop-Limited Converter

System or Sub-system Frame

Producer

Consumer

Miscellaneous Box

Switching Actions, On-Off Processes

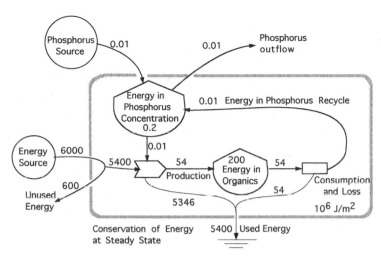

FIGURE 3.4.
Energy systems diagram of an ecosystem in steady state with numbers representing energy flows and storages and the conservation of energy.

represent the principle of conservation of energy, then numbers on pathways to represent energy inflowing must equal those going into storage and those outflowing. At steady state, values in tanks are constant, and flows in and out of each storage tank must balance. Figure 3.4 is an example of numbers for energy flows in a steady state. Even though systems vary daily, seasonally, and from year to year, long-term averages of flows may balance and be used to make a useful budget diagram.

Where a storage tank represents pure energy storage (example: heat, kinetic energy, sound) without accompanying matter, depreciation is represented by a pathway direct from tank to heat sink. If the storage also contains matter, then the depreciation pathway has two branches, one to the heat sink and one for the materials going elsewhere. In Figure 1.1 the depreciation loss is combined with the energy dispersal of the outflow pathway.

◆

3.3 DIAGRAMS WITH NUMBERS OF MORE THAN ONE KIND

Numbers representing more than one kind of unit are often placed on the same diagram (Figure 3.5). For calibration pur-

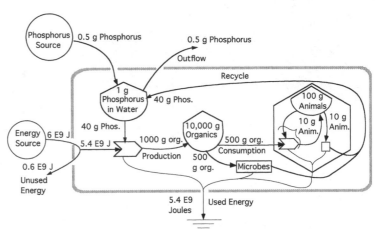

Numbers on Pathways are Flows per Square Meter per Year

FIGURE 3.5.
Energy systems diagram with several kinds of numbers as is appropriate for flows in or out of each storage.

poses, the units (mass, volume, numbers, energy, etc.) used for the cluster of flows around one storage tank can be different from those used for another. This is the kind of numerical diagram used to calibrate equations for simulation in Chapter 7.

However, those flows that join or branch have to be of the same kind (as already explained), and those flows that enter or leave a storage have to be of the same kind as the quantity stored. In Figure 3.5 the numbers on pathways in and out of the storage tank on the left are flow rates for phosphorus. The numbers in and out of the organics storage are rates of flow of organic matter. The numbers in and out of the animal storage are flow rates for animal production and losses. The diagram is composed of several clusters of numbers, each of a different kind. To avoid confusion on systems diagrams with numbers, write the units (joules, grams, liters, individuals, etc.) as in Figure 3.5. For the pure energy flows (sunlight and heat sink flows in Figure 3.5), an energy unit has to be used.

ZONES SEPARATED BY INTERACTIONS

The clusters of pathways and storage, each of a different type, join through interaction symbols. As explained with Figure 2.6, production processes usually use two or more inputs of different kinds, generating a product of a third kind, and heat sink flow of a fourth type. When different kinds of units are used in the same diagram as in Figure 3.5, the interaction symbols separate the clusters of numbers of different kind. In Figure 3.5 phosphorus units are in the upper left, organics in the middle and consumers to the right. Energy values are shown on the pure energy pathways (sunlight and heat sink outflows).

◆

3.4 NUMBERS FOR FLOWS BETWEEN TWO STORAGES

Normally, a flow between two storages has some kind of energy transformation, and thus the pathway should show the transformation process on the pathway. For example, a small box is used in Figure 3.1. Several kinds of tank-to-tank pathways are included in Figure 3.6 with examples of pathway numbers.

If numbers are being assigned for one kind of material (Section 3.1), then the value coming out of the first tank and that going into the second tank can be the same (Figure 3.6a). If numbers are being assigned for energy only (Section 3.2), then the number outflow from the first tank branches with energy going into the next tank and into the heat sink (Figure 3.4). If numbers are being assigned separately for each tank (preparing for calibration, Section 3.3), then the numbers for tank outflow can be different from those for inflow to the next tank (Figure 3.6b).

A simple line with a heat sink is appropriate for a frictional or diffusion type of process (Figure 3.6a). For example, with water flowing in a pipe, the energy from its driving pressure is dispersed by friction into dispersed heat.

In Figure 3.6b the simple box represents the transformation of organic matter being decomposed, yielding output of inorganic nutrients to the next tank and used energy to the heat sink. The simple box (without interaction of other inputs) is appropriate for a simple model where other factors are constant.

The interaction in Figure 3.6c represents the decomposition pathway, or organic matter interacting with oxygen to produce nutrients and dispersed heat.

A common error is drawing a pathway from one storage tank to another without indicating the usual transformation process (pathway box or interaction) and the heat sink. There is always

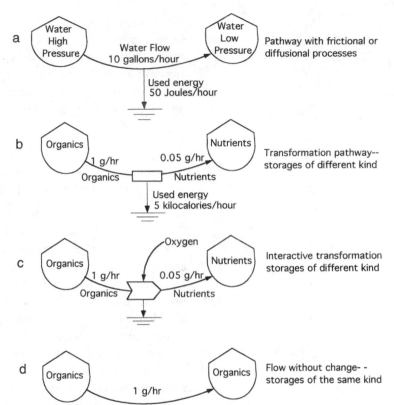

FIGURE 3.6.
Pathways between two storages. (a) A flow in which available energy is degraded as with friction or diffusion; (b) a flow with an energy transformation along the way; (c) a flow with an interaction transformation along the way; (d) a transfer of something unchanged from one storage to another (where energy dispersal is negligible).

some processing such as movement in space. For purposes of simplification it is correct to use the simple pathway (Figure 3.6d) if you are making the assumption that the magnitude of the transformation is negligible relative to other values in the model. That assumption requires the flow to pass the content of one storage tank to the other without change and the two storages to contain something of the same kind. For example, very little energy is involved when dead leaves stored on a tree fall to the ground to become dead leaves stored as litter.

PART TWO

◆

SIMULATION

Part II introduces ways of simulating systems models after they are diagrammed and made quantitative. Simulation shows what happens as a result of network designs and numerical values. Simulation plots graphs that show system behavior.

Chapters 4 and 5 simulate systems models with preprogrammed symbol blocks for the program EXTEND. Connecting the icons on the computer screen automatically arranges mathematical operations. Chapter 4 uses pictorial blocks. Chapter 5 uses general systems blocks. Appendix A explains how to program additional blocks.

We explain how to write equations from energy systems diagrams in Chapter 6 and calculate the coefficients (calibration) in Chapter 7. Calibrated equations are used to simulate models with several methods in Chapters 8, 9, and 10. Simulation is with spreadsheets in Chapter 8, BASIC programming in Chapter 9, and STELLA in Chapter 10. Chapter 11 introduces emergy (spelled with an "m") and its simulation.

Chapter Four

SIMULATION WITH PICTURE BLOCKS FOR EXTEND

S imulation makes systems models come alive. With the computer program EXTEND,* models are simulated by arranging icon blocks on the screen and connecting them with a mouse. The program does the rest, plotting a graph of properties with time. This chapter uses *pictorial icons* in which all the equations and most of the numbers are hidden in the preprogrammed blocks. Simulation is introduced by running prearranged minimodels. These models, the library of symbol blocks, and a version of EXTEND are included on the disk provided with this book.

◆

4.1   EXTEND

EXTEND is a user-friendly computer program for simulating systems models for PC and Macintosh computers. It uses preprogrammed blocks that are connected by the user to form systems models on the computer screen (examples: Figure 1.3a and Figure 4.1b). Each block is represented on screen by an icon (small picture). Groups of related blocks are stored together in computer files called *libraries*. After EXTEND is loaded on your computer, you can use its menu to load a library of icon blocks. Then you can pull down blocks from the library to the screen and use a mouse to connect the blocks to form a model. You connect each block you want to plot to the plotter icon. The model can be given a name and saved on a disk. Double clicking the mouse on each icon causes its *dialog box* to appear, a small box for typing in appropriate numbers. Then you use the simulation menus to run the simulation. A window appears on the screen plotting graphs. Graphs can be printed or the whole screen saved to the computer "clipboard" for printing later.

The usual way to use EXTEND is to represent equations by

* EXTEND 4 is available from Imagine That, 6830 Via del Oro, Suite 230, San Jose, CA, 95119-1353. Telephone: 408.365.0305. Email: extend@ imaginethatinc.com·Web Site:http://www.imaginethatinc.com

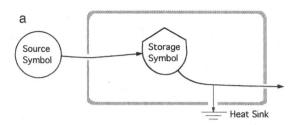

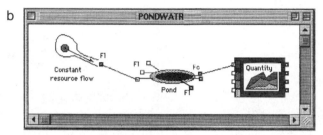

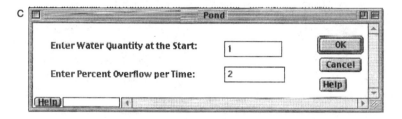

FIGURE 4.1.
Model of simple storage
arranged for EXTEND.
(a) Energy systems diagram;
(b) screen of EXTEND with
picture version of the model
(PONDWATR); (c) dialog box
for the storage symbol;
(d) simulation graph.

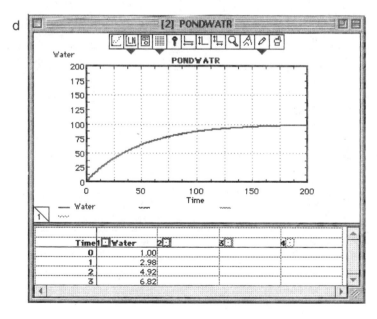

connecting blocks from the icon block libraries supplied by Imagine That, the EXTEND company in San Jose, California (see Chapter 6). However, the pictorial icon blocks used in this chapter and the next were specially programmed by the authors to introduce simulation without the user being concerned with equations.

◆

4.2 SIMULATING PREASSEMBLED MINIMODELS

To begin, let's use the book disk to load EXTEND, our library of pictorial blocks called PICTLIB, and several prearranged minimodels (in the IntrModl folder). On PC computers, library files have the extension "lix" (example: PICTLIB.lix); prearranged model files have the extension "mox" (example: TANK.mox). Follow the procedure in Table 4.1, which lists the steps for loading the program, icon block library, and minimodels; setting numerical values; and running the simulation.

Consider first the fundamental process of storage, its energy

TABLE 4.1
Procedures for Running EXTEND Models

1. Place the book disk in your CD drive. *Open the EXTEND program* by clicking on the EXTEND icon with the mouse.
2. Use the LIBRARY menu to *open the library* that has the blocks for the models you plan to simulate. If the computer asks you to find the appropriate library, locate the file in the folder for EXTEND libraries.
3. Use the FILE menu to *select and open a prearranged model* (or a new screen if you are arranging a new model—see Table 4.2). Double click on the model you want to run. A picture screen will appear.
   If you have not loaded the library of icon blocks, opening the model usually causes the appropriate library to open (step 2).
4. To change *numerical values,* double click on the icons. Dialog boxes appear in which you can type new numbers.
5. To *run the simulation,* use the RUN menu. One or more screens appear on which those quantities are plotted that were connected to the plotter on screen.
6. To set the *time of the run,* use the setup screen in the RUN menu.
7. To *read quantities from a graph,* move the mouse line across the graph to the point you want to read. The time and various quantities appear across the top of the table below the graph.
8. To *add new icons* to the model, see Table 4.2.
9. To *print the model or plotter screens,* use the Print option in the FILE menu.

TABLE 4.2
Procedures for Arranging EXTEND Models

1. Place the book disk in your CD drive. *Open the EXTEND program* by clicking on the EXTEND icon with the mouse.
2. Use the LIBRARY menu to *open the library* that has the blocks for the model you plan to assemble. If the computer asks you to find the appropriate library, locate the file in the folder for EXTEND libraries.
3. Use the FILE menu to *open a new model.* A blank window will appear.
4. Pull down the LIBRARY menu, *select the library,* and by dragging the mouse to the right of its name, select Open Library Window. A long narrow strip appears on the left showing the icons and names of the blocks in that library.
5. Move the blank screen to the right of the icon strip. Use the mouse to click and *drag icon blocks from the strip to the blank screen,* starting with a plotter.
6. You can also access blocks directly from the LIBRARY menu. Use the LIBRARY menu to select the needed library. While still holding the mouse down, move the arrow to the right and down the list to the icon you want and let it go. The icon block will appear on your screen.
7. To arrange blocks for your model, move the mouse cursor over a block until the arrow becomes a hand. Then click and hold the cursor on the block to move it around the screen.
8. To send data from one block to another, put the cursor over the output connector of one block. When it becomes a pen, hold it down to draw a line connecting the flow from that block to another. When the line is correctly connected, the line becomes thick. A bad connection remains dashed.
9. Connect flow (Fl) outputs only to flow (Fl) inputs, and force (Fc) outputs only to force (Fc) inputs.
10. To *erase a line,* move the point of the mouse arrow to the line and click. When the line becomes thick, press the delete key on the keyboard.
11. To *erase a block,* move the mouse over it until the mouse cursor becomes a hand. Click once to highlight it. Then press the delete key on the keyboard.
12. *Connect the blocks* to match your energy systems diagram. Place the plotter on the right, and connect outputs of the blocks to the plotter input.
13. Double click on each icon to bring up its dialog box. *Type numbers in the dialog box of each block.*
14. To set the length of time for the simulation, pull down the Simulation Setup option from the RUN menu. This brings up the simulation time dialog page where you can type in the time span.
15. Also use the simulation setup in the RUN menu to enter the time interval (dt) to be covered in each computer iteration. Use 1 unless the plots show scrambled and blended lines indicating a chaotic artifact. In that case try smaller values (0.1, 0.05, 0.01, etc.).
16. When you have a plot on the screen, set the vertical scales. Use the mouse to click at the top of the scale, which opens a small box in which you type the upper value. Click at the bottom of the scale to get another box where you can enter a zero. You can type one scale on the left and a different scale on the right end of the plot.
17. To *change the scale* for a plot from the left y axis to the right y axis, click on the ▢ tool in the plot window. Click the ▢ tool and it will shift its assignment to the opposite side. ▢ indicates that the quantity will be plotted on the right axis.
18. To *add titles* in the data table shown under a graph, click on the ▢ tool in the plot window, and type labels in the boxes that appear.
19. To *save your model,* use Save As in the FILE menu, which provides a box for you to type in a file name and indicate folder location.

systems diagram in Figure 4.1a, and a pondwater example containing an inflow, a storage, and a quantity-dependent outflow. Look at the model and run the simulation. Then run "What If" experiments. Try to predict what will happen before you run the program.

PONDWATR

Using EXTEND, click FILE menu to open the model POND-WATR (in the IntrModl folder), a process that also loads the library PICTLIB. Figure 4.1b shows the computer screen with three icons (Spring, Pond, and Plotter). The icons on screen show water flowing from a spring into a pond. Tiny squares are connectors: outputs black and inputs clear. Built within the pond icon is a hidden outflow process. This pictorial model has the storage and pathways of the energy systems diagram above it (Figure 4.1a). Double click on the spring and pond icons to bring up dialog boxes to see how conditions are already set for simulation. The box for the spring has space to type the flow rate. Figure 4.1c is the dialog box for the pond icon. Simulate the model by using the Run instruction on the RUN menu. The resulting plot grows and levels (Figure 4.1d).

"What If" Experimental Problems for the EXTEND Program PONDWATR

4.2–1 What happens with a run if the flow rate from the water source is reduced to a small value?

4.2–2 What happens with a run if the outflow rate is set to zero? (Double click on dialog box of the pond.)

4.2–3 What happens if the pond is started with a high storage, 200?

PEXPO

Figure 2.13 introduced a system of autocatalytic growth on an unlimited energy source (design with feedback to multiply by the input force). This system has accelerating growth. Load and run PEXPO, a pictorial version of this exponential growth model. When loaded to the screen (Figure 4.2a), the model has an icon

Outside Source
of Inflows

Storage

Heat Sink,
Pathway of
Used Energy

Interaction,
Production
with Two
Inputs

Exchange

Constant
Gain
Amplifier

Loop-Limited
Converter

System or
Sub–system
Frame

Producer

Consumer

Miscellaneous
Box

Switching
Actions,
On–Off
Processes

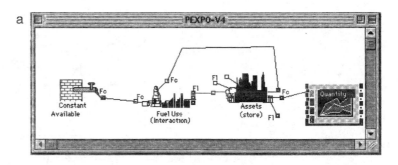

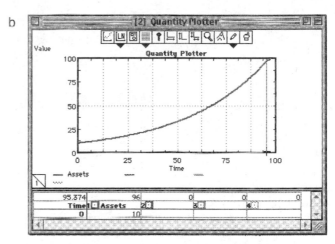

FIGURE 4.2.
Model of exponential growth
of a city using pictorial icons.
(a) Systems diagram using
pictorial icon blocks; (b) graph
of simulation.

for a city (Assets) feeding its services back to an icon for production (Fuel Use) and use of fossil fuels. The more assets in the city, the more fuel is pumped from the icon Constant Available. The graph resulting from simulation (Figure 4.2b) has steeply accelerating growth.

"What If" Experimental Problems for the EXTEND Program PEXPO

4.2–3 Triple the starting value of assets in the dialog box of the city assets block. How does the curve compare with the one with the smaller starting value?

4.2–4 What happens if the concentration of available energy in the dialog box of the source is reduced to 10,000?

EARTHPIC

A model of the earth metabolism for EXTEND is provided in energy systems language (Figure 4.3a) and with pictorial icons

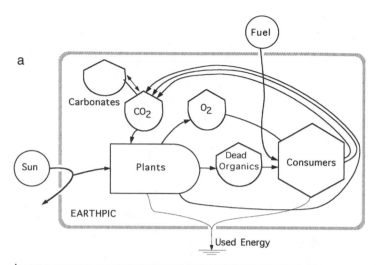

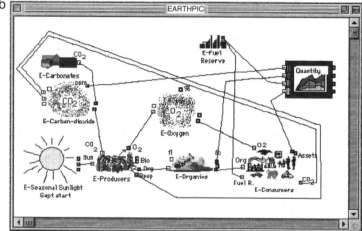

FIGURE 4.3.
EARTHPIC model of the earth's biosphere. (a) Energy systems diagram; (b) model on screen of EXTEND showing arrangement of picture icon blocks; (c) 10-year simulation with carbon dioxide in oscillating steady state; (d) 100-year simulation with rise and decline of fossil fuel use.

(Figure 4.3b). The producers, plants of the earth use sunlight and carbon dioxide to make plant biomass (organic matter) and oxygen. The consumers of the earth including animals, microbes, and cities use the organics and oxygen to operate their growth while releasing carbon dioxide as a by-product. The organic matter on the surface of the earth is increased by a flow of fossil fuels pumped from underground. Some of the carbon dioxide flows from the air into the ocean sink, reacting with carbonates there.

Open and double click on the model EARTHPIC (in the Earthsys folder). The program loads the library ESLIB, and the system diagram with pictorial icons appears (Figure 4.3b). The fossil fuel reserve is not connected. The model uses a time unit of days.

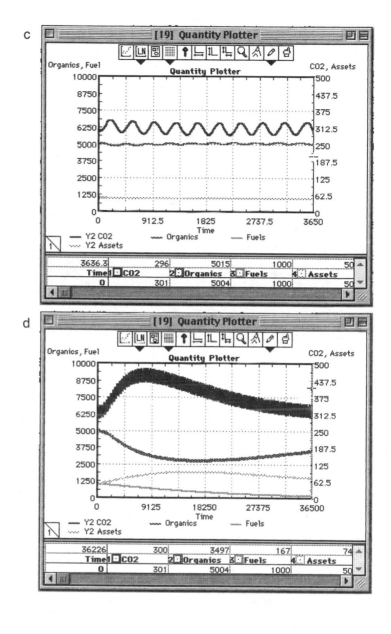

FIGURE 4.3.
(*Continued*)

The simulation is set to run for 10 years (3650 days). Look at the simulation graph (Figure 4.3c). Because the sunlight is supplied with a sinusoidal curve representing the seasonal changes, the organic matter and the carbon dioxide go up and down each year also. However, as examined over a period of years, the storages remain fairly constant, indicating a balance between the rates of production of carbon dioxide by consumers and the

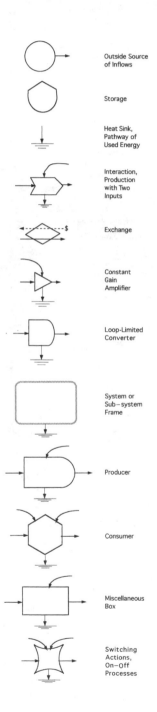

Outside Source
of Inflows

Storage

Heat Sink,
Pathway of
Used Energy

Interaction,
Production
with Two
Inputs

Exchange

Constant
Gain
Amplifier

Loop-Limited
Converter

System or
Sub−system
Frame

Producer

Consumer

Miscellaneous
Box

Switching
Actions,
On−Off
Processes

removal of carbon dioxide by the plants and the exchange with the carbonates.

Next, use the mouse to connect a pathway from the Fuel Reserve source to the box labeled Fuel R on the consumer icon (Figure 4.3b). The dialog box of the fuel reserve should be set to 1000. Run the simulation and examine the plot. Now the carbon dioxide concentration is increasing, the consumer assets are increasing, and environmental resources (organics) are decreasing.

Change the simulation time to 100 years (36,500 days) and run again. The result (Figure 4.3d) shows the carbon dioxide reaching high values, increasing the greenhouse effect on climate. By then the fossil fuel reserve has been reduced, decreasing the consumption that is generating carbon dioxide. Consumer assets begin to decline again and environmental storages increase.

"What If" Experimental Problems for the EXTEND Model EARTHPIC

4.2–5 What happens if the flow into the ocean carbonate sink is disconnected while the fossil fuel source is being used? Click on the pathway with the mouse and press the Delete key.

4.2–6 What happens if the fossil fuel reserve is doubled?

4.2–7 Without the fossil fuel connection, what happens if something reduces the sunlight intensity by half?

4.3 CREATE AND SIMULATE A PICTURE MODEL

You can make your own models by combining icons on a blank screen. Figure 4.4 shows the icons available in the pictorial library (PICTLIB in the IntrModl folder) assembled on screen without connections. If the library has been loaded, you can bring these icons to screen view on a strip by pulling down the LIBRARY menu to PICTLIB and Open Library Window. Choose Open New Model and drag icons from the library window to make a new model. Use the steps in Table 4.2. Make your first one simple.

Include a "quantity plotter" to which you connect outputs

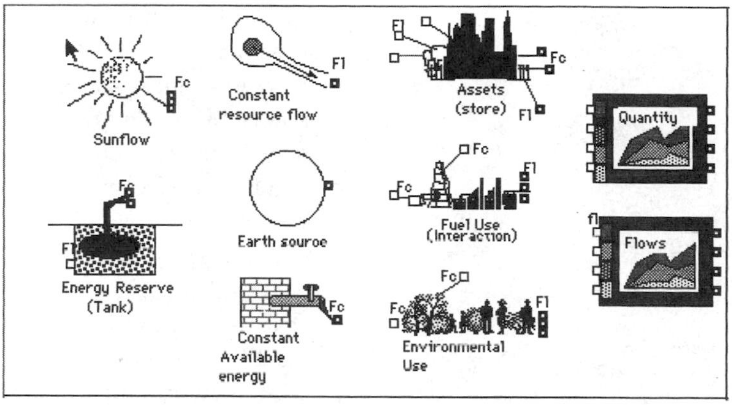

FIGURE 4.4.
Computer screen with the picture symbols of the EXTEND library PICTLIB.

marked Fc (force) in order for the program to plot a graph of the stored quantity. As explained in Chapter 2, some system parts deliver "forces" to pathways, and some deliver "flows." Output connectors marked with Fl deliver flows. Output connectors marked with Fc deliver forces. Input connectors marked Fl require flow inputs. Input connectors marked Fc require force inputs. Correct use requires that the mouse connect from the black output of one block to the clear input of another block.

One of the icon blocks in the PICTLIB library is the *flow plotter*. You can use it to plot flows by connecting the black output connectors marked Fl (flow) to one of the plotter's input connectors. For example, after opening the PONDWATR model, use the LIBRARY menu and select the Flow Plotter option. Then if you click the mouse on the model screen, the flow plotter appears. Use the mouse to drag it into a position on the right. The outflow from the lake in the PONDWATR model will be plotted if you connect the output of the lake icon to one of the input connectors of the flow plotter and rerun.

◆

4.4 BIOQUEST MODELS

Using EXTEND, a set of exercises introducing system simulation was prepared for a software package named BioQUEST, at Beloit College, Wisconsin. These exercises are available from Academic Press and are included in the EXTEND folder of this book's disk, each with an instruction manual in WORD. After loading the program EXTEND, load a model and the appropriate library. All folders and files are also listed in Appendix E.

Folder EDM (Environmental Decision Making) contains pre-arranged models including BQFISH, a model of a fishing food chain; BQGRASS, a model of grassland and fire; BQLOG, a model of forest logging and sale; and FISHSALE, a model of fishing and sales. The first three models use pictorial icons and also general systems blocks. Figure 4.5 shows the pictorial icons used in these models (file BIOICON in the IntrModl folder). These EDM models use the library BIOQULIB.lix. From the EDM folder, call up the Instruction Manual (EDManual, in WORD), which leads you through the simulation exercises.

In the Biosphr2 folder is a model SIMBIO2, which simulates the atmosphere in Biosphere 2, the giant experimental glass chamber in Arizona. The model in picture icons is SIMBIO2P, and the model in general systems blocks is SIMBIO2G. These use the library SB2LIB. To use these models, load the Instruction Manual (SBManual, in WORD) and follow its directions.

◆

4.5   SPECIALIZED ICON BLOCKS FOR EXTEND

The introductory picture blocks in this chapter (PICTLIB library in the IntrModl folder) are simple units prepared for intro-

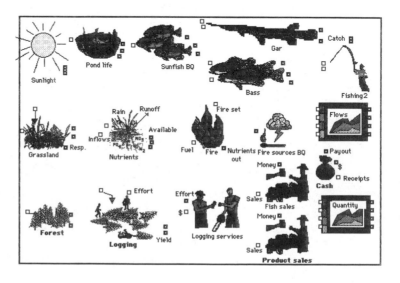

FIGURE 4.5.
Computer screen with the picture symbols used in the BioQUEST models for EXTEND.

ductory teaching. Most of the data and calibrations that make the blocks quantitative are hidden within the programs. Clicking on the icons only causes simple dialog boxes to appear with blanks for only one or two values to be typed. Not having many numbers to consider makes the first experience in simulation an easy one.

However, specialized picture blocks can be made of a particular species, chemical relationship, city, etc., with many details, and quantitatively calibrated so that the block can represent what that unit does in the real world. For example, developing species blocks for simulation of ecological systems is a research opportunity for the future. Appendix A has instructions for making new blocks or changing the program script within blocks.

Chapter Five

◆

SIMULATION WITH GENERAL SYSTEMS BLOCKS FOR EXTEND

$\mathbf{W}$hereas *pictorial icon blocks* were used with EXTEND to simulate models in Chapter 4, Chapter 5 introduces *general icon blocks* for simulating any system. Preprogrammed blocks were made for each of the symbols of the energy systems language (GNSYSLIB library). Simulation is arranged by connecting icon blocks on the screen to match an energy systems diagram. Programs are made quantitative by copying numerical values from the systems diagrams into dialog boxes. Like the networks of picture icons, networks of general symbols can generate graphs without involving the user in the equations. Simulation modeling with general systems symbols helps compare similar processes on many scales.

5.1 ENERGY SYSTEMS SYMBOLS PREPROGRAMMED AS ICON BLOCKS FOR EXTEND

With procedures detailed in Appendix A, icons were drawn and program scripts written for the symbols of the energy systems language and saved for use in simulation. Whereas the quantitative properties of picture symbols are built in and hidden in the object-oriented programs, the general symbols have dialog boxes for the user to type in numerical values. The program uses these values of flow and storage to calibrate the coefficients that make the model quantitative.

GENSYS ICON BLOCKS

The icon blocks made for the energy language symbols are shown together on a computer screen (without connections) in

66

Figure 5.1. These blocks are in the GNSYSLIB file of the disk supplied with this book. After opening EXTEND, use the LIBRARY menu to load the GNSYSLIB library (Gensymb folder). Then appropriate blocks can be assembled on the screen, or a model with prearranged symbols can be opened and simulated.

In Figure 5.1 sources are on the left, and storage units and production modules of several types are next. In the middle are interactions and switch junctions. To the right are several types of consumer units and units for economic exchange.

EXTEND PLOTTERS FOR ENERGY SYSTEMS MODELS

The right side of Figure 5.1 shows several kinds of plotter blocks. The *quantity plotter* records stored quantities and the

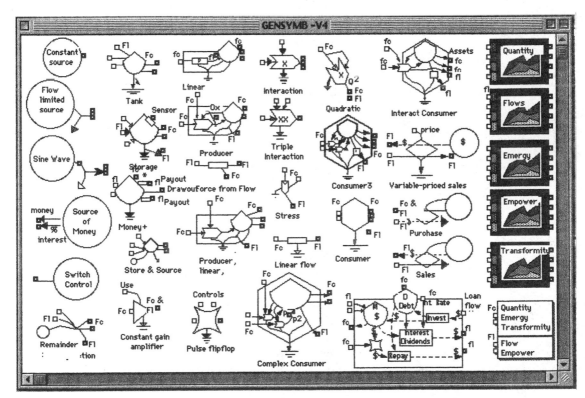

FIGURE 5.1.
Screen with energy language blocks for EXTEND from library GNSYSLIB.

forces delivered by sources. The *flow plotter* plots the outputs of interactions and production processes. Three plotters are available to graph *transformity, emergy,* and *empower,* concepts explained in Chapter 11. The EMERGY plotter multiplies the force received, representing quantity times the transformity received. The EMPOWER plotter multiplies the transformity received times the flow received. The plotter in the lower right is a compound plotter that arranges for all five plotters to operate simultaneously (five screen windows).

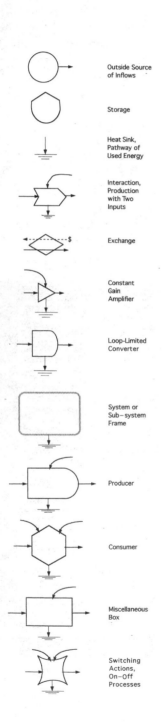

Outside Source
of Inflows

Storage

Heat Sink,
Pathway of
Used Energy

Interaction,
Production
with Two
Inputs

Exchange

Constant
Gain
Amplifier

Loop-Limited
Converter

System or
Sub–system
Frame

Producer

Consumer

Miscellaneous
Box

Switching
Actions,
On–Off
Processes

5.2 EXTEND MODELS USING GENERAL SYSTEMS BLOCKS

Several prearranged EXTEND models that correspond to simple energy systems diagrams are supplied on the book disk. Instructions for loading and running already-assembled models are given in Table 4.1.

TANK MODEL OF STORAGE USING EXTEND

In the first chapter the simple storage model TANK was shown as an energy systems diagram (Figure 1.1b) and with the equivalent EXTEND model, using general icon blocks in Figure 1.3a. The simulation result in Figure 1.3b shows the growth when storage starts with a low value. Its program is similar to the pictorial PONDWATR model.

After opening the EXTEND model, use the FILE menu to open the prearranged TANK model from the EXTEND folder (in the IntrModl folder) of the book disk. Run it.

*"What If" Experimental Problems for the
EXTEND Model TANK*

5.2–1 What happens if there is a large stored quantity at the start of the simulation and the source flow is 2? Double click on the tank symbol to get the dialog box where you can type in a large quantity (e.g., 100). Then use the menu to run again.

5.2–2 With the starting storage set at 100, what happens if the inflow is set to 0 (or the pathway is disconnected)?

5.2–3 With the starting value 1 and source flow 2, what happens if the rate of outflow is increased? Change the dialog box so that the rate of outflow is 10%. ("When the storage is set to 100, the outflow is 10.")

ACCELERATING GROWTH MODEL: EXPO

When there is a constant (unlimited) energy source driving an autocatalytic process (Figure 2.13), growth accelerates. In Chapter 9 this pattern will be shown to be "exponential," hence the name "expo." A pictorial version of this model (PEXPO in the IntrModl folder) was simulated in Chapter 4 (Figure 4.2a).

Using EXTEND, load the prearranged model EXPO (in the Gensymb folder) and note that the system design is like that of the energy systems diagram (Figure 2.13). Run its simulation. An accelerating growth curve similar to that in Figure 4.2b is observed.

"What If" Experimental Problem for the EXTEND Model EXPO

5.2–4 What changes can you make in the numbers in the dialog boxes so that the storage drains away toward zero?

PC&CYCLE: A MODEL OF PRODUCTION, CONSUMPTION, AND RECYCLE

An ecological minimodel with some of the main processes of ecosystems is PC&CYCLE, shown as an energy systems diagram in Figure 5.2a. Nutrient materials from outside flow into a storage and some flow out in proportion to it. Interaction between an energy flow and necessary nutrient material generates production of organic matter that goes into a storage. Organics go to a consumer represented by a box and to a depreciation flow (small box). Both of these pathways release materials that recycle to the nutrient storage tank.

Using EXTEND, open the model PC&CYCLE (in the Gensymb folder) and observe the icon blocks prearranged on the screen (Figure 5.2b) to represent the energy systems diagram

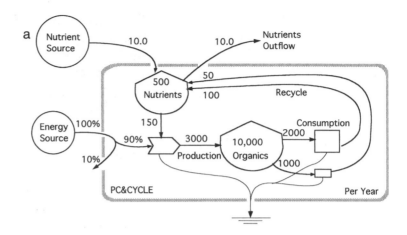

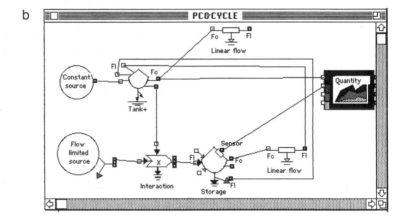

FIGURE 5.2.
Use of EXTEND to calibrate
and simulate program
PC&CYCLE starting with an
energy systems diagram with
numbers. (a) Energy systems
diagram and numerical values
of flows and storages; (b) screen
of EXTEND with model
assembled; (c–f) dialog boxes
for system components with
calibration numbers entered;
(g–i) dialog boxes of sources
with numbers entered for
simulation run.

(Figure 5.2a). Run the simulation and observe the storages reaching a steady state.

◆

5.3   ENTERING NUMBERS IN DIALOG BOXES OF GENSYS BLOCKS (MODEL PC&CYCLE)

So that they can be used for any system or scale, the general systems icon blocks of this chapter (GNSYSLIB library in the Gensymb folder) have no built-in numerical values (no precalibration). Instead, dialog boxes are provided for copying flow values and storage values from an evaluated energy systems diagram like that of Figure 5.2a. The program uses these inputs to

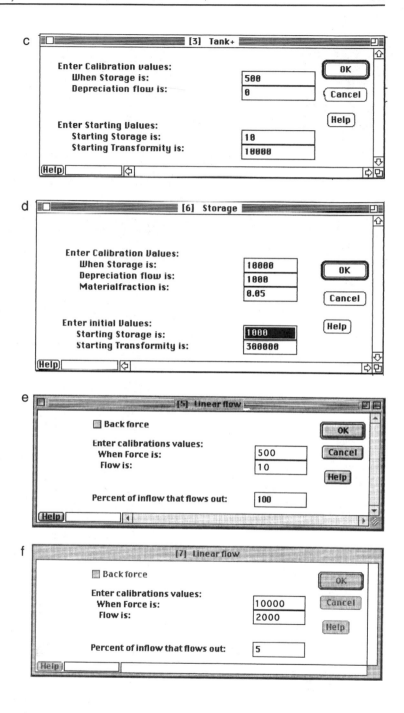

FIGURE 5.2.
(*Continued*)

calculate automatically the coefficients of the hidden equations that are set up when blocks are assembled and joined.

Placing numbers on diagrams was introduced in Chapter 3.

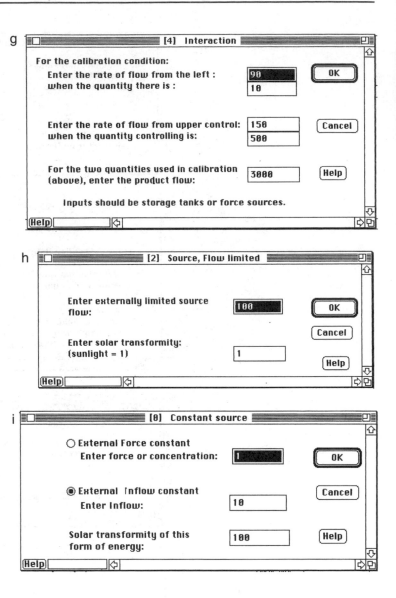

FIGURE 5.2.
(*Continued*)

Care is used at that stage to properly represent units (i.e., grams, dollars, joules, etc.). Flows and values for the same storage have to be in the same units. All flows have to use the same time unit. After that, copying of these numbers on diagrams to the dialog boxes can be made without further consideration of units.

For example, to enter numbers from the PC&CYCLE model in Figure 5.2a to the EXTEND version in Figure 5.2b, copy the numbers to the appropriate dialog boxes for the symbols. These appear when you double click on the icons. To illustrate the process, the values copied from the energy systems diagram are shown in the dialog boxes as follows (Figures 5.2c–h):

In the dialog box in Figure 5.2c, set *depreciation from nutrient storage* to zero, because concentrations are too low to diffuse and disperse further.

In the dialog box in Figure 5.2d, enter *depreciation and material recycle from organic storage* with a flow 1000/time when storage is 10,000.

In the dialog box in Figure 5.2e, enter the *outflow from the nutrient storage* with a flow of 10/time when the storage is 500.

For the *linear flow from organic storage to the consumer block* (dialog box in Figure 5.2f), enter a flow of 2000 and a storage force of 10,000. Consumption releases nutrients. For the pathway from the box enter a flow of 100 by entering 5% (of the inflow of 2000).

For the *interaction* dialog box (Figure 5.2g), enter numbers for inflows and output. Set *flow from the left* to 90 driven by the unused energy inflow 10%; set *flow from above* to 150 when the quantity above is 500; set the *output production* at 3000.

Enter numbers for input sources: 100 for energy source inflow (dialog box in Figure 5.2h) and 10 for nutrient source inflow (dialog box in Figure 5.2i).

◆

5.4   ENTERING NUMBERS FOR THE START OF SIMULATION (PC&CYCLE)

Dialog boxes of the sources and storages have places to enter the starting values, not to be confused with the entries made earlier for the program to calculate coefficients (calibration). Starting values (initial conditions) need not be the same as those used for calibration.

For example, conditions for a run of the PC&CYCLE simulation can be set lower than the calibration values so that we can watch the system grow: Set organic storages in Figure 5.2d to 100 (1% of the value used for calibration). Similarly, set the starting condition for nutrients in Figure 5.2c at 10.

Next, use the Set Up option in the RUN menu to set the time

interval of iteration (DT). Normally this value can be 1. To
determine if the interval is small enough to avoid error, see if
simulations with smaller values are identical. If the flows per
iteration are a large proportion of the storages, the simulation
can go into artificial chaos with values jumping up and down,
often leaving the screen graph with vertical lines and solid areas.
In such a case, a smaller DT is required. With EXTEND, small
DTs may require that more memory be assigned to the program operation.

5.5   GRAPHS FROM SIMULATING PC&CYCLE WITH EXTEND

In the simulation run in Figure 5.3, nutrients (plotted on the
right side scale Y2) reached a steady state in 75 years, and the
organic matter storage (left scale) leveled in 300 years.

"What If" Experimental Problems for the EXTEND Program PC&CYCLE

5.5–1 Simulate the model in Figure 5.2b with a lower nutrient
flow (5.0/time). How long does it take for the system to

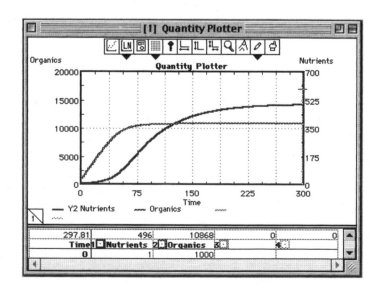

FIGURE 5.3.
Plotter screen of EXTEND
with results of simulation of
PC&CYCLE model in
Figure 5.2.

reach steady state because of the smaller inflow of nutrients? What happens to the organic matter?

5.6   SIMULATING THE EARTH SYSTEM MODEL WITH ENERGY SYSTEM BLOCKS (EARTHGS)

The earth system model in Figure 4.3b of Chapter 4 was simulated with EXTEND using the picture block icons (ESLIB in the Earthsys folder), thus introducing system ideas about the biosphere. However, the observer has to look at the corresponding energy systems diagram (Figure 4.3a) to know what the symbols represent. By using the general systems blocks, it is easier to remember what the blocks do and thus comprehend model networks better. In Figure 5.4b the earth system model was arranged using general systems icon blocks for simulation with EXTEND. Numbers for flows and storages written on the energy systems diagram in Figure 5.4a were used for calibration. EARTHGS (abbreviation for "Earth in General Systems Symbols") is the simulation program for EXTEND on the book disk. If you run the program (Earthsys folder), observe simulation graphs similar to those in Figure 4.3. The CO_2 graph is somewhat different, because carbon dioxide is expressed in grams per square meter of surface area, rather than parts per million as used in Figure 4.3.

5.7   BUILDING NEW MODELS

General systems icon blocks (GNSYSLIB library) can be arranged and connected to represent most energy systems diagrams. (Table 4.2 has instructions for arranging new EXTEND models.) As the first example, assemble the simple TANK model. Using EXTEND, open a blank screen (New Model). Call up a *constant source* icon, setting the buttons on its dialog box to external flow. Call up a *tank icon* and a *quantity plotter*. Connect the

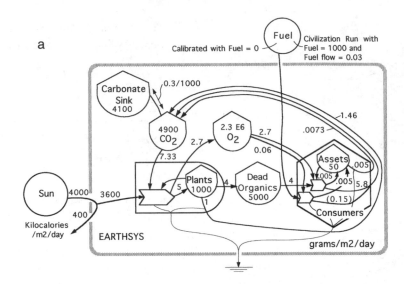

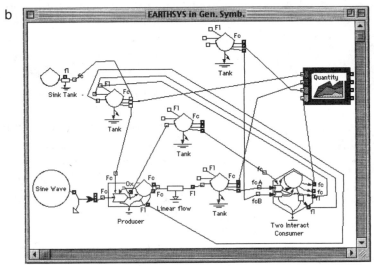

FIGURE 5.4.
General systems blocks
arranged to simulate the earth
system model with EXTEND.
(a) Numbers on the energy
systems diagram; (b) model
(EARTHGS) on computer
screen.

icons, enter values for source flow and initial storage, and run
the program. Compare the model and simulation graph with
Figure 1.3. If the blocks in the library do not have the function
you need, Appendix A has instructions for making new blocks
or changing the program scripts.

Chapter 6

◆

EQUATIONS FROM DIAGRAMS

$\mathbf{B}$ecause of the way the energy language was devised, the act of making diagrams defines a set of mathematical equations for that system. Writing the equations is an automatic translation process that does not require mathematical training. Derived in this way, equations are consistent with the principles of energy, materials, and information for that system. Equations are necessary for communication with those who think and represent systems with mathematics. Simple equations can be manipulated to show useful properties not seen from words or diagrams alone. The equations are also needed for simulation procedures in chapters to follow. You should be able to write equations from diagrams and draw diagrams that translate the equations. With the models of later chapters, equations are included with the systems diagrams. A method of simulation is explained that arranges the parts of an equation on screen using math blocks for EXTEND.

6.1 EQUATIONS FOR A SINGLE STORAGE

From Chapter 1, Figure 1.1, let's start with the simple model of a storage with two pathways (Figure 6.1). As already discussed in Chapter 3, the model says that the rate of change of the quantity in the storage is equal to the rate of inflow minus the rate of outflow. For example, the rate of increase of water in a storage tank is equal to the liters of water per hour inflowing minus the liters per hour outflowing. After putting a storage and its pathway connections in words, we write the mathematical equation with one expression for the "rate of change" and one expression for each pathway carrying inflows or outflows.

DIFFERENTIAL EQUATION

In Figure 6.1 the inflow per time on the left input pathway is represented by J. Drawing a line pathway (Figure 2.2) out of a

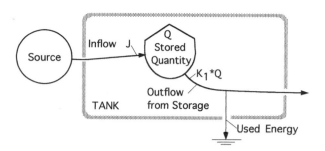

Differential Equation:

Rate of Change of Quantity Q with Time: $dQ/dT = \dot{Q}$

$$\dot{Q} = J - K_1{*}Q$$

Difference Equation:

Quantity Q after Time Interval DT Changed from Q_T at Earlier Time T

$$Q_{T+DT} = Q_T + DQ{*}DT$$

Change Rate: $DQ = J - K_1{*}Q$

In BASIC Programs on Each Iteration:

A Value at a Memory Location Q is Replaced with a New Value:

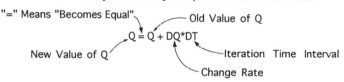

FIGURE 6.1.

Kinds of equations for the simple storage model TANK. See introductory diagram (Figure 1.1) and simulation with BASIC in Figure 9.1 (program TANK.bas in Tables 9.1 through 9.5). Simulation with other methods is shown in Figures 4.1 and 10.3.

storage normally means that the flow is in proportion to the "force" exerted by that storage (such as physical force, chemical concentration, or other property to which the flow is proportional). In other words, the outflow is proportional to the quantity stored (Q in Figure 6.1). When a flow is proportional to a quantity, it means that increasing the quantity increases the flow. A pathway (and its mathematical expression) that is proportional to one other quantity is said to be "linear." For example, water flows through sand in proportion to the quantity of headwater storage (the stored water causes pressure that pushes the water flow).

The amount that the flow increases is indicated by a *constant* (also called a *coefficient*), usually obtained from observed data. It is called a constant because its value does not change as the stored quantities (variables) go up and down. While using one set of data for calibration and another set for experiments, a constant is not changed for the study of a model's behavior over

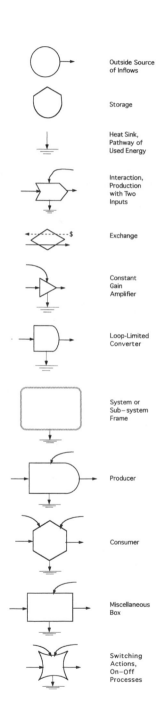

Outside Source of Inflows

Storage

Heat Sink, Pathway of Used Energy

Interaction, Production with Two Inputs

Exchange

Constant Gain Amplifier

Loop-Limited Converter

System or Sub–system Frame

Producer

Consumer

Miscellaneous Box

Switching Actions, On–Off Processes

time. Then, to see the sensitivity of changing in that relationship, you can change it and observe the effect. In this book most of the constants are given the letter "K" (some are L's) and each is given a different subscript. For example, the outflow from Q in Figure 6.1 is the constant K_1 multiplied times Q. This is sometimes written $K_1 \times Q$ where $\times$ means to multiply. Sometimes multiplication is written simply as KQ where multiplication is implied when two letters are closed up side by side.

However, with computers multiplication is indicated by the asterisk (*), so that the flow in Figure 6.1 is written $K_1 * Q$. Separating the items multiplied in this way is useful not only so the computer can understand the equation, but also to avoid confusion when a quantity is represented by more than one letter and/or number.

Verbalizing what is shown in the two-pathway storage model of Figure 6.1, we state:

The rate of change of storage with time is the difference between the inflow J and the outflow $K_1 * Q$.

The most common notation for "rate of change of a quantity with time" is dQ/dT. Another notation is Q with a dot over it. Usually the plus (+) is omitted if the first expression in an equation is plus. Without words, the equation is:

$$dQ/dt = J - K_1 * Q.$$

A rate equation is sometimes called a *differential equation.* Its equation states the rate of change in general terms, without actually using numbers. For a particular case, you can find out what the inflow rate is numerically and what the outflow constant is numerically. For example, Figure 3.3b has data observed for flows of water in and out of a tank in steady state. For the model in Figure 6.1 the outflow is 100 liters per hour. Therefore, the outflow expression $K_1 * Q$ represents 100 liters per hour. ("Per" means divided by. Division is indicated with a slash, /.)

$$K_1 * Q = 100 \text{ liters/hr.}$$

Dividing both sides by Q:

$$K_1 = 100/Q.$$

When the flow was 100, Q was 1000. Therefore,

$$K_1 = 100/1000 = 0.1/\text{hr}.$$

More on using data to calibrate equations is given in Chapter 7. To encourage visual images of real systems, we use the words *storage* and *tank* for the locations of accumulation. In mathematical teaching of systems these are called *state variables*.

6.2   EQUATIONS FOR STEP SIMULATION

Another way of handling changes of storages with time is to write an equation for step changes over a specified interval of time. Referring to the TANK model in Figure 6.1, we verbalize the changes:

Quantity Q after the next time interval will equal the quantity now plus the change rate times the time interval:
(New quantity) = (old quantity) + (change rate) * (time interval).

Computers simulate by making repetitive calculations of step changes over small time intervals. This book uses the letters used in the computer programs as follows: *Time interval is DT. Change rate is DQ*, which is the sum of flow rates in and out of storage. With this notation the equation in the simulations is:

$$Q_{new} = Q_{old} + DQ*DT.$$

When the computer makes a calculation during simulation, the "=" sign means "will equal." It responds to the equation by replacing the present value of Q (a memory location in the computer) with a new value of Q calculated as the old value plus the change for that interval of time (DQ*DT). The equation given to the computer simulation is simply:

$$Q = Q + DQ*DT.$$

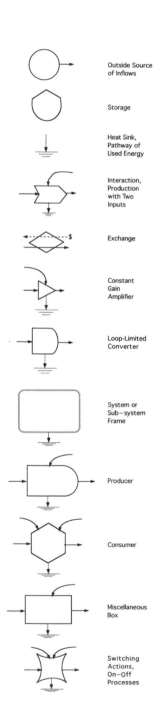

Outside Source
of Inflows

Storage

Heat Sink,
Pathway of
Used Energy

Interaction,
Production
with Two
Inputs

Exchange

Constant
Gain
Amplifier

Loop-Limited
Converter

System or
Sub–system
Frame

Producer

Consumer

Miscellaneous
Box

Switching
Actions,
On–Off
Processes

When the computer attention gets to the equation during simulation it recalculates what the change rate DQ is at that time, multiplies by the setting for time interval (iteration interval DT), and adds it to the old Q. It prints, plots, or saves the number before repeating the calculation for the next time step. It repeats the calculation over and over for all the storages in the model, plotting graphs of the way all aspects of the system change with time.

Graphs can be compared with expectations to see if the model is the right one to show what happens in the real world. People who have verbal models about systems are often surprised to find that the simulation behavior is not what they expected from their qualitative thinking.

STEP SIMULATION, RAMP EXAMPLE

Figure 6.2 is an example of stepwise summing of changes according to a simulation equation. The model is a storage with only one inflow pathway. The change rate is constant: DQ = J, and J = 10 units per hour. The quantity in the tank at each hour will be what it was at the start of the hour plus the amount added. In words:

New Q will equal old Q plus DQ for one time step:
Q = Q + J*DT.

Adding 10 at each time interval produces an upward sloping line. This model is appropriately called a RAMP (Figure 6.2).

◆

6.3 STEP SIMULATION AND CONTINUOUS PROCESSES

For a process that operates in steps, the RAMP model with stepwise digital simulation generates a realistic graph. For example, a pump turning on briefly every hour to pump in 10 gallons of water would produce a record of water level steps like that shown in Figure 6.2b.

Digital simulation usually simulates continuous processes even

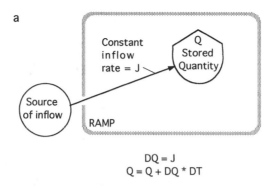

$$DQ = J$$
$$Q = Q + DQ * DT$$

FIGURE 6.2.
Simulation by summing the step additions for a time interval according to a very simple rate change equation. (a) Energy systems diagram of a single storage model without an outflow (RAMP) and its equation; (b) graphical result of simulation.

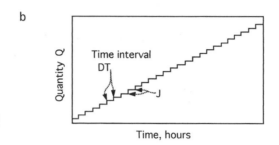

though its iterations occur in steps. A pump operating all the time to send 10 gallons of water into the tank (Figure 6.2a) is a continuous process. Even though the change rate DQ = 10 is correct, simulating it with repeated digital recalculations (Figure 6.2b) is an approximation. The more you reduce the time interval (DT) between calculations (iterations), the closer the simulation approaches the correct graph for continuous pumping, in this case a straight line. However, the smaller the time step is, the more iterations the computer has to make and the longer it takes for the program to run.

Because the flow rates are usually changing during a time step, calculating change with the rates at the start of the time interval is not correct for the whole time step. The error is large and cumulative if rates are changing rapidly and the time step is large (example: steeply accelerating growth curves). Several kinds of mathematical calculations can be added to simulation programs to base the changes on the average rates during each time step. Swartzman and Kaluzny (1987) summarize Runge-Kutta methods and mathematics of numerical integration. Many commercial software packages for simulation include several options for increasing accuracy (Chapter 22).

As the time step approaches zero, the errors approach zero. In this book we reduce the time step DT until no change is observed in the simulation curves, and the error is judged to be negligible. This method is called *Euler numerical integration.*

Outside Source
of Inflows

Storage

Heat Sink,
Pathway of
Used Energy

Interaction,
Production
with Two
Inputs

Exchange

Constant
Gain
Amplifier

Loop-Limited
Converter

System or
Sub–system
Frame

Producer

Consumer

Miscellaneous
Box

Switching
Actions,
On–Off
Processes

6.4   EQUATIONS FROM SYSTEMS DIAGRAMS

For each model we need rate equations to put into the simulation programs. In this book, for a storage labeled Q, *DQ is shorthand for the rate of change with time: dQ/dt.* Most of this chapter explains details for writing rate equations for energy systems diagrams. Figure 6.3 shows the mathematical expressions for many kinds of pathways. With a little practice it gets easy to visualize the *network* and the *mathematical terms* together. Depending on previous experience and aptitude, some people will think first in one while deriving the other.

FIVE-PATHWAY MODEL FOR ONE STORAGE

Figure 6.4 is another example with five different kinds of pathways, two flowing in and three flowing out. The rate of change of the quantity in the tank is the difference between the plus contributions of the inflows minus those subtracted by the outflows. There is one mathematical term (expression) for each pathway. We write the equation by setting DQ equal to two plus terms and three minus terms. The first term J is for an inflow whose value is determined from outside the system by whatever is varying J or making J constant. The second inflow is coming from a storage of reserve energy E by the linear pathway whose expression is $+K_0*E$.

The topmost outflow is in proportion to an interaction, which means the product of the storage and an outside source of active pumping F calibrated with the constant K_3. The middle pathway flows out in proportion to the self-interaction of thequantity Q calibrated with the constant K_2, which equals K_2*Q*Q (Q times Q is Q squared $= Q^2$). For example, crowding may cause a

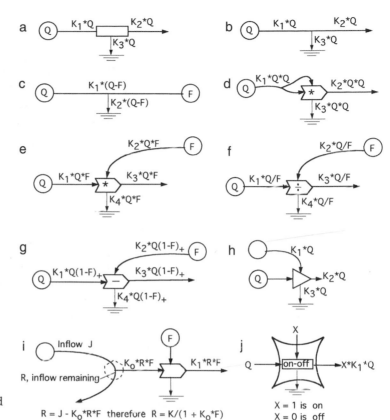

FIGURE 6.3.
Mathematical expressions for common pathways of systems models. (a) Linear transformation; (b) linear frictional process; (c) reversible, linear flow with backforce; (d) quadratic flow; (e) product interaction; (f) dividing interaction; (g) subtracting interaction; (h) constant gain amplifier (gain = K_2); (i) pumping from source limited flow; (j) on/off switch.

population to drain energies negatively in proportion to the frequency of self-interactions. The lower pathway drains storage linearly as required by the second energy law. The term here is $-K_1{}^*Q$, like that of Figure 6.1.

EQUATIONS FOR SYSTEMS WITH MANY STORAGES

The examples used so far have one rate equation for one storage. Most models are more complex than one storage, and therefore a set of several equations is used to represent the system.

Let's write the equations for the production–consumption–recycle model (PC&CYCLE) given in Figure 3.4. First, copy that model to the top of a page (Figure 6.5a) and add some letters

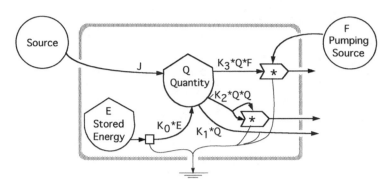

$$DQ = J + K_0{*}E - K_1{*}Q - K_2{*}Q{*}Q - K_3{*}Q{*}F$$

Rate of change of quantity Q with time equals:

inflow rate	$+ J$
plus storage inflow	$+ K_0{*}E$
minus outflow	$- K_1{*}Q$
minus outflow	$- K_2{*}Q{*}Q$
minus outflow	$- K_3{*}Q{*}F$

FIGURE 6.4.
Writing the equation for a storage model with five kinds of pathways.

representing the sources and storages. Then assign a "K" for the constant coefficient on each pathway (Figure 6.5a). Next we suggest you cover up the rest of the page and try to write the mathematical terms for each pathway. Compare your result with Figure 6.5b. Then write the equation for each tank as the sum of the plus inputs and minus outputs. Compare your results with those given in the figure.

FLOW USE JUNCTION

A special design is appropriate where pumping is from a flow-through source with a fast turnover. The energy source on the left in Figures 3.4 and 6.5 sends a flow into the system (as defined by the frame). Components within the system can interact with that flow using part or most of it in production processes. That which is unused flows out. An example is the flow of sunlight into a forest, most of which is captured and used, with some being reflected out. Another example is the flow of water in a stream used by a village. It can pump from whatever water remains until that flowing out is too small. A special equation

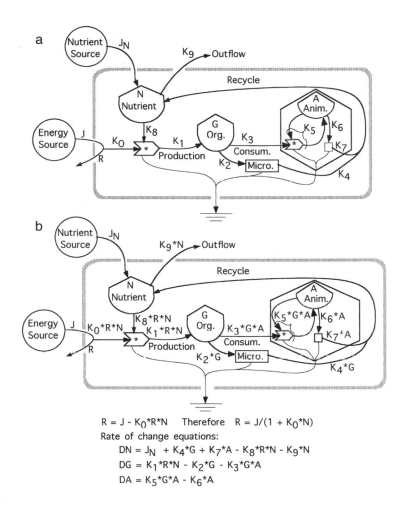

FIGURE 6.5.

Equations for a more complex model (given with numbers in Figure 3.4). (a) Shown with constant coefficients for each pathway; (b) shown with the complete math terms for each pathway.

$R = J - K_0*R*N$ Therefore $R = J/(1 + K_0*N)$
Rate of change equations:
$DN = J_N + K_4*G + K_7*A - K_8*R*N - K_9*N$
$DG = K_1*R*N - K_2*G - K_3*G*A$
$DA = K_5*G*A - K_6*A$

is used for this configuration that is diagrammed in Figure 6.3i. The unused remainder R is the inflow from outside minus that used.

The productive interaction that uses that flow-through resource does so in proportion to the unused remainder. The use and the productive output are modeled as the product of the interacting force (F in Figure 6.3i) and the available remainder R.

This mathematical expression is useful where the inflow of the resource is much faster than any storage of that resource within the system. For example, light energy flowing thousands of miles per second almost instantaneously replaces the quantity of light energy that is within the system at one time. It would

Outside Source of Inflows

Storage

Heat Sink, Pathway of Used Energy

Interaction, Production with Two Inputs

Exchange

Constant Gain Amplifier

Loop-Limited Converter

System or Sub−system Frame

Producer

Consumer

Miscellaneous Box

Switching Actions, On−Off Processes

not be possible to simulate light storage because the turnover time is too fast compared to the scale of the rest of the system. Where turnover times of some quantity are much higher than the other storages of a system, show the resource as a flow from which other units can pump according to the unused remainder. Then write the mathematics as shown in Figure 6.3i.

6.5 TRANSLATING EQUATIONS INTO ENERGY SYSTEMS DIAGRAMS

Whereas previously we wrote math for systems networks, we next do the reverse. Looking at a set of equations, even by those mathematically trained, does not immediately provide a systems overview. But an energy systems diagram provides an immediate understanding of parts, processes, and overall relationships. Getting understanding from the raw math is tedious thinking. The easiest way is to translate the equations into energy systems language.

Writing equations for an energy systems diagram is rigorous and not ambiguous. However, for the reverse process, drawing diagrams for a set of equations produces more than one alternative. This is because the equations do not indicate which letters are sources, where the energy and material constraints are, nor which is the author's intent out of a possible number of different configurations with the same mathematics. We have discovered 18 different systems models that produce logistic mathematics when the equations are manipulated into a common form. As Oster and Auslander (1971), among others, have explained, typical mathematics practices are too general to represent systems accurately.

EXAMPLE OF DIAGRAMMING A SET OF SYSTEMS EQUATIONS

With the difficulties in mind, let's illustrate the process of translating equations with Figure 6.6. A set of equations is given at the top of the figure. Cover up the rest of the figure and try to visualize what the equations mean without diagramming—pretty

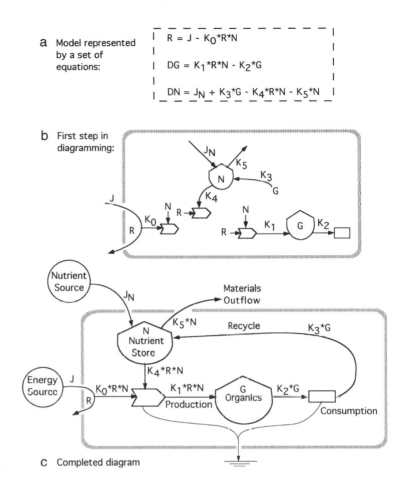

a Model represented by a set of equations:

$$R = J - K_0*R*N$$

$$DG = K_1*R*N - K_2*G$$

$$DN = J_N + K_3*G - K_4*R*N - K_5*N$$

b First step in diagramming:

c Completed diagram

FIGURE 6.6.
Diagramming a model that is in the form of a set of systems equations. (a) Equation set; (b) start of diagram; (c) completed diagram.

obscure. You can keep the lower part of the figure covered as you try the following procedure.

Next, place a storage symbol on the paper for each rate equation. Connect one pathway for each term in that equation. For each pathway, fix its structure appropriate to that expression (see pathway examples in Figure 6.3). Figure 6.6b shows the first stage. Try to make the connections of one storage to another consistent according to terms in both equations.

Next, try to identify and draw the plausible energy sources and sinks. Estimate the order of items on the transformity scale so as to position units on the diagram in that order. These provide a reasonable, correct solution, which may be only one among several possible. To complete a diagram of someone else's model equations, you need that person's collaboration to obtain other information and intent.

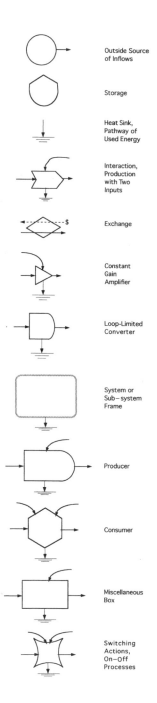

Outside Source of Inflows

Storage

Heat Sink, Pathway of Used Energy

Interaction, Production with Two Inputs

Exchange

Constant Gain Amplifier

Loop-Limited Converter

System or Sub-system Frame

Producer

Consumer

Miscellaneous Box

Switching Actions, On–Off Processes

6.6   MATHEMATICAL CHARACTERISTICS OF UNITS

Whereas most models are simulated to determine what they do, the equations for some simple models can be manipulated mathematically to find the graphs of storages with time, calculate steady states, and learn other special properties. The mathematical characteristics of simple units are helpful in their use to build more complex models. The mathematics of energy systems models was given in another book (Odum, 1983, 1994). Properties of one unit are given next to explain exponential processes.

EXPONENTIAL DRAINING STORAGE MODEL (DRAIN)

A storage with one linear draining pathway is about the simplest model that you can draw (Figure 6.7). Because the outflow is in proportion to the remaining storage, its *percent outflow per time is constant.* For example, water stored in a temporary pond percolating into the ground drains out—quickly at first and then more slowly as the pressure driving it decreases. A constant fraction K of what is remaining in storage Q flows out in each unit of time.

A differential equation is integrated to get an equation for storage over time (Figure 6.7). Time T in the resulting expression for quantity Q is a negative exponent classifying the equation as a declining exponential (Figure 6.7). In Chapter 9, Section 9.8, simulation of this model generates the same curve. DRAIN is a general systems model because it fits similar processes on all scales. Remembering the equations that go with a draining storage helps you recognize when it is an appropriate model. Data fit when they follow a straight line of Q plotted against T on semi-log paper.

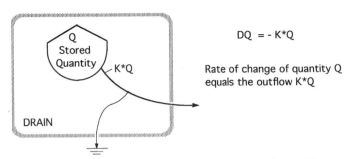

$$DQ = -K*Q$$

Rate of change of quantity Q
equals the outflow K*Q

Mathematical manipulations to see other properties of the model:

In differential equation form:
$$\frac{dQ}{dT} = -K*Q$$

Dividing through by Q:
$$\frac{dQ}{Q*dT} = \frac{-K*\cancel{Q}}{\cancel{Q}}$$

Equation now says that the
fractional loss in Q with time
is constant:
$$\frac{dQ}{Q*dT} = -K$$

Multiplying both sides by dT:
$$\frac{dQ*\cancel{dT}}{Q*\cancel{dT}} = -K*dT$$

Add Integral symbols $\int$ on both sides
to indicate integration to follow:
$$\int \frac{dQ}{Q} = \int -K*dT$$

Logarithmic equation
after integration
where Q_0 is the
storage at the start:
$$Log_e \ Q/Q_0 = -K*T$$

Exponential form of
equation follows from
definition of log as
an exponent:
$$Q/Q_0 = e^{-K*T}$$

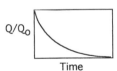

FIGURE 6.7.
Model (DRAIN) of a single tank
with one linear outflow
pathway with mathematical
manipulations to show the form
of its plot on a semi-logarithmic
plot with time and its
exponential form on an
ordinary plot with time.

◆

6.7 MATHEMATICS AND NETWORKS

In translating diagrams, we write a rate equation for each
storage. A model with four storages will have four rate equations.
If these storages are connected with each other through at least
one mutual pathway, then the letters for each storage will occur
in the equation for at least one other tank, and together directly

or indirectly they all share terms. Although there is not much
point for complex systems, each equation could be combined
with the others to form one giant equation. This is another way
of recognizing that everything is connected to everything else
directly or indirectly. In practice, for simplicity, we keep each
storage rate equation separate but simulate them together as a
system set.

In common mathematics, operations terms are combined, fac-
tored, and manipulated in various ways to get results in a compact
and more easily understood form. However, this is the same as
changing the systems diagram, and it results in differences in
meaning regarding pathways. For example, Figure 6.8 shows
some equation changes that most people would regard as simplifying without changing the meaning, but when translated into
network form, different configurations result. For example, in
Figure 6.8a factoring changed the number of pathways entering
storage. In Figure 6.8b it eliminated an outflow pathway. The
more complex the model, the more possibilities there are in ener-
getics and configurations for the same equations.

For this book the equations are used mainly as an intermediate
stage in setting up programs for computer simulation. Because
rate equations are a widely understood language for representing
system structures and processes, but are ambiguous in represent-

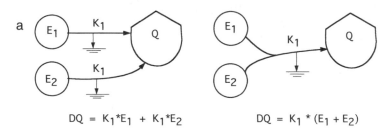

$$DQ = K_1{}^*E_1 + K_1{}^*E_2 \qquad\qquad DQ = K_1 * (E_1 + E_2)$$

FIGURE 6.8.
Changes in diagrams implied
by manipulating equations.
Factoring K_1 implies a
combination of pathways.
(a) Pathways into storage;
(b) pathways out of storage.

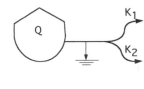

$$DQ = - K_1{}^*Q - K_2{}^*Q \qquad\qquad DQ = - (K_1 + K_2){}^*Q$$

ing networks without a diagram, this book presents all models with rate equations *and* energy systems diagrams.

◆

6.8   SIMULATING EQUATIONS WITH EXTEND

Although we use EXTEND in a special way to simulate without writing equations in Chapters 4 and 5, a manual and other libraries for simulating equations come with the EXTEND program. There are blocks for adding, subtracting, dividing, integrating, etc. Equations can be calculated by connecting math blocks on screen. Most of these blocks are not compatible with our special blocks because of their array structure (Appendix A). Figure 6.9 is an example of how EXTEND is used to generate system graphs by manipulating equations with the math icon blocks from the "generic library." In the library there are also blocks where simulation is accomplished just by typing in equations.

The differential equation derived from the energy systems diagram for the TANK model (Figure 6.1) is $dQ/dT = J - K*Q$. Integrating both sides produces an integral equation for Q (Figure 6.9a). The math steps in the integral equation are also diagrammed in Figure 6.9a in the form used to explain negative feedback (a feedback that subtracts) in classical engineering control systems.

After opening EXTEND and the generic library, mathematical icons are arranged on screen to match (Figure 6.9b). The place in the network that computes Q is connected to a plotter from the plotter library (Figure 6.9b). A graph of the storage Q results (Figure 6.9c). You need merely know what integration is to use this approach to simulation, because the integration is done by the blocks. Integration is finding the equation for a state (storage) from the rate equation.

TIME LAG

Most of the models in this book have the time lag inherent in storages. Because storage tanks require time to fill, a lag occurs before the inputs to the tank affect the outputs. Another kind of

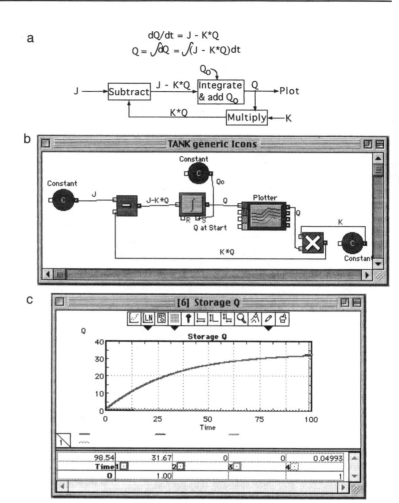

FIGURE 6.9.
Simulating the model TANK
(Figure 6.1) with EXTEND by
arranging math blocks of the
generic library to represent its
integral equation. (a) Equation
and its diagram; (b) math
blocks connected on computer
screen; (c) simulation graph.

time lag occurs when a backforce is generated in opposition to
the input. Examples are electrical current flows through coils of
electrical wire (electromagnets) and the resistance of people to
new ideas. The equations for this "impedance" are found in
electrical engineering texts. Another kind of time lag is digital
time delay, when an input is delayed for a time interval before
being passed to its destination. Examples are delay between food
inputs to an animal and production of offspring and delay in
production on an assembly line due to a holiday. When the time
delay is added to a system that usually grows into a steady state
(example: LOGISTIC in Chapter 13), the system oscillates.

Chapter Seven

◆

CALIBRATING MODELS

Calibration is the process of fitting a model with numbers. Numbers are used to calculate the constant coefficients of equations to be used in simulation. The numbers may be observed data from measurements in nature, or they may be numbers assumed to investigate how the model works. Chapter 7 explains how to calculate coefficients after numbers are placed on energy systems diagrams (Chapter 3).

Coefficients are constant values of the K's in the program that indicate how much flow there is on a pathway in terms of causal forces or concentrations. After numerical values of flows and storages are placed on an energy systems diagram, calibration is accomplished by calculating a table of coefficients from these numbers, a simple task for a small minimodel. For larger models, calculations should be made with a computer "spreadsheet," which makes subsequent changes easy. For some programs, the calibration calculations are included within the simulation program, and you merely have to enter the flows and storage values from the evaluated systems diagram. After equations are calibrated, you are ready to simulate with one of the programs explained in this book (QUICKBASIC, EXTEND, STELLA, EXCEL).

◆

7.1 PROCEDURE FOR CALIBRATING A MODEL

The numbers used for calibration are flows (quantities processed per time) and storages at the time. The data may be for a particular moment or, more often, average data. With the first calibration of a model, getting approximate numbers is sufficient. Later when the model is running, calibration can be refined using knowledge of how the model performed.

USING STEADY-STATE VALUES

If the system is one that tends to level its growth with storages in steady state (inflows balancing outflows), you may use the steady-state conditions for the calibration. If you have some of the flows but not all, the missing values for flows can be assumed as the values necessary to make the inflows and outflows equal.

Calibration with steady-state values helps you find errors when simulation models are run for the first time. If the model is not an oscillator, if the starting values are the same as the calibration values, and if there are no errors, then the graphs are straight horizontal lines. In other words, the model is started and run at steady state. Error is recognized if the lines are not straight.

SELECTING UNITS

First, decide what the unit of time will be for the whole model (example: hours, days, years, etc.). It better to select a time unit that is much smaller than the span of interest of your planned simulation. For example, if your interest is in a model's behavior over many years, use a unit of days, weeks, or months. You can use a large unit on the same scale as the span of time of the simulation, but then the iteration interval (DT) will need to be very small.

Decide what the spatial units will be. Whereas the units for a storage and all the flows into it have to be the same (one equation), the different storages can have different units (example: Figure 3.5). However, it may be simpler to make the spatial units the same for all the storages.

Whereas you could use the area of the whole system as the spatial unit, such numbers are large and hard to visualize. Instead, you can select a commonly used unit with which people have had experience. For example, flows and storages can be given per square meter even though the area of the system is much larger. Whatever is chosen, place the units on the energy systems diagram to help you keep them straight (example: Figure 3.5).

Then, for the complex diagrams, use a colored pencil to write the numbers for the storages in the tanks and for the flows on the pathways in and out of the tanks (Figure 7.1). Everything that flows in and out of the same tank must have the same units. For example, for a tank representing population, the flows in and out must be numbers of people per unit of time. For a tank

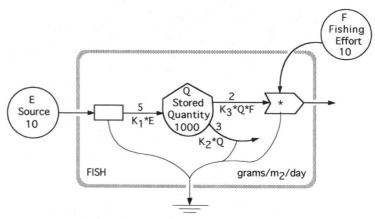

FIGURE 7.1.
Minimodel FISH with three
evaluated pathways to show
how to calibrate equations.
Table 7.1 shows the calibration
result.

$DQ = K_1*E - K_2*Q - K_3*Q*F$
Rate of change of fish stock DQ equals production K_1*E minus
   mortality K_2*Q minus fishing harvest rate K_3*Q*F.
Fish caught is in proportion to stock Q and to fishing effort F.

representing biomass, the flows must all use a consistent weight
unit, such as grams per time.

The flows in and out of an interaction will be in different
units. At this symbol the input forces or concentrations multiply,
divide, or involve other functions, but they do not add to generate
the production output. Even if two flows to an interaction have
the same units (grams, calories, etc.), they represent different
qualities of energy and should not be added.

CALCULATING COEFFICIENTS

After the flows and storage numbers have been placed on the
diagram (Figure 7.1), calculate the coefficients as shown in Table
7.1. Usually there is one coefficient for each pathway. In the first
column list the mathematical terms for each of the pathways to
a storage (the terms in the rate equation for that storage). Using
these numbers, divide through by the values of the causal variables (sources and storages) in each expression (columns 2, 3,
and 4). Values for the coefficients result in the last column. For
example, in the first line of the table, $K_1*E = 5$ when $E = 10$.

TABLE 7.1
Calibration of Constants for the Model FISH Using Numbers on the Diagram in Figure 7.1[a]

Expression and value	Coefficient expression	Calculation	Coefficient value
K1*E = 5	K1 = 5/E	K1 = 5/10	K1 = 0.5
K2*Q = 3	K2 = 3/Q	K2 = 3/1000	K2 = 0.003
K3*Q*F = 2	K3 = 2/(Q*F)	K3 = 2/(1000*10)	K3 = 0.0002

[a] Sources: E = 10, F = 5, and storage, Q = 1000. Values for sources and storages used for calibration may or may not be those used for simulation.

Therefore, $K_1 = 5/E$; substituting the value of E = 10, $K_1 = 5/10 = 0.5$.

◆

7.2 CALIBRATING A MODEL WITH A SPREADSHEET PROGRAM

Especially if the table of constants is long (many coefficients), it is convenient to set up the calibration table using a spreadsheet program (examples: LOTUS, QUATRO, EXCEL). Because items occur in several places in a set of equations, each may be part of several calculations. A common error occurs when you change something in one place, but forget to change it in all the places it occurs. Spreadsheets respond to changes by automatically changing all the relevant calculations. Formulas for automatic calculation are built into the cells for the coefficient values.

After the spreadsheet program is loaded, a screen grid appears. You can enter captions and labels similar to those used for the hand calculation (Table 7.1). Figure 7.2 shows a spreadsheet with letters for column headings, and numbers labeling the rows. The numbers used in Figure 7.2 are for the FISH model shown in Figure 7.1. Each cell in the grid has an address according to its column and row. For example, the source value 10 is in cell C4 and the inflow value 5 is in cell C9.

When you type something into a cell, the program treats

	A	B	C	D	E	F
3						
4	Sources:	E =	10			
5		F =	10			
6	Storages:	Q =	1000			
7						
8	Coefficient:					
9	Inflow	K1*E =	5	Therefore	K1 =	0.5
10	Loss	K2*Q =	3	Therefore	K2 =	0.003
11	Outflow	K3*Q*F =	2	Therefore	K3 =	0.0002

These are formulas for manipulating the contents of other cells which are built into the calculation of these cells

+C9/C4
+C10/C6
+C11/(C6*C5)

FIGURE 7.2.
Numbers for the model FISH in Figure 7.1 demonstrating the use of spreadsheets for calibrating coefficients.

an entry that starts with a letter as a label; whereas those that start with a numerical item such as "+" are processed as arithmetic calculations. As set up in Figure 7.2, entries in columns A, B, D, and E are labels, whereas entries in columns C and F are numerical. When the cursor is on a cell, its contents appears in a line at the top of the screen, where you can type a change.

For cells in the final column, enter the formulas for calculating the constants (K's) in terms of the addresses of the cells containing numbers to be multiplied or divided. For example, for the expression K1* E = 5 (line 9 in Figure 7.2), the formula entered in the program for cell F9 is the ratio of the addresses of the numbers to be divided 5/10. A plus is added to indicate that these are not labels, but numbers to be processed.

$$+C9/C4.$$

The program automatically takes the value in cell C9 and divides it by the value in cell C4. The advantage of this means of calculation is that when you change a number anywhere in the table, the program instantly recalculates every coefficient affected by the change. A printout of the EXCEL spreadsheet for this model is shown in Figure 7.3. Vertical and horizontal lines to show the addresses and cell limits are optional in the printout.

The rate of change equation for the FISH model in Figure 7.1 is:

$$DQ = K_1{}^*E - K_2{}^*Q - K_3{}^*Q^*F.$$

If the evaluated constants from Figure 7.2 or 7.3 are substituted for the K's in the equation, the calibrated equation is:

$$DQ = 0.5^*E - 0.003^*Q - 0.0002^*Q^*F.$$

◆

7.3   CALIBRATING MORE COMPLEX MODELS

For a more complex model than that just discussed, the process is the same, except there are coefficients for several equations (for several storages). The numbers on the energy systems diagram of model PC&CYCLE used in previous chapters (Figures 5.2 and 5.3) are shown again in Figure 7.4a and used to calculate coefficients in Figure 7.4b. These coefficients are used for the simulation in BASIC in Chapter 9. See more complex examples in Chapter 21.

FIGURE 7.3.
Printout of EXCEL spreadsheet (file FISHNUMB.wk1) containing the calibration table for the model FISH in Figure 7.2.

Spreadsheet Calibration Table for Model FISH					
Calibration Quantities:					
Sources:	E =	10			
	F =	10			
Storages:	Q =	1000			
Calculation of Coefficients:					
	K1*E =	5	Therefore,	K1 =	0.5
	K2*Q =	3	Therefore,	K2 =	0.003
	K3*Q*F =	2	Therefore,	K3 =	0.0002

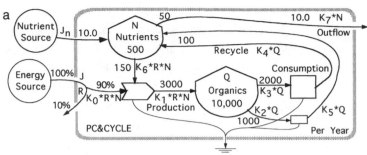

a

b   Calculation of Coefficients for the Model PC&CYCLE

Sources and Storages used for Calibration:
  N = 500 grams per square meter
  Q = 10,000 grams per square meter
  R = 10 %

Calculation of Coefficients:

$K_0*R*N = 90$	Therefore	$K_0 = 90/10/500$	$= 0.018$
$K_1*R*N = 3000$	Therefore	$K_1 = 3000/10/500$	$= 0.6$
$K_2*Q \quad = 1000$	Therefore	$K_2 = 1000/10,000$	$= 0.1$
$K_3*Q \quad = 2000$	Therefore	$K_3 = 2000/10,000$	$= 0.2$
$K_4*Q \quad = 100$	Therefore	$K_4 = 100/10,000$	$= 0.01$
$K_5*Q \quad = 50$	Therefore	$K_5 = 50/10,000$	$= 0.005$
$K_6*R*N = 150$	Therefore	$K_6 = 150/10/500$	$= 0.03$
$K_7*N \quad = 10$	Therefore	$K_7 = 10/500$	$= 0.02$

FIGURE 7.4.
Calibration of the model
PC&CYCLE. (a) Numbers
on energy systems diagram;
(b) coefficient calculations.

SPREADSHEET USE

By now spreadsheets are part of general education of most
students. Details on their use can be found in the manuals for
those programs. The tables can be converted into text tables or
transferred to other computers. The spreadsheet FISHNUMB.
wk1 used in this chapter is included on the disk supplied with
this book.

Chapter Eight

◆

SIMULATING WITH SPREADSHEET

S preadsheet computer programs process data in tables. They provide a grid of "cells" (boxes) on the computer screen in which numbers and labels are typed. Then cells are given formulas for calculating quantities from numbers in other cells. Chapter 8 explains how to use spreadsheet programs to simulate systems models. The simple model FISH is used as an example.

◆

8.1 PROCEDURE FOR SIMULATION WITH A SPREADSHEET

After rate equations have been written for a model (Chapter 6) and its coefficients calibrated (Chapter 7), a spreadsheet program such as LOTUS, EXCEL, QUATRO, or one of many others is selected and loaded. The program brings an empty grid to the screen showing the cells formed by the crossing lines (Figure 7.2). The vertical columns are labeled with letters and the horizontal rows with numbers. Cells are identified by the column's letter and the row's number (examples: B1, H3, etc., in Figure 8.1b).

Type in the numbers for coefficients and other constant values in cells at the top of the screen. Below, set up the main simulation table by entering the terms of the model equations in a row as headings for the columns (Figure 8.1b) including sources, storages, math terms for pathway flows, and change rates.

When you type labels, numbers, or calculation formulas into the cells, the program treats each of these three kinds of information differently. Normally, something typed into a cell is treated as a label if it starts with a letter; it is treated as a number if the first character is a number. You can set the number of decimal places. However, if a + or = sign is used first (example: = G9), the program treats it as an address but uses its contents as a number. Figure 7.2 shows examples of formulas used in a spreadsheet for calibration.

After a formula is written into a cell it automatically calculates

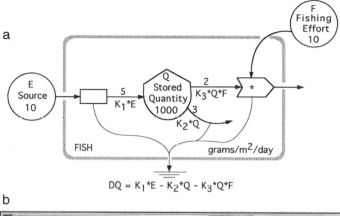

a

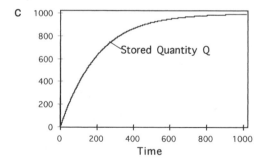

b

	A	B	C	D	E	F	G	H
							FISH.wk1	
1					Simulation of the Model Fish			
2					DQ = K1*E - K2*Q - K3*Q*F			
3								
4		E = 10.00		K1 =	0.50	K2 =	0.003	K3 =	0.0002
5								
6		Days	Source	Fishing	Quantity	Deprec-	Yield	Change in
7			flow	Effort	Storage	iation	outflow	storage
8		T:	k1*E	F:	Q:	k2*Q:	k3*Q*F:	DQ:
9								
10		0	5	1.00E+01	1.00E+00	3.00E-03	2.00E-03	5.00E+00
11		1			6.00E+00	1.80E-02	1.20E-02	4.97E+00
12		2			1.10E+01	3.29E-02	2.19E-02	4.95E+00
13		3			1.59E+01	4.77E-02	3.18E-02	4.92E+00
14		4			2.08E+01	6.25E-02	4.17E-02	4.90E+00
15		5			2.57E+01	7.72E-02	5.15E-02	4.87E+00
16		6			3.06E+01	9.18E-02	6.12E-02	4.85E+00
17		7			3.54E+01	1.06E-01	7.09E-02	4.82E+00
18		8			4.03E+01	1.21E-01	8.05E-02	4.80E+00
19		9			4.51E+01	1.35E-01	9.01E-02	4.77E+00
20		10			4.98E+01	1.50E-01	9.97E-02	4.75E+00
21		11			5.46E+01	1.64E-01	1.09E-01	4.73E+00
22		12			5.93E+01	1.78E-01	1.19E-01	4.70E+00
23		13			6.40E+01	1.92E-01	1.28E-01	4.68E+00

c

FIGURE 8.1.
Simulating the model FISH with a spreadsheet. (a) Model with calibration numbers and equation; (b) spreadsheet simulation table (formulas are given in Table 8.1.); (c) simulation graph derived from the table.

a number for that cell using numbers in other cells. It does this by adding, subtracting, multiplying, and otherwise manipulating the numbers at the cell addresses. For example, the formula =

G9*E6 means multiply the number in cell G9 by the number in cell E6.

Each horizontal row represents the values at one time. In the first row, enter zero for the time and the starting value for storage. Attach formulas to calculate the other numbers. For example, the formula for the change rate DQ is the sum of the cells that have the terms of the rate equation.

For the second line down (the next time step), put in formulas that calculate values from the first line. The formulas for this one line are duplicated on the third line and beyond with the FILL DOWN command that automatically increases the line numbers of the cell addresses by one. In this way, every calculation is repeated on the next line for the next interval of time, similarly on the next, and then the next, etc. Those numbers that need to be held constant from line to line in the fill down process are given the $ symbol in front of each character. For example, if address F4 contains a constant, it is kept from changing when copied down by entering it as $F$4.

The result of copying the calculation line to successive positions below is a long table that tabulates the storages and flows as they change with time. The vertical columns of numbers for time, for storage, and other quantities are convenient for plotting graphs.

◆

8.2   SIMULATING THE MODEL FISH

An example of simulation is shown in Figure 8.1 for the systems model FISH previously used in Figure 7.1 to illustrate calibration. The energy systems diagram (Figure 8.1a) shows the storage Q to be a balance between the production of young fish from the left K_1*E the recruitment in fisheries, and the losses along two pathways: the normal mortality and the mortality due to interaction with the fishing effort F. Figure 8.1b shows the top of the simulation table of the spreadsheet.

Values of the constants E, K1, K2, and K3 were entered in line 4 (Figure 8.1b). Then in row 10, time was set at 0 and initial conditions of storage Q set at 1. The formulas shown in Table 8.1 calculated the source inflow, the fishing effort, the depreciation loss, the yield outflow, and the storage change rate from

Outside Source of Inflows

Storage

Heat Sink, Pathway of Used Energy

Interaction, Production with Two Inputs

Exchange

Constant Gain Amplifier

Loop-Limited Converter

System or Sub–system Frame

Producer

Consumer

Miscellaneous Box

Switching Actions, On–Off Processes

TABLE 8.1
Formulas Used in Spreadsheet for Model FISH in Figure 8.1

Terms in equation		Formulas in cells
K1*E	Line 10	=$D$4*$B$4
K2*Q	Line 10	=$F$4*E10
K3*Q*F	Line 10	=$H$4*E10*$D$10
DQ	Line 10	=$C$10 − F10 − G10
Q	Line 11	=E10 + H10
T	Line 11	=B10 + 1

numbers in the other boxes. In line 11 and all those below, the storage and time were calculated as the storage plus the change in storage from the previous time (row above). Table 8.1 has the formulas typed in (invisibly attached) to cells in these columns. When the rectangular cursor is on a cell with a formula, it shows in the edit line at the top of the program. See if you can derive them for yourself just by looking at rows 10 and 11 in Figure 8.1b.

If you make a change, the program automatically recalculates numbers throughout the table without being instructed further. Figure 8.1c shows the growth of fish obtained by graphing the column of 1000 numbers in the Q column as a function of the column of numbers in the T column, each representing a day. The simulation table was saved on disk memory as template file FISH.wk1. The wk1 extension indicates a spreadsheet template.

PRACTICE EXERCISE

The spreadsheet template for Figure 8.1b (FISH.wk1 in the IntrModl folder) is included on this book's disk. After loading your spreadsheet program, bring the template to screen using the program menu. Load the template. The contents of Figure 8.1b appear on screen. The grid lines can be turned on or off. Try varying source values and fishing pressures and observing the changes in the same way as with other "What If" exercises in this book. For more details on use of the spreadsheet program refer to the appropriate program manuals.

SAVING LARGE TEMPLATES

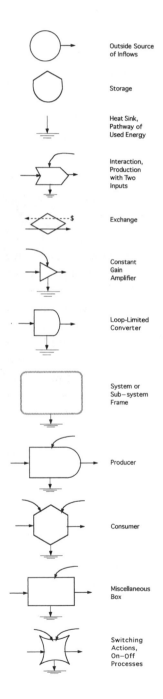

Outside Source
of Inflows

Storage

Heat Sink,
Pathway of
Used Energy

Interaction,
Production
with Two
Inputs

Exchange

Constant
Gain
Amplifier

Loop-Limited
Converter

System or
Sub–system
Frame

Producer

Consumer

Miscellaneous
Box

Switching
Actions,
On–Off
Processes

If you have a complex model to simulate, the headings and first calculation line may be very wide. Saving the whole table may use too much disk memory. It is not necessary to save all the repetitive lines below the first calculation lines. If the lower part of the table is not saved it can be restored after the abbreviated template is returned by using the FILL DOWN command. For example, suppose only the first 11 lines of the spreadsheet in Figure 8.1b were saved. After the template is reloaded from disk, highlight lines 11 and below to include 1000 lines. From the menu activate the FILL DOWN command. The formulas in line 11 are transferred to all the lines below and the values calculated. Thus, the whole table is immediately restored on screen.

People already familiar with spreadsheets can simulate with equations for models without learning a special simulation program. Larger models may reach the limits of some computers' memory. Because spreadsheet programs are common in business and are becoming part of everyday knowledge, a full explanation of their use is left to other books and manuals.

Chapter Nine

◆

PROGRAMMING IN BASIC

In this chapter simulation programming is introduced using the simple language BASIC. Learning to visualize processes as a sequential list of instructions, that is, as a program, provides special insights about systems. These are tested for consistency when the program is run on computer. An essential part of education, learning some programming, permits students to simulate their own models, testing their own ideas in a special way. For these purposes, people should do their own programming.

For minimodels, which we use to overview systems, a simple language may be sufficient. BASIC is the best known simple programming language, written by J. Kemeny at Dartmouth in the early days of computing. A generation of college-bound students learned BASIC in high school. Learning BASIC is often recommended as a first step that makes it easy to learn more complex programming languages later.

In this book a BASIC program is provided for each minimodel along with the diagram of the system in energy language, equations, and a typical sample of the graphic output from running the program. There are 46 BASIC programs in Parts III and IV. Although some modern versions of BASIC are elaborate, only a few commands are needed for simulating systems. The reader needs to learn enough of the BASIC commands to read programs, understand what they do, run them, and make changes. Some readers may want to write programs for new minimodels. The easiest way is to load a somewhat similar minimodel program and make changes.

In any system there are storages such as wood, money, materials, or water. Let's start with a program of a simple model with one storage called TANK, which was introduced in Figures 1.1 and 6.1. Instructions and explanations given in the TANK example apply to most of the models in this book. More advanced concepts are introduced with examples in Part III and programming details in Appendix B. To aid the comparison of methods, models simulated with BASIC in Chapter 9 (TANK, EXPO, DRAIN, and PC&CYCLE) are the same ones used in the previous chapters on model building, spreadsheets, and EXTEND, and in the next chapter on STELLA.

9.1 RUNNING A SIMULATION MODEL WITH QUICKBASIC

To run a simulation program with QUICKBASIC, you should have on hand a computer with a monitor, a disk drive, access to a printer, and a graphics "screen dump" program, which instructs the printer to record the the output graphs on paper. Directions for putting in disks, loading programs and running simulations with QUICKBASIC, saving programs, and printing out graphs are given in Table 9.1. More details and notes on running models with other programs and computers are given in Appendix B.

MODEL TANK

Try the various operations in Table 9.1 for the simple model TANK. It is a storage with an inflow and an outflow (Figure

TABLE 9.1
Steps in Running a Simulation Program with QUICKBASIC[a]

1. With the computer on and running, load QBASIC.exe from PC windows files. Each of the words that appears across the top of the screen is a pulldown menu.
2. Load the CD-ROM that comes with this book and locate the BASIC simulation programs in the file Basicprg.
3. If QUICKBASIC for Macintosh is not available, Macintosh users can load Chipmunk Basic from the CD using it to do the book exercises. Follow instructions in Appendix B.
4. Use the QBASIC file menu to load a simulation program and observe the program listing that appears on the screen.
5. In the RUN menu, select the Run option and observe the plotted graph that appears on the screen.
6. You can make changes and run the program again. To save the revised program, select Save As in the FILE menu. When the screen provides open dialog boxes, type a new name and select a folder on a floppy disk or hard disk as the destination for the new file.
7. To print the program listing, select the Print option in the FILE menu.
8. To save a simulation graph, use a screen-dump procedure to save the screen control to the clipboard. (For PCs using windows, use the Print Screen key; for Macintoshes, use Command-Shift-3).
9. To print a simulation graph, load a program that can read graphics (SIMPLETEXT, CANVAS, COREL DRAW, WORD, etc.). Use the menus in that program to load the file with the saved graph. On Macintoshes the screen dumps are on the hard disk labeled "Picture 1," "Picture 2," etc. On Windows PCs the graphics are on the clipboard ready to be pasted into a drawing program. Use features of the drawing program to label and make changes.
10. Use the FILE menu of the drawing program to save the graph. Print the graph as a final illustration.

[a] Details and notes for use with other programs are given in Appendix B.

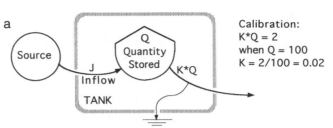

a

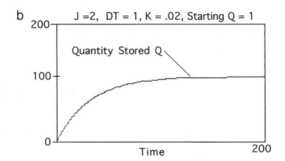

Calibration:
K*Q = 2
when Q = 100
K = 2/100 = 0.02

Rate of Change of Quantity Stored: DQ = J - K*Q
Stored Quantity: Q = Q + DQ*DT

FIGURE 9.1.
Model of growth of storage
with inflow and outflow
(program TANK.bas).
(a) Systems diagram, equations,
and calibration of the coefficient
K; (b) graph of growth to
steady state. Also see equations
in Figure 6.1 and hand
simulation in Figure 9.2.

9.1a). As the water flows in and out, the total quantity of water in the basin will increase and then level off. It levels off when the amount of water inflow equals the amount of water outflow. Water continues to flow in and out, but the level stays the same. The graph in Figure 9.1b shows these changes. The change in time is graphed on the horizontal axis and the change in quantity of water on the vertical axis. As you can see, the line which represents quantity of water goes up steeply at first, and then more gradually, until it finally reaches a point at which it levels off.

◆

9.2   SIMULATION WITH CALCULATIONS
BY HAND

Simulation is a sequence of calculations, and the list of instructions used to make the calculations is a *program*. First, let's simulate the model TANK by hand using its rate equation: $DQ = J - K*Q$ (Figure 9.1a). The instructions follow:

1. First list the numbers to be used from Figure 9.1b: The rate of inflow J = 2, the starting storage quantity Q = 1, and the constant coefficient for the outflow K = 0.02.
2. Make a simulation table with time T, the storage value Q, and the terms in the rate equation as headings of the columns (Figure 9.2). Each line in the table represents a time step.
3. Enter the starting values in the first line: The source J = 2; time is 0, and storage Q is the inital value 1. After multiplying K = 0.02 times the inital storage 1, put the result (0.020) in the column K*Q.
4. Now calculate what the change in Q will be during the next time interval: DQ = J − K*Q = 2 − (0.02)(1) = 2 − 0.020 = 1.98.
5. On the next line put time as 1, and Q = Q + DQ from the first line Q = 1 + 1.98 = 2.98; enter the other numbers. The

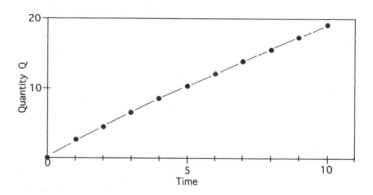

FIGURE 9.2.
Hand simulation of the model TANK (calculating and graphing without computer). Graph plots the points for Q with time calculated in the table.

Hand Simulation of Tank Model   DT = 1; J = 2; K = 0.02

| Time | Source | Storage | Outflow | Change |
T + DT	J	Q = Q + DQ	K*Q	DQ = J - K*Q
0	2	1	0.020	1.98
1	2	2.98	0.060	1.94
2	2	4.92	0.099	1.90
3	2	6.82	0.136	1.86
4	2	8.64	0.173	1.83
5	2	10.51	0.210	1.79
6	2	12.30	0.246	1.75
7	2	14.05	0.281	1.72
8	2	15.77	0.315	1.69
9	2	17.46	0.349	1.65
10	2	19.12	0.382	1.62
Etc.				

change for the next time step is DQ = J − K*Q = 2 − (0.02)(2.98) = 2 − 0.060 = 1.94.

6. On the third line time is 2 and Q = Q + DQ = 2.98 + 1.94 = 4.92.

7. Continue calculating down the table. Next, if you plot the series of Q's on a graph with time, it should resemble Figure 9.2.

You can see how tedious hand simulation is. The computer can calculate and graph thousands of points in a second. The simulation using a spreadsheet in Chapter 8 let the computer calculate a simulation table.

ROUNDING-OFF ERRORS

The calculations in Figure 9.2 are made to two or three decimal places. With these calculations, you will notice that it is necessary to round off figures, thus introducing a small error. However, the error can accumulate. When the computer does the calculations, they may be accurate to 40 decimal places or more, but there is still a rounding-off error. However, for most minimodel simulations great precision is not required.

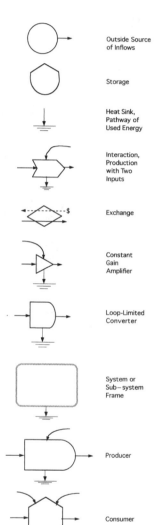

Outside Source of Inflows

Storage

Heat Sink, Pathway of Used Energy

Interaction, Production with Two Inputs

Exchange

Constant Gain Amplifier

Loop-Limited Converter

System or Sub–system Frame

Producer

Consumer

Miscellaneous Box

Switching Actions, On–Off Processes

9.3   WRITING A BASIC PROGRAM FOR THE TANK MODEL

To write a program for the computer, you use the same instructions used in the hand simulation, but you have to use a language of short words and phrases to which the computer will respond. In this chapter we use the words understood by BASIC programs such as QUICKBASIC and BASICA. You only need to learn two dozen BASIC commands to simulate most minimodels.

A computer program is like a cookbook recipe with the quantities listed and then the directions given. In BASIC each instruction statement is given a number. You usually start using 10, 20, 30, etc., so that you can later add statements between your original statements. When all the statements are written and you type RUN and press Return, the computer will start at the first number and follow your directions in sequence. In QUICKBASIC the line numbers are optional.

From a PC Windows file call up QBASIC, and in the FILE menu type NEW to get a blank screen ready for your program. Giving each a line number, type and *learn the commands* in Table 9.2 (PC) or Table 9.3 (Macintosh). Then use the RUN menu to start the program. If there are no errors, the same graph should appear that you observed when running the TANK file from the book disk. If something different results, search for bugs (errors in typing, items left out, wrong punctuation, erroneous thinking). Spaces are required between items in a line, but the amount of space is not critical. Use the instructions given in Table 9.1 to save and/or print this result of your first program. Compare it with Tables 1.1 and 9.1 or 9.2.

EXAMPLES THAT FIT THE STORAGE MODEL TANK

An ecological example that fits the TANK model is the change in quantity of leaves on the forest floor when there is a steady leaf-fall and creatures decomposing them at a constant fraction of

TABLE 9.2
PC Program for TANK with Explanations

Program	Explanations
10 REM IBM: TANK	REM means remark; does not affect the program
20 CLS	Clears the screen
30 SCREEN 1,0:COLOR0,0	Sets medium-resolution graphics and color
40 LINE (0,0)-(320,180),1,B	Plots a border around the graph
50 J = 2	Enters the quantity for J
60 Q = 1	Enters the beginning quantity for Q
70 K = .02	Enters the quantity for K
80 DT = 1	Time step between recalculations
100 REM Start of Iteration	Start of recalculation loop
110 PSET (T,180 − Q),2	Tells the computer to plot a point for T and a point for Q (180 − Q is necessary to get a graph with the normal starting 0 point in the lower left corner)
120 DQ = J − K*Q	Equation for the change in Q over time
130 Q = Q + DQ*DT	Equation for Q after the change
140 T = T + DT	Time is increased by one unit
150 IF T < 240 GOTO 100	If the graph is not at the end of the screen (240 points), it will run again starting at statement 100

TABLE 9.3
Macintosh Program for TANK with Explanations

Program	Explanation
10 'TANK' or REM TANK	Means remark; does not affect the program
20 CLS	Clears the screen
40 LINE (0,0)-(280,240),,B	Draws a border around the graph
50 J = 2	Enters the quantity for J
60 Q = 1	Enters first quantity for Q
70 K = .02	Enters the quantity for K
80 DT = 1	Time step between recalculations
100 REM Start of Iteration	Start of recalculation loop
110 PSET (T,180 − Q)	Tells the computer to plot a point for T and a point for Q (180 − Q is necessary to get a graph with the normal starting 0 point in the lower left corner)
120 DQ = J − K*Q	Equation for change in Q over time
130 Q = Q + DQ*DT	Equation for Q after change
140 T = T + DT	Time is increased by one unit
150 IF T < 240 GOTO 100	If the graph is not at the end of the screen (240 points), it will run again starting at statement 100

what is stored per unit time. An economic example of the TANK model is a bank account that receives a steady income and has expenditures in proportion to the money in the account (program CASH.bas, Figure 9.3 and Table 9.4). In this case money is inflowing as income. The tendency for more money to cause higher spending might be called a "first law of economics." Spending with time is set to be a proportion of the stored money. Because work is required by those processing the money, heat sinks are shown dispersing energy, but the quantities in the program are money.

Simulation models are like controlled experiments, holding everything the same while you change one variable to see the effect. After running a program, you can change one thing in the program to see what would happen. For each program we suggest some "What If" experiments. No doubt you have others to consider. Sometimes you systematically test the effects of all the sources and other factors in a model, one at a time, to see what responds more or less. This is called a *sensitivity analysis.*

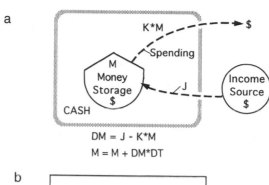

$$DM = J - K*M$$
$$M = M + DM*DT$$

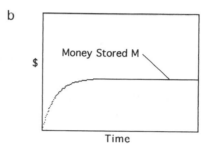

FIGURE 9.3.
Model of growth of storage of money (program CASH.bas). (a) Systems diagram and equations; (b) typical simulation of growth in cash money to steady state.

TABLE 9.4
Program of Money Storage
(CASH.bas) (Figure 9.3)

```
10    REM PC: CASH (money storage)
15    CLS
20    SCREEN 1,0
30    COLOR0,0
40    LINE (0,0) - (240,180),3,B
50    J=4
60    M=1
70    K=.05
80    DT=1
90    T0=1
95    M0=1
100   PSET(T/T0,180-M/M0),3
110   DM=J-K*M
120   M=M+DM*DT
130   T=T+DT
140   IFT/T0<240GOTO100
```

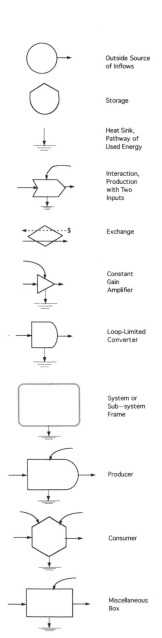

Outside Source
of Inflows

Storage

Heat Sink,
Pathway of
Used Energy

Interaction,
Production
with Two
Inputs

Exchange

Constant
Gain
Amplifier

Loop-Limited
Converter

System or
Sub–system
Frame

Producer

Consumer

Miscellaneous
Box

Switching
Actions,
On–Off
Processes

◆

9.4   ITERATION TIME

At the bottom of the program there is a loop with the calculations of time and quantity and plotting of points repeated over and over (statements 100–150 in Table 9.4). DT is the iteration time interval, the time step, the time represented by each passage around the loop. The way values are added in successive steps is illustrated in Figure 6.2.

DT is part of the statement that advances the time:

140   $T = T + DT$ New time equals old time plus the time interval for iteration.

DT, the interval, is multiplied by the change rate (DQ) for each interval

130   $Q = Q + DQ*DT$ New quantity equals the old quantity plus the change for the time interval.

DT was put in the scaling factor section of the program so that it can be easily changed.

83   $DT = 1$ Start with 1, but adjust after testing for accuracy.

By making DT smaller than 1, the program goes through the loop more often for the same passage of time. More points are plotted. The simulation mathematics is more accurate, but it takes longer for the run to cover the same time. If you make DT large, the simulation is fast but the graph it generates may be in error. You can find out if it is erroneous by setting DT smaller and running again. The large value is OK if the smaller runs with the smaller value are identical. Varying DT does not change the time scale (amount of time represented by the horizontal axis). Time span and scale represented on the horizontal axis of the plot are controlled by T0 (T zero).

When the DT is too large, the flow into a storage in one iteration may be so large that the storage jumps up much higher than its true value; then on the next iteration, with high value the outflows are too large and the storage may discharge too much and go to low values, even zero or minus values. This erroneous jumping back and forth is called *artificial chaos* and shows up on the screen as scattered points or vertical solid lines, often stopping the program and causing overload statements to appear.

◆

9.5   FACTORS FOR KEEPING GRAPHS ON SCALE

The simulation graph in Figure 9.1, using the TANK program in Table 1.1, had values that were within the scale of the y and x axis of the graph. However, a different set of initial conditions and coefficients might have caused the graph to go off scale (too large) or be so small that its plots are not distinuishable from the zero line. Plots that go way off scale may stop the program with an error message. Thus, a good general procedure to practice is to include scaling factors on the plot statements so that it is easy to increase or decrease the amplitude of the quantity being plotted. In Table 9.5, for example, Q0 is included in line 90 and 110:

$$90 \quad Q0 = 1$$
$$110 \quad PSET \ (T, \ 180\text{-}Q/Q0), \ 3$$

TABLE 9.5
Program for a Stored Quantity (TANK2.bas) (Figure 9.1)

```
10   REM PC: TANK2 (Storage with scaling)
15   Cls
20   SCREEN 1,0
30   COLOR0,0
40   LINE (0,0) - (240,180),3,B
50   J=2
60   Q=1
70   K=.02
80   REM Scaling Factors
83   DT=1:                    REM Time interval for each loop
86   T0=1:                    REM Factor scaling Time horizontal axis
90   Q0=1:                    REM Factor for scaling vertical axis
100                           REM Start of Loop
110  PSET(T,180-Q/Q0),3:      REM Q0 factor controlling Q scale
120  DQ=J-K*Q
130  Q=Q+DQ*DT:               REM Q Added depends on DT
140  T=T+DT:                  REM DT is time added in each iteration
150  IF T/T0<240 GOTO 100:    REM Use of T0 to control time scale
```

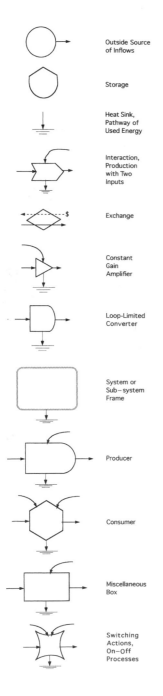

Outside Source
of Inflows

Storage

Heat Sink,
Pathway of
Used Energy

Interaction,
Production
with Two
Inputs

Exchange

Constant
Gain
Amplifier

Loop-Limited
Converter

System or
Sub–system
Frame

Producer

Consumer

Miscellaneous
Box

Switching
Actions,
On–Off
Processes

The scale of Q is increased or decreased by changing Q0.

If one of the quantities on the graph goes off scale, a statement can be put in to limit it:

$$115 \quad IFQ/Q0 > 180 \text{ THEN } Q/Q0 = 180$$

Time on the horizontal axis is scaled by putting T0 (0 is a zero) in the plot statements. T0 also has to be included in the last statement that sends the calculation back around the loop so that the simulation will stop at the end of the space in the frame. In Table 9.5, for example:

$$86 \quad T0 = 1$$
$$110 \quad PSET (T/T0, 180 - Q/Q0),3$$
$$150 \quad IF T/T0 < 320 \text{ GOTO } 100$$

Scaling factors (Q0, T0, etc.) may be used as divisors as in the TANK2.bas program in Table 9.5, or they may be used as multipliers. If scaling factors are used as multipliers, these lines are revised as follows:

$$115 \quad IF Q*Q0 > 180 \text{ THEN } Q*Q0 = 180$$
$$110 \quad PSET (T*T0, 180 - Q*Q0),3$$
$$150 \quad IF T*T0 < 320 \text{ GOTO } 100l$$

"What If" Experimental Problems for the TANK2.bas Program in Table 9.5

9.5–1 What would the simulation graph look like if you started with 100 liters of water instead of 1 liter? First, make an educated guess (an hypothesis). Then change Q to 100 by typing: 60 Q = 100. The computer will substitute this new statement 60 for the original. Type RUN and press Return; the new simulation graph will appear. Was your hypothesis right? To make another change you need to return the program to the original by retyping statement 60: 60 Q = 1.

9.5–2 What would happen if you cut off the water inflow after 96 hours? What statement would you add that tells the computer to change J when the time is 96 hours? Type:

$$115 \quad IF T = 96 \text{ THEN } J = 0$$

9.5–3 For another experiment study the effect of the constant outflow fraction per time. What if leaves on the forest floor were exposed to higher temperatures, causing faster decomposition? Change K to 0.1. Predict the effect of a faster outflow rate.

9.6 PROGRAM FORMAT AND USE OF REMARKS IN PROGRAMS

Although each program has been chosen because it illustrates a new concept or different system, the sequence of lines in all the programs in this book is similar to that in the TANK program.

Before the Iteration (Repeating Calculations):

Remark with name of program and whether PC or Macintosh
Statement to clear the screen
Scaling factors and iteration interval
List of constant coefficients
Values of sources
Starting values (initial conditions)

Start of Iteration:

Graphic plot statements
Ordinary equations
Change rate equations
Calculation of new values for storages (state variables)
Limit statements
Statement returning the program to the start of iteration
Program stops when there are no statements left

The last four categories of items entered before the iteration may be in any order. Within the iteration, statements that use other values must follow the lines where those values were put into memory. To be correct, all of the change equations (example: DQ) have to precede the calculations of new state variables (example: Q = Q + DQ).

The program will stop if a denominator (X) of a quotient

becomes zero. This can be avoided by including a statement preventing the value from being zero.

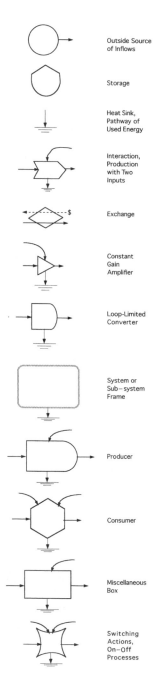

$$\text{If } X = 0 \text{ THEN } X = 0.00000001$$

In BASIC, anything in a program line that follows REM is ignored by the program as a remark. Normally programmers put extensive remarks on each line so that later people can see what that line does and why. However, for simple minimodels that have the same structure, the remarks can distract the user from visualizing the system's structure and function in overview. In this book some remarks are used, but not for the regular features common to all the listings.

Some programmers avoid using the GOTO commmand in large, complex programs lest it cause tortuous pathways that are hard to follow or become confused when iterations involve nested loops within loops. Instead, the FOR and NEXT commands are suggested. However, GOTO is more easily understood with the simple, standard structure of programs in this book.

Whereas most of the statements of our programs will fit most versions of BASIC, the plotting statements that make the computer draw a graph of the results are different.

◆

9.7   SIMULATING EXPONENTIAL GROWTH WITH PROGRAM EXPO

A model of autocatalytic growth on constantly available energy, EXPO, was introduced in Figure 2.13 and is shown with its equations for growth in Figure 9.4a. For example, a population of mice that has all the food it needs will increase faster and faster because the more individuals, the more reproduction, the more food consumed, and the more growth of the whole population.

EXPLANATION OF EQUATIONS

In the equations of Figure 9.4, E is the source that stays at a constant concentration no matter how much is used from it; Q

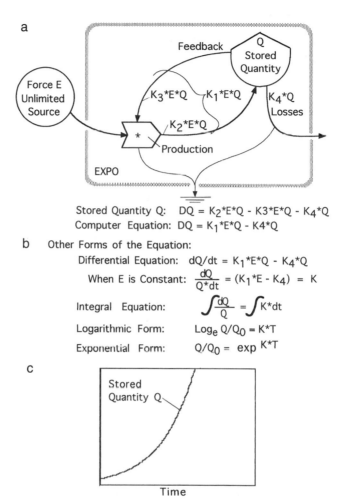

a

Stored Quantity Q: $DQ = K_2*E*Q - K_3*E*Q - K_4*Q$
Computer Equation: $DQ = K_1*E*Q - K_4*Q$

b   Other Forms of the Equation:

Differential Equation: $dQ/dt = K_1*E*Q - K_4*Q$

When E is Constant: $\dfrac{dQ}{Q*dt} = (K_1*E - K_4) = K$

Integral Equation: $\int \dfrac{dQ}{Q} = \int K*dt$

Logarithmic Form: $Log_e\ Q/Q_0 = K*T$

Exponential Form: $Q/Q_0 = exp\ K*T$

c

FIGURE 9.4.
Model of exponential growth on a constant available energy source (program EXPO.bas) including equations. (a) Systems diagram; (b) equations in other forms; (c) typical simulation graph.

is the quantity of animals using source E. The increase in quantity of mice per time K_1*E*Q is proportional to the source E and the population Q. The interaction symbol shows that the mice are eating the food to produce more mice. The term K_1 is a combination of two coefficients, K_2 and K_3. The increase in quantity of mice depends on their growth and reproduction (K_2*E*Q) minus the effort they expend in getting their food and water (K_3*E*Q). The net growth is K_1*E*Q ($K_1 = K_2 - K_3$); K_4 is the death rate coefficient, the proportion of Q which dies; and K_4*Q is the number of mice that die each week, the death rate.

To understand the process, divide both sides of the rate of change equation (dQ/dt) (Figure 9.4b) by Q to get the growth

Chapter Nine   Programming in BASIC

Outside Source
of Inflows

Storage

Heat Sink,
Pathway of
Used Energy

Interaction,
Production
with Two
Inputs

Exchange

Constant
Gain
Amplifier

Loop-Limited
Converter

System or
Sub–system
Frame

Producer

Consumer

Miscellaneous
Box

Switching
Actions,
On–Off
Processes

rate per individual $[dQ/(Q*dt)]$. Combining terms when E is constant shows the growth rate per individual to be a constant K. Exponential growth has a constant percent increase in each time period.

Integrating both sides of the equation produces a logarithmic equation, which says that the log of the population is a constant function of the time. In other words, the semi-log graph (log Q/Q_0 versus T) is a rising straight line.

If the logarithm of population (an exponent) is KT, then the population Q relative to that at the start, Q_0, is the number "e" with the exponent KT. Thus, the growth is exponential. Graphing the exponential equation generates Figure 9.4c, which is also produced by simulating the change equation. Running the BASIC program EXPO.bas listed in Table 9.6 plots the graph in Figure 9.4c. The change in quantity of mice over time DQ is the increase (K_1*E*Q) minus the decrease (K_4*Q): $DQ = K_1*E*Q - K_4*Q$. The quantity of mice Q after a week is the number at the start plus the change: $Q = Q + DQ*DT$.

EXAMPLES Of EXPONENTIAL MODELS

During the early states of population growth, when the demand for food is small compared to the amount available, almost

TABLE 9.6
Program of Exponential Growth
(EXPO.bas) (Figure 9.4)

```
10   REM PC: EXPO (Exponential growth)
15   CLS
20   SCREEN 1,0
30   COLOR0,0
40   LINE (0,0)-(240,180),3,B
50   Q=10
60   E=1
70   K1=.07
80   K4=.05
85   DT=1
90   T0=1
95   Q0=1
100  PSET(T/T0,180-Q/Q0),3
105  DQ=K1*E*Q-K4*Q
110  Q=Q+DQ*DT
120  T=T+DT
130  IFT/T0<240GOTO100
```

any population of microbes, plants, or animals can grow exponentially. Because an unlimited food supply is not possible indefinitely, eventually the population of mice would stop growing so fast. Then we would have to use a different model to fit the new situation of a limited food supply.

Industries, like oil and mining, have grown exponentially when new oil fields and sources of gold were found. The economy of the United States, from the early 1800s to the mid-1900s was growing exponentially using the great abundance of natural resources and newly discovered fossil fuels.

GROWTH WITH QUADRATIC FEEDBACK

If the autocatalytic growth is proportional to the square of the storage Q^2, the quadratic growth rate is less at first but accelerates to grow faster than the simpler EXPO model. You can test this by changing the equation line in Table 9.6 to read: 105 DQ = K1*E*Q*Q − K4*Q. Over much of its history the growth of the United States economy was quadratic, faster than exponential. This means that growth was accelerated by self-interactions. In other words, growth processes were cooperative.

"What If" Experimental Problems for the EXPO.bas Program in Table 9.6

9.7–1 Predict what will happen to the growth of the mice population if you double the concentration of food. Retype line 60 as E = 2. Then press Return and type RUN. Next cut the concentration of food (E) in half (as if each food pellet has half the nutrition). What happens to the mouse population?

9.7–2 What will the graph of Q look like for another kind of mice that eats more efficiently, increasing its growth rate per individual? Type 70 K1 = 0.08. Change statement 70 again, making K1 less than the original 0.07. What does that graph look like?

9.7–3 What would happen to the growth of the mice population if the mice caught a virus, increasing the death rate? To test your hypothesis, would you increase or decrease K4? Show what happens to the growth of the population.

9.7–4 If E is 1 and you make K1 equal to K4, what will happen to the population? Try it.

9.8   SIMULATING DISPERSAL FROM STORAGE WITH MODEL DRAIN

Figure 9.5 (and Figure 6.7) shows a storage with only one pathway, an outflow. The dispersal model and its BASIC program in Table 9.7 are useful for showing the mathematics of exponential decline. This model is a version of storage like TANK but without inflow. When the run starts, the storage drains rapidly and then more slowly because a constant fraction K of what is remaining of Q flows out in each unit of time. The mathematics in Figure 6.7 is like that for exponential growth except with a minus sign. The result (Figure 9.5b) is a declining exponential.

EXAMPLES OF THE DRAIN MODEL

Water draining from a tank with an outlet or a lake with a stream flowing out are examples of this model. Another example is the decay of radioactivity of elements like uranium. The same proportion of what remains decays in each time period. If half is lost in one period of time, then half of what remains is lost in the next, and half of that in the third period, and so on. We

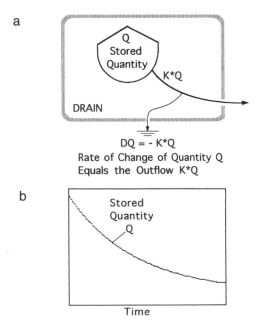

a

$DQ = -K*Q$
Rate of Change of Quantity Q
Equals the Outflow K*Q

b

FIGURE 9.5.
Model of a storage with one outflow pathway proportional to the storage remaining (program DRAIN.bas). See Figure 6.7 for equations explaining exponential decay. (a) Systems diagram and change equation; (b) typical simulation graph.

TABLE 9.7
Program of Draining Tank
(DRAIN.bas) (Figure 9.5)

```
10   REM PC: DRAIN  (Draining tank)
15   CLS
20   SCREEN 1,0
30   COLOR0,0
35   LINE (0,0)-(240,180),3,B
50   Q=150
60   K = .01
65   Q0=1
70   T0 = 1
75   DT=1
80   PSET (T,160 - Q/Q0),3
90   DQ = - K * Q
100   Q=Q+DQ*DT
110   T = T + DT
120   IFT/T0<240GOTO80
```

describe decay systems by their *half-lives,* the time to lose half. Uranium 238 has a half-life of 4 billion years. That means that in 2 billion years half of a sample has decayed to lead. In the next 2 billion years only half of that will decay.

Spending money at a rate dependent on the money left is an economic example. You are given a bank account from which you spend a certain percent every month. Your account goes down exponentially.

Decomposition of a pile of leaves by microbes often fits this model. Another example is the survival curve of those populations of organisms in which deaths are proportional to the number of individuals remaining. The survival curve is the fraction of those born at the same time that are still living at a later time.

In pollution studies, a rough but widely used method of measuring organic matter in waters is the measurement of BOD (biochemical oxygen demand). When water is put into a bottle and then kept in the dark, decomposition (respiration) occurs in proportion to the amount of organic matter available. The oxygen remaining after 5 days is used as a comparative measure of the organic matter originally present. The change in oxygen in the bottle often follows this graph.

"What If" Experimental Problems for the DRAIN.bas Program in Table 9.7

9.8–1 If you started with a larger pile of leaves in the forest, but used the same rate of decomposition, would the pile be gone sooner, at the same time, or later? Change Q to 160 and run both graphs.

9.8–2 Double the rate of your spending of your bank account (change K to 0.02). How long does your money last?

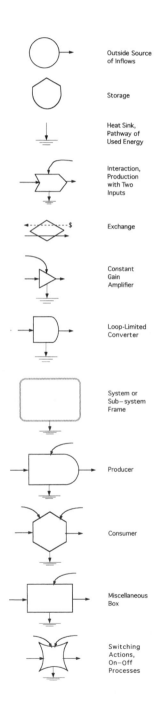

Outside Source of Inflows

Storage

Heat Sink, Pathway of Used Energy

Interaction, Production with Two Inputs

Exchange

Constant Gain Amplifier

Loop-Limited Converter

System or Sub–system Frame

Producer

Consumer

Miscellaneous Box

Switching Actions, On–Off Processes

9.9 SIMULATING PRODUCTION, CONSUMPTION, AND RECYCLE WITH MODEL PC&CYCLE

A model with the essence of production, consumption, and recycle of materials was simulated with EXTEND in Chapter 5 (Figure 5.2). It is simulated again here using BASIC. Using numbers for flow and storage on the energy systems diagram (Figure 7.4a) and procedures from Chapter 7, coefficients were calculated in Figure 7.4b. With the equations in Figure 9.6a a BASIC program was written (Table 9.8). A simulation run with this program started with low initial values for the two storages, showing the pattern of growth, which takes some years for organic and nutrient storages to reach steady state (Figure 9.6c).

We verbalize the energy systems diagram (Figure 9.6a) as follows. Nutrients flow into and out of the soil. Production by plants uses the energy source of the sun, rain, and the soil nutrients to make organics in plant and soil biomass. Consumers including animals and microbes use the organics and release nutrients to the environmental storage again. Included in the lower pathway is depreciation of the organic matter by chemical and physical processes— the degradation described by the second law of thermodynamics.

EXPLANATION OF EQUATIONS

Available energy R is the remaining difference between the inflow J and that drawn into plant production K_0*R*N. Nutrients flow into the system J_n and flow out K_7*N in proportion to nutrient storage N. Production K_1*R*N is proportional to available nutrients N and available energy R. Organics storage Q is a balance between production and losses by the animal–microbe consumption pathway K_2*Q and a pathway of depreciation and

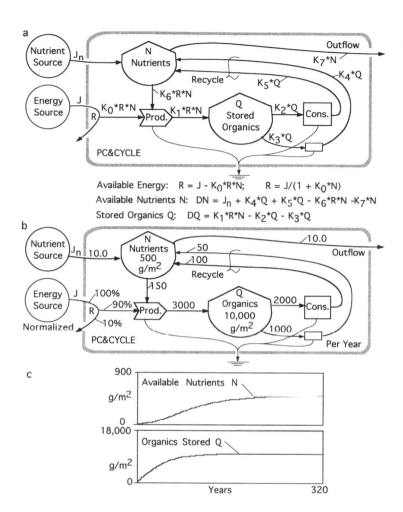

FIGURE 9.6.
Model of production, consumption, and recycle typical of ecosystems (program PC&CYCLE.bas). (a) Systems diagram with equations and math terms on the pathways; (b) diagram with values of flows and storage used for calibration; (c) typical simulation result. See simulation with other methods in Figures 5.2 and 10.10.

other losses K_3*Q, both in proportion to the storage Q. Nutrients are released back to the soil by the pathway, K_4*Q and K_5*Q.

EXAMPLES OF PC&CYCLE

Examples of this model are ecosystems on land such as deserts or grassland, and in water such as a pond or the ocean. Because there is no outflow from the system, it would have to be changed to use it for a farm or other yield system.

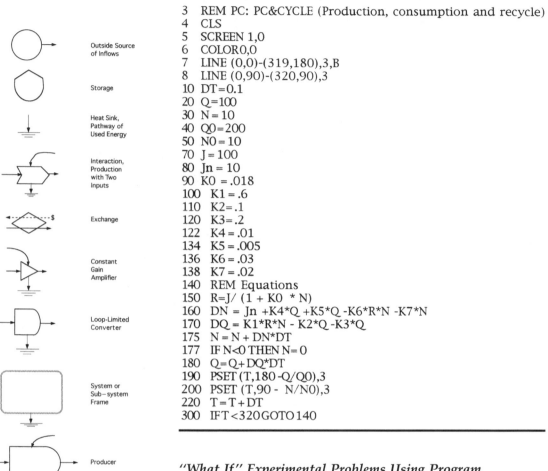

TABLE 9.8
Program of Production, Consumption, and Recycle
(PC&CYCLE.bas) (Figure 9.6)

```
3    REM PC: PC&CYCLE (Production, consumption and recycle)
4    CLS
5    SCREEN 1,0
6    COLOR0,0
7    LINE (0,0)-(319,180),3,B
8    LINE (0,90)-(320,90),3
10   DT=0.1
20   Q=100
30   N = 10
40   Q0=200
50   N0 = 10
70   J=100
80   Jn = 10
90   K0 =.018
100  K1 =.6
110  K2=.1
120  K3=.2
122  K4=.01
134  K5=.005
136  K6=.03
138  K7=.02
140  REM Equations
150  R=J/ (1 + K0 * N)
160  DN = Jn +K4*Q +K5*Q -K6*R*N -K7*N
170  DQ = K1*R*N - K2*Q -K3*Q
175  N = N + DN*DT
177  IF N<0 THEN N= 0
180  Q=Q+DQ*DT
190  PSET (T,180 -Q/Q0),3
200  PSET (T,90 - N/N0),3
220  T = T + DT
300  IFT<320GOTO 140
```

"What If" Experimental Problems Using Program PC&CYCLE.bas in Table 9.8

9.9–1 What would happen to the growth of organics and nutrients in the soil if there were more consumers? Make a prediction and then increase K3 to 0.25 and run the program. Explain the results.

9.9–2 If there were less sunlight and rain (energy resource), how would it affect the system? Change J to 50 and run the program. Explain.

9.9–3 If the soil started richer in nutrients, how would that affect the growth? Change N to 100. Predict the results. Explain why the nutrients' line leveled off at the same place as in the original graph.

TABLE 9.9
Location of Useful Modeling and Programming Mechanisms

Item	Model	Location
Entering tabular data with READ, DATA	WETLAND	Table 16.1
Modeling effect of temperature	MUSMOD	Figure 21.3a
Plotting with expressions to get a straight line	SPECAREA	Table 16.4
Integrating to derive equations for graphs	EXPO	Figure 9.4
	DRAIN	Figure 6.8
Steady state by setting change to equal zero	INTLIMIT	Table 12.3
Conservation of recycling materials (nutrients)	AUTOCYCL	Table 12.6
Conservation of recycling land area	ROTATION	Table 18.4
	RESERVE	Table 17.6
Conservation of recycling money	ECONP&C	Table 17.4
Values per area and per volume together	LAGOONII	Figure 21.2a
Representing calibration on isolated units	LAGOONII	Figure 21.2c
Generating a family of curves with time	LOGISTIC	Table 13.5
Generating a family of curves relating variables	RESERVE	Table 17.6
Plotting the steady-state values	SPECAREA	Table 16.4
Plotting rates and net changes	CLIMAX	Table 16.2
Plotting both with time and other variables	CHAOS	Table 15.6
	PREYPRED	Table 15.1
Change of inputs after a time	DESTRUCT	Table 15.5
Plotting lines between plotted points	FIRE	Table 15.3
	NETPROD	Table 12.1
Use of sine wave to represent day-night inputs	DAYP&C	Table 12.4
Turning on pathways by changing $Z = 0$ to $Z = 1$	TRADE	Table 19.1
Setting upper limit to curves going off scale	EXPO	Table 14.2
Plotting production rates	FACTORS	Table 12.2
"Force" proportional to purchases	TANKSALE	Table 17.2
Representing light with "flow remainder"	Most models in Pt III	
Scarce materials with "flow remainder"	EARTHGEO	Table 20.3
One model driving another	STATECON	Table 20.4

9.9–4 If you increase or decrease the inflow of nutrients, how will this change the system? Change Jn to 1 and run the program; then change it to 20 and make another run. What do these graphs tell you about the effect of nutrients? What *is* the limiting factor in this system?

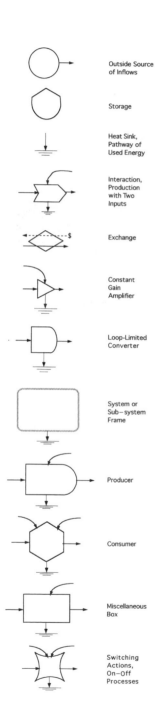

Outside Source of Inflows

Storage

Heat Sink, Pathway of Used Energy

Interaction, Production with Two Inputs

Exchange

Constant Gain Amplifier

Loop-Limited Converter

System or Sub-system Frame

Producer

Consumer

Miscellaneous Box

Switching Actions, On–Off Processes

9.10   USEFUL MODELING AND PROGRAMMING MECHANISMS

Whereas this chapter concentrates on the essential BASIC commands for programming energy systems minimodels, the examples in Parts III and IV show other techniques and algorithms. Table 9.9 lists where some of these special methods are used.

CHIPMUNK BASIC FOR MACINTOSH

Since the older QUICKBASIC for Macintosh could not be included on the book disk, a similar freeware program *Chipmunk BASIC* for Macintosh developed by Ronald H. Nicholson was included on the CD-ROM along with manuals and tutorials. This form of BASIC is similar to BASICA and QUICKBASIC except for slight differences in statements for graphics and plotting as explained in Appendix B. Appropriate substitutions of graphics lines were made in many of this book's BASIC programs and included in the book disk so that Macintosh users can do the exercises in this book with Chipmunk BASIC.

Chapter Ten

◆

SIMULATING WITH STELLA

Jay F. Martin* and David R. Tilley**

*Department of Oceanography and Coastal Sciences,
Louisiana State University, Coastal Ecology Institute,
Baton Rouge, Louisiana 70803 and **Environmental Engineering
Sciences and Center for Wetlands, University of Florida,
Gainesville, Florida 32611

133

$\mathbf{A}$nother tool for modeling dynamic systems with minimal mathematical sophistication and little programming knowledge is the iconographic modeling software STELLA. By not requiring the user to develop computer programming code this software allows more time for analyzing model behavior. This chapter shows how to convert models from the energy systems language into STELLA models by creating, calibrating, simulating, graphing, and playing "What If" with three minimodels: a single storage model, TANK; an exponential growth model, EXPO; and a production, consumption and recycle model, PC&CYCLE. Each minimodel is diagrammed with energy systems language symbols and STELLA icons, and PC&CYCLE is also diagrammed with Forrester symbols, in order to compare the three systems languages. Figure 10.1 shows the equivalency of each language's symbols. The icons and modeling language of STELLA are based on the system dynamic modeling methodology developed by Jay Forrester (1961). The programming language DYNAMO was introduced at that time also. DeWit and Goudriaan (1978) provided an introduction to simulation using DYNAMO.

To create a simulation model in STELLA, symbol icons representing storages, flows, and system variables are pulled down from the toolbar menu and placed on a worksheet. The symbols are connected and mathematical equations written to relate the model components. The four basic STELLA icons are displayed and labeled in Figure 10.2. The explanation in this chapter is applicable to both the Windows PC and Macintosh versions of STELLA II.

$\blacklozenge$

10.1 SIMULATING THE TANK MODEL
USING STELLA

The simulation of a single storage with one inflow and one outflow is well suited for introducing the STELLA modeling

STELLA is available from High Performance Systems, Inc., 45 Lyme Rd., Hanover, NH 03755.

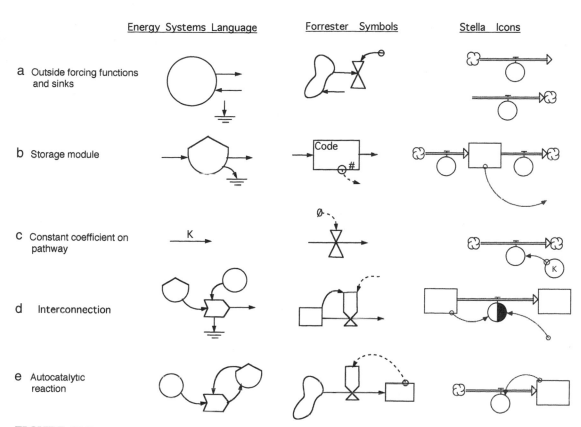

Energy Systems Language	Forrester Symbols	Stella Icons

a Outside forcing functions and sinks

b Storage module

c Constant coefficient on pathway

d Interconnection

e Autocatalytic reaction

FIGURE 10.1.

Comparison of representations of basic modeling concepts using energy systems language, Forrester symbols, and STELLA icons.

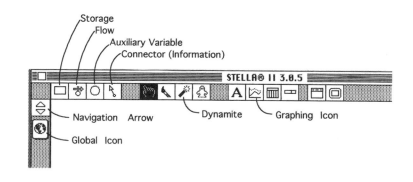

FIGURE 10.2.

The four basic icons used in STELLA.

software. The TANK model could represent the flow of water entering and leaving a kitchen sink, deposits and withdrawals from a bank account, or refrigerators entering and leaving a warehouse. A step-by-step example of how to create such a model using STELLA software follows. For comparison, the model is presented using the energy systems language and the STELLA icons (Figure 10.3).

CONSTRUCTION OF TANK MODEL

1. After entering Windows, double click on the STELLA icon. This opens the STELLA program group, where the STELLA icon should be double clicked again.

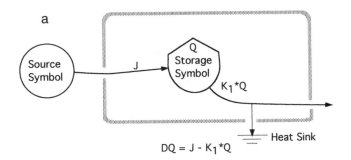

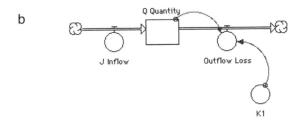

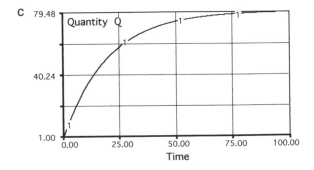

FIGURE 10.3.
Diagrams of TANK model using (a) energy systems language, (b) STELLA icons, and (c) output graph.

2. The software will open into the Model Construction Layer. If using a version of STELLA earlier than 5.0, the Mapping Layer should be switched to the Model Construction Layer by clicking once on the navigation arrow (Figure 10.2) on the left of the screen.

3. In the Model Construction Layer, use the hand icon to select a storage icon from the toolbar and place it onto the worksheet. This is accomplished by moving the hand to the rectangular storage icon in the toolbar, clicking once, positioning the storage on the worksheet, and clicking again to deposit the icon. The storage will be shaded with the highlighted text NONAME I appearing above. With the storage symbol highlighted, rename it Q_QUANTITY by typing from the keyboard. Press Enter.

4. To add the J_INFLOW pathway to the worksheet use the hand icon to select the flow icon from the toolbar by clicking on it once. Place the flow icon on the worksheet by moving the mouse slightly to the left of Q_QUANTITY, press the mouse button down and hold it while dragging the flow icon to the right until Q_QUANTITY becomes shaded. Release the mouse button to establish the connection between J_INFLOW and Q_QUANTITY. The flow icon will be shaded with the highlighted text NONAME I appearing below. With the flow icon symbol highlighted, rename it J_INFLOW by typing from the keyboard. Press Enter.

5. To add the outflow loss pathway to the model, obtain another flow icon from the toolbar by clicking on it once with the hand icon. Place the flow icon on the worksheet by moving the mouse over Q_QUANTITY and press the mouse button down and hold it while dragging the flow icon to the right. Release the mouse button when the flow icon is sufficiently long (refer to Figure 10.3b). Rename this flow icon from NONAME I to OUT-FLOW LOSS following the method described in step 4.

Note: When a mistake is made placing an icon onto the worksheet, it may be deleted using the dynamite icon (Figure 10.2). After obtaining the dynamite from the toolbar, position it above the icon to be deleted and click once.

6. The coefficient regulating the OUTFLOW LOSS from Q_QUANTITY should now be added to the model. After clicking once on the circle icon from the toolbar, move the mouse below the VALVE ON OUTFLOW LOSS and click once to position it. Rename this circle icon from NONAME I to K1 following the method described in step 4.

7. Information about Q_QUANTITY and K1 are required at the OUTFLOW LOSS VALVE for controlling the flow rate from Q_QUANTITY. The red arrow connector icon, symbolizing information flow, is used to do this. Select a red arrow connector icon from the toolbar by clicking once on it with the hand. Position the red arrow connector icon over Q_QUANTITY, depress the mouse button and drag the arrow icon to the OUTFLOW LOSS VALVE, releasing the mouse button when OUTFLOW LOSS becomes shaded.

8. The outflow coefficient, K1, and OUTFLOW LOSS VALVE must also be connected with a red arrow connector icon. Select a red arrow connector icon from the toolbar by clicking once on it with the hand. Position the red arrow connector icon over the K1, depress the mouse button and drag the arrow icon to the OUTFLOW LOSS VALVE, releasing the mouse button when OUTFLOW LOSS becomes shaded.

CALIBRATION OF TANK MODEL

1. Click on the global icon located on the left of the screen to switch from the Mapping Mode to the Modeling Mode, which is represented by X^2. Question marks will now appear in the model where calibration information is needed.

2. Double click on Q_QUANTITY to bring up the dialog box shown in Figure 10.4. Enter the initial value of 1.0 in the

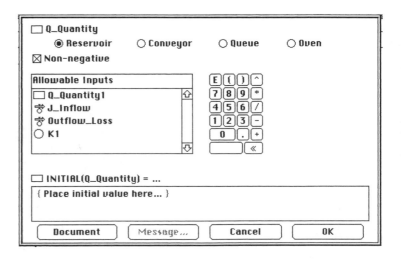

FIGURE 10.4.
Dialog box used to enter initial value for storage in the TANK model.

highlighted Initial Value box at the bottom of the dialog box. This can be entered from the keyboard, or using the mouse and digit pad in the dialog box. Click OK to return to the model. The question mark should have disappeared from Q_ QUANTITY.

3. To calibrate the inflow pathway, double click on J_INFLOW to reveal a Flow dialog box (Figure 10.5). In the box located below the text labeled Inflow_J = . . . enter 4.0. Click OK. This means that 4.0 units will enter Q_QUANTITY for each time step.

4. To set the value of K1, double click on it to open a dialog box. Enter a value of 0.05. This is the proportion to Q_ QUANTITY that flows out each time step.

5. To add the equation which controls the outflow loss pathway, double click on the OUTFLOW LOSS VALVE to open the dialog box displayed in Figure 10.6. The model components (K1 and Q_QUANTITY), which determine the outflow loss via the red arrow connectors, are listed in the Required Inputs box. With the mouse, select the outflow coefficient K1 to make it appear in the bottom box. Follow this by clicking on the multiplication symbol (*) on the number keypad, and Q_QUANTITY in the Required Inputs box to establish the equation OUTFLOW LOSS = K1*Q_QUANTITY. Click OK.

6. The model is now calibrated and ready to run but should be saved first. From the menu bar select FILE and then the Save As . . . option to save the model as TANK.stm. Click OK.

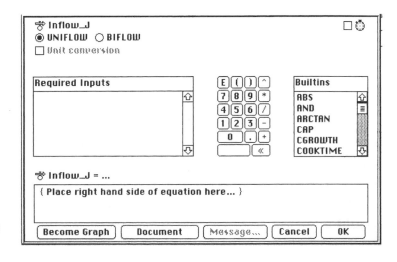

FIGURE 10.5.
Dialog box used to calibrate the inflow in the TANK model.

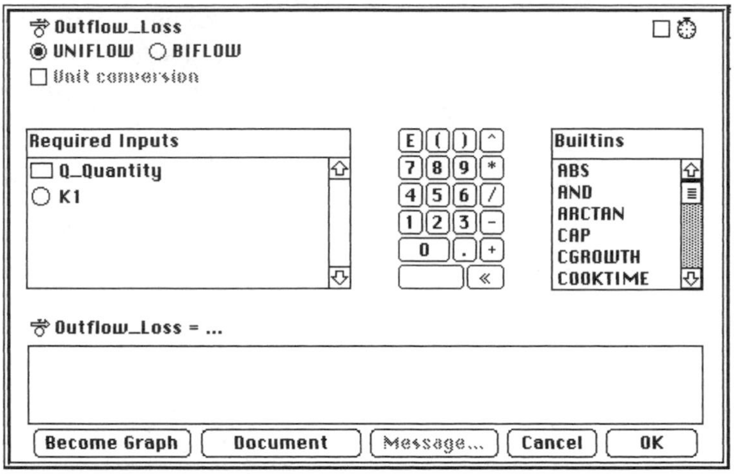

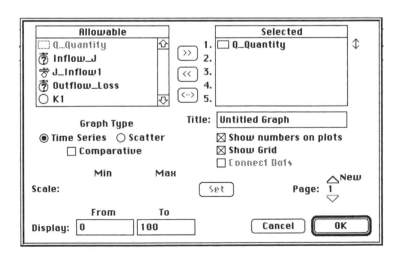

FIGURE 10.6.
Dialog box used to calibrate
outflow in the TANK model.

RUNNING SIMULATION AND
GRAPHING OUTPUT

1. The graph icon (Figure 10.2) must be added to the model
   worksheet in order to plot the simulation output. Click once
   on the graph icon from the toolbar. Position the mouse hand
   anywhere on the worksheet and click the mouse button once
   again to add the graph icon. A blank graph will appear when
   the icon is released.
2. To open the graphics dialog box (Figure 10.7), double click
   anywhere on the blank graph. In the Allowable box double

FIGURE 10.7.
Graphics dialog box.

click on Q_QUANTITY. This places Q_QUANTITY in the Selected box, allowing the model to plot its value as a function of time for the simulation run. Click OK. Close the graph screen by clicking on the Control Menu button located in the upper left corner of the screen.

3. Before running the simulation you must first set the RUN specifications. From the menu bar select RUN and then Time Specs . . . to open a dialog box (Figure 10.8). The length of the simulation should be changed from the default value of 12 time steps to 100. Do this by entering 100 in the To: box. Click OK to return to the model.

4. To run a simulation, select RUN RUN from the menu bar.

5. View the graphical output (Figure 10.3c) by double clicking on the graph icon in the worksheet. To view the graph during the simulation, open the graph screen before running the model.

6. If you need to view the code for this model, click on the downward navigator arrow (triangle located to the left of the worksheet).

"What If" Experiments

We will show how to change the characteristics of the TANK model so that "What If" questions can be explored.

10.1–1. What happens if the model is started with a higher amount in storage? If this TANK represented a sink,

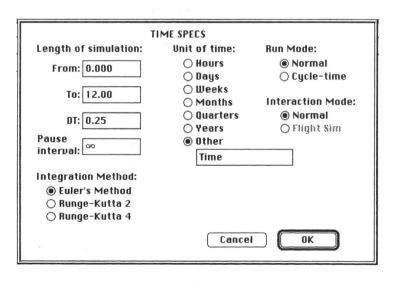

FIGURE 10.8.
Dialog box to set simulation run specifications.

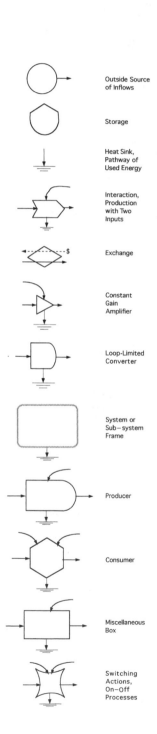

Outside Source
of Inflows

Storage

Heat Sink,
Pathway of
Used Energy

Interaction,
Production
with Two
Inputs

Exchange

Constant
Gain
Amplifier

Loop-Limited
Converter

System or
Sub–system
Frame

Producer

Consumer

Miscellaneous
Box

Switching
Actions,
On–Off
Processes

imagine starting with 100 liters of water instead of 1 liter. Change the initial value of Q_QUANTITY by double clicking on it and replacing 1.0 with 100.0. Click OK. Run the model by selecting RUN RUN from the menu bar. Double click on the graph icon to reveal the results (note the change in the scale of the vertical axis). After completing this experiment return Q_QUANTITY to 1.0.

10.1–2. What happens if the outflow coefficient K1 is doubled? Using the sink analogy again, imagine doubling the size of the drain pipe. Double click on K1 and change the value from 0.05 to 0.10, then run the model and view the results.

◆

10.2   SIMULATING THE EXPO MODEL USING STELLA

This section demonstrates how to change the TANK model to EXPO, a model of exponential growth. It is assumed that the TANK model was created, calibrated, and simulated in the previous section. The following is a list of the steps necessary for converting the TANK model to the EXPO model. The EXPO model is diagrammed with the energy systems language and STELLA icons in Figure 10.9 (refer to energy systems representation of this model (Chapter 9.7)). The EXPO model has as its main differences with the TANK model a force source rather than a flow, and an autocatalytic feedback (pathway from Q_QUANTITY to interaction symbol).

1. Open STELLA by double clicking on its icon. Once STELLA is open it is necessary to first close the model file that opens when STELLA does. Do this by selection FILE and then Close Model from the menu bar. Open the TANK model by selecting from the FILE menu Open, highlighting TANK.stm in File Name box, and clicking OK. Make sure you are in the Modeling Mode of STELLA (the X^2 symbol should be showing on the left of the screen).

2. Double click on the text K1 below the outflow pathway. Rename this coefficient K4. This will allow the use of K4 for an outflow coefficient in the EXPO model.

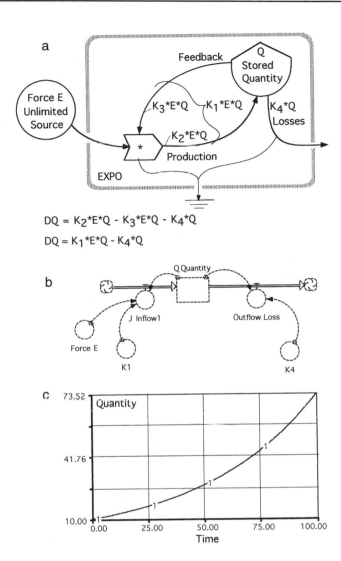

$$DQ = K_2{*}E{*}Q - K_3{*}E{*}Q - K_4{*}Q$$

$$DQ = K_1{*}E{*}Q - K_4{*}Q$$

FIGURE 10.9.
Diagrams of EXPO model using
(a) energy systems language,
(b) STELLA icons, and
(c) output graph.

3. Double click on the Q_QUANTITY storage icon and enter a new initial value of 10.

4. To change J_INFLOW from a constant flow source to a flow using a constant force source and to add an autocatalytic feedback pathway it is necessary to add two more variables (Force E and K1) and three connector pathways. To add the constant force source Force E, add a circle icon from the toolbar to the worksheet, placing it above and slightly to the left of J-Inflow. Name this variable Force E once it is placed on the worksheet.

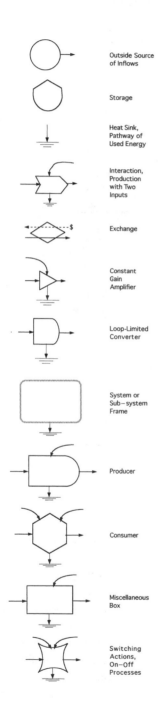

Outside Source
of Inflows

Storage

Heat Sink,
Pathway of
Used Energy

Interaction,
Production
with Two
Inputs

Exchange

Constant
Gain
Amplifier

Loop-Limited
Converter

System or
Sub–system
Frame

Producer

Consumer

Miscellaneous
Box

Switching
Actions,
On–Off
Processes

5. Add another variable K1 following the same steps (refer to Figure 10.9b for placement).

6. These added variables must now be connected to the J_ Inflow1 valve. Select the red arrow connector from the icon bar. Hold mouse over K1. Click and drag the arrow connector to the J_Inflow valve until it becomes shaded, and release mouse button. Repeat for Force E.

7. A red arrow connector must also be drawn from Q_ QUANTITY to the J_Inflow1 valve. Refer to Step 6 for instructions.

8. There should now be a "?" in K1, Force E, and the J_Inflow1 valve indicating that calibration information is needed. Double click on K1 to open the dialog box so that the initial value 0.07 can be set. Double click on Force E to set its value to 1.0. Double click on J_Inflow to open its dialog box. In the highlighted box enter FORCE E*K1*Q_QUANTITY to set the inflow equation. This should be done by clicking on these terms in the Required Inputs box. The TANK model is now converted to the EXPO model and is ready to run.

9. Run the simulation by selecting RUN RUN from the menu bar. (The output should resemble Figure 10.9c.)

"What If" Experiments and Sensitivity Analysis

10.2–1. Suppose the energy available for growth were doubled. Change E from 1.0 to 2.0. What happens to Q_ QUANTITY over the simulation run?

10.2–2. STELLA has the ability to run multiple "what if's" at the same time. This is called *sensitivity analysis*. This allows the user to specify changes of the model parameters and run simulations consecutively, plotting the results from each run on one graph for easy comparison. To try this, open the RUN menu and choose Sensi Specs. . . . Enter the number of simulation runs as 3 in the # of Runs: box. Select the FORCE E variable by double clicking on it in the Allowable box. Highlight FORCE E in the Selected (Value) box by clicking on it. This will open the Start: and End: boxes in the lower middle of the dialog box. Then define a beginning value of 1.0 and an ending value of 12.0 in these boxes. Click Set. The values for E to be used in the sensitivity analysis will appear in the Run # Value box. Click OK when finished.

Double click on the graph icon, then double click anywhere on the graph page so that the graph type can be modified. In the graph dialog box under Graph Type click once in the Comparative box. This allows STELLA to plot multiple simulations on one graph. Note that only one parameter can be plotted in the comparative model. Click OK. Now run the simulation (select RUN, S-Run from the menu bar).

10.3 PRODUCTION–CONSUMPTION–RECYCLE MODEL (PC&CYCLE)

Multiple storage and flow icons can be linked together to represent more complex models in STELLA. Figure 10.10 displays a production–consumption–recycle model, PC&CYCLE. This model can demonstrate the dynamic relationship between energy sources, nutrient inputs, and organic matter storages. The flow-limited source used in the energy systems language is shown as a combination of auxiliary variables (Energy Source, R, K0) in the STELLA model.

Both the interaction and linear flow symbols of the energy systems language have heat sinks indicating that a transformation process is occurring. To represent a transformation process in STELLA, the Unit Conversion feature is employed. This feature must be used to transform a flow entering a flow valve from the left into a new product exiting to the right. The Unit Conversion option is found at the top of the Flow dialog box (Figure 10.11). Clicking on the Unit conversion activation box will allow the selection of a UC auxiliary variable from the Required Inputs box. Half of the circle on the flow icon will become shaded as a result of activating Unit conversion.

Using the Unit Conversion feature to represent a transformation process in STELLA requires a modification to the method presented in the energy systems diagram. In Figure 9.6a the flow leaving Q Organics and entering the Consumption box ($K2*Q = 2000$) has a different value from the flow leaving the Consumption box ($K5*Q = 100$) due to the transformation of

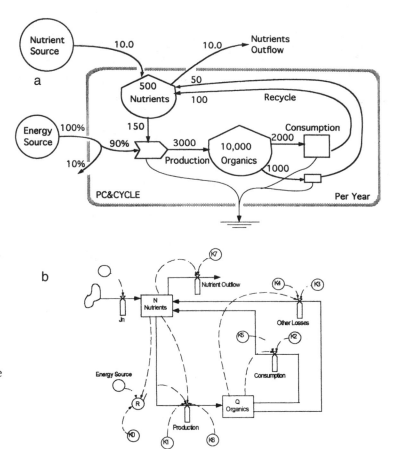

FIGURE 10.10.
Diagrams of
production–consumption–recycle
model PC&CYCLE using
(a) energy systems language,
(b) Forrester symbols,
(c) STELLA icons, and
(d) simulation with STELLA.

organic matter into nutrients. The Consumption Process could represent microbial decomposition of leaves into nutrients. To accomplish this transformation in STELLA the flow (K2*Q) entering Consumption is multiplied by UC2, a variable not present on the energy systems diagram. The UC2 variable equals K5/K2, which is the ratio of the coefficient of outflow to the coefficient of inflow. When UC2 is multiplied times K2*Q the resulting equation is K5*Q, the outflow from Consumption. The production and other loss pathways are treated similarly.

Referring to Figure 10.1c, Table 10.1, the two previous examples given in this chapter, and the PC&CYCLE model description given in Chapter 9.9, it should be possible to construct this model. The graphical output should resemble Figure 10.10d.

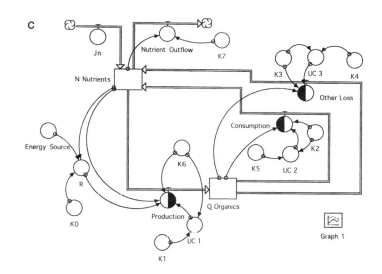

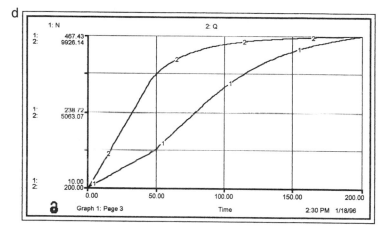

FIGURE 10.10.
(*Continued*)

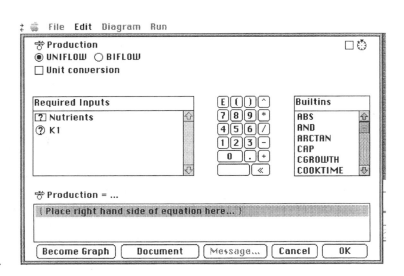

FIGURE 10.11.
Dialog box of flow icon
showing unit conversion option.

TABLE 10.1

Description and Quantification of Terms for PC&CYCLE Model

Model parameters	Description	Value/equation
N Nutrients	Initial value	10000
Q Organics	"	500
Energy Source	Constant	100
Jn	Constant inflow	10
K0	Coefficient	0.018
K1	"	0.6
K2	"	0.1
K3	"	0.2
K4	"	0.01
K5	"	0.005
K6	"	0.03
K7	"	0.02
UC 1	Equation	=K1/K6
UC 2	"	=K5/K2
UC 3	"	=K4/K3
R	"	=Energy Source/ (1 + K0*N)
Production		
Nutrients used	Flow equation	=K6*R*N Nutrients
Organics produced	"	=K6*R*N Nutrients *UC 1
Consumption		
Organics consumed	Flow equation	=K2*Q Organics
Nutrients produced	"	=K2*Q Organics *UC 2
Other Loss		
Organics consumed	Flow equation	=K3*Q Organics
Nutrients produced	"	=K3*Q Organics *UC 3
Nutrient Outflow	Flow equation	=K7*N Nutrients

Outside Source of Inflows

Storage

Heat Sink, Pathway of Used Energy

Interaction, Production with Two Inputs

Exchange

Constant Gain Amplifier

Loop-Limited Converter

System or Sub–system Frame

Producer

Consumer

Miscellaneous Box

Switching Actions, On–Off Processes

"What If" Experiments

10.3–1. What if the microbes are able to produce a larger amount of nutrients from the same amount of organic input? Simulate this by changing K4 from 0.01 to 0.05. What happens?

10.3–2. What if a similar system is modeled at a lower latitude where the intensity of sunlight is double the present amount? Simulate this by increasing the auxiliary variable Energy Source from 100 to 200. What happens to the level of organics? Nutrients?

10.4 COMPARISON OF SYSTEMS LANGUAGES

Converting from an energy systems diagram to a STELLA model can be accomplished, however fundamental differences between the two modeling languages should be recognized. A key difference involves the visual representation of mathematical relationships. The energy systems language shows the mathematical relationships explicitly. For example, the EXPO model demonstrated in this chapter (Figure 10.9a) contains an interaction symbol which shows that the production of Q_QUANTITY equals the energy source E times Q_QUANTITY (via the feedback). The STELLA representation of this same interaction (Figure 10.9b) does not reveal the mathematics involved in producing more Q_QUANTITY.

Inherent in the energy systems language is the flow of energy to the heat sink in any transforming or storing process due to the second law of thermodynamics. The Forrester and STELLA languages do not incorporate this energetic principle in their symbols. The contrast between the STELLA and energy systems versions of the PC&CYCLE model illuminates this important difference between the two languages (Figure 10.10). The energy systems diagram shows the consumption of organics transforming a flow of 2000 to 100. In this transformation a different product results and some of the available energy is lost to the heat sink. In STELLA the Unit Conversion feature located within the Flow dialog box is used to simulate transformations. If the Unit Conversion option is not chosen then the amount that exits a storage through a flow icon equals the amount which enters the downstream storage, as when material flow is conserved. The energy systems representation clearly shows the loss of available energy to the heat sink, whereas the STELLA representation does not.

Another difference concerns the representation of an autocatalytic or exponential system. The energy system representation of the EXPO model in Figure 10.9a shows feedback from the storage interacting with the outside energy source to produce more of

the storage. This feedback flow is an actual flow from storage
necessary to obtain more of the outside energy source. In the
STELLA representation of this same model (Figure 10.9b), the
connector between the storage and the inflow represents only
information, not material. This approach results in the same
equations for inflow to the storage, $K1*E*Q$. However, in the
energy systems language $K1 = K2 - K3$, while in STELLA the
feedback flow from the storage is not incorporated into $K1$.

EXERCISES

More instructions and practice with STELLA can be found in
books by Hannon and Ruth (1994) and Ruth and Hannon
(1997). These books have interesting minimodels. They use the
STELLA icons for their network diagrams. As an exercise to
relate to concepts in this book, we suggest translating some of
these into energy systems diagrams.

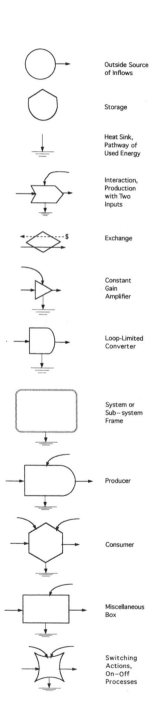

Outside Source
of Inflows

Storage

Heat Sink,
Pathway of
Used Energy

Interaction,
Production
with Two
Inputs

Exchange

Constant
Gain
Amplifier

Loop-Limited
Converter

System or
Sub—system
Frame

Producer

Consumer

Miscellaneous
Box

Switching
Actions,
On—Off
Processes

Chapter Eleven

◆

SIMULATING EMERGY AND TRANSFORMITY

Chapter 11 provides procedures for simulating emergy (spelled with an "m") and transformity. Emergy measures previous work that produced a product or service. Transformity is the emergy per unit energy. These are energy-based measures of value, useful for putting environment and economy on the same basis. A full explanation is provided in a recent book (Odum, 1996).

In this chapter we show how to add equations for emergy and transformity to models so that simulations can include graphs of emergy per time (empower), emergy storages, and transformities. When equations are added to a minimodel to calculate emergy and transformity, the outputs plot or print emergy flows, emergy ratios and transformities.

◆

11.1   CONCEPTS

In Chapter 1, Figure 1.5, we explained the way all phenomena fit into a hierarchy of energy in which processes that require more work have higher emergy, operate on a larger scale, and have a higher transformity. Expressing everything as emergy or its economic equivalent, emdollars, helps simulation models generate public policy. According to the maximum power principle, systems that reinforce emergy production and use prevail and generate the most benefit.

DEFINITIONS

Emergy is the energy of one type required directly and indirectly to make a product. Its unit is the emjoule. *Transformity* is the energy of one type required to make one joule of another type. Because energies of different kinds are not equal in their work contributions, calculating emergy is a way to put everything on a common emjoule basis. For example, a joule of human effort contributes more to real wealth than a joule of tree growth, and this contributes more than a joule of sunlight. Transformities indicate the position in the energy hierarchy that we recognize in energy systems diagrams by placing units from left to right (Chapter 1). Transformity increases with scale (Figure 1.6).

EMERGY OF FLOWS

To calculate emergy flow from sources or storages, multiply an energy flow by its transformity. For example, a block of wood containing 1 million joules (1 E6 J) of wood energy has a solar emergy content of 4 E9 solar emjoules obtained by multiplying the energy content (1 E6 joules) by the solar transformity of wood (4E3 solar emjoules per joule):

(Solar emergy) = (energy) * (solar transformity)
(1 E6 joules wood) * (4E3 solar emjoules/joule) =
 4 E9 solar emjoules (4E9 is computer notation for 4×10^9.)

You can also evaluate emergy flow from the flow of matter or dollars. If data are in grams (rather than energy), energy quality may be expressed as emergy per unit mass. If data for goods and services are expressed in dollars, an emergy/money ratio results and may be expressed as solar emjoules per dollar. Emergy/ money ratios are obtained from a nation's annual emergy budget divided by its gross economic product. There are many published values for transformity, emergy/mass, and emergy/money, many of which are cited in a summary book on environmental accounting (Odum, 1996). Where two independent pathways join in a production interaction, the emergies of both pathways are added (providing the sources of the two are independent and not from a closed loop feedback).

EMERGY OF STORAGES

Rate equations of simulation models are first calibrated in units of energy, matter, and other units. These quantities follow laws of matter and energy conservation: Matter inflowing must equal that stored and outflowing; energy inflowing must equal that stored and outflowing. Emergy, however, is a tabulation of what was previously processed in making the product. It is not reduced by the drain pathways necessary for dispersal of energy according to the second energy law. It is a memory evaluation. You can tally the emergy for each storage and calculate transformities as needed to compute emergy storages, flows, and emergy indices at each transformation step of the model's simulation. Transformities and emergy should be of only one energy type, such as the solar transformities used in the examples.

11.2 PROCEDURE FOR SIMULATING EMERGY AND TRANSFORMITY

To include emergy and transformity calculations in a simulation program, first draw the systems diagram, derive the equations, calibrate the constants, complete the simulation equations, run the program, and debug as may be necessary to generate graphs in the usual way. Then add emergy flow equations to the diagrams and to the program and print the statements.

NOTATION

Emergy flows and storages are designated with E as the first letter. E_S is flow of emergy from the source; E_Q is emergy in the storage. Transformities are designated with T_r for the first letters. For example, T_{rQ} is the transformity of stored quantity Q; T_{rF} is the transformity of fishing pressure F from the economy (Figure 11.1).

After drawing an energy systems diagram with numbers and rate equations, prepare a second diagram on which emergy flows, storages, and transformities are designated (example: Figure 11.1b). Use that diagram to write equations for emergy flow and storage. Terms from the first diagram are multiplied by transformities to get terms for emergy flow. Three equations are required: one for the period of growth, one for time of constant storage, and one for declining storage. We supply three examples in the following sections to illustrate the methods.

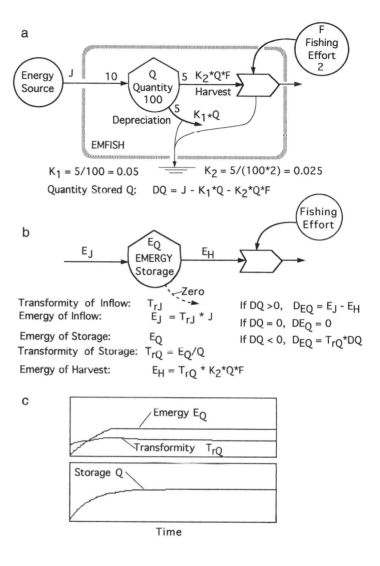

FIGURE 11.1.
Model EMFISH and equations for simulating emergy. (a) Energy system diagram and equation; (b) notation for simulating transformity and emergy; (c) simulation using BASIC (program EMFISH.bas in Table 11.1).

♦

11.3   SIMULATION MODEL EMFISH

A demonstration of emergy and transformity definitions and behavior is provided with the model EMFISH in Figure 11.1 and the BASIC program EMFISH.bas (Table 11.1). Figure 11.1a shows the rate of change equation for a storage of any quantity with one inflow, one outflow, and one depreciation pathway. The simulation uses these to calculate flows of energy in and out

TABLE 11.1
Program of Emergy Storage and Use (EMFISH.bas) (Figure 11.1)

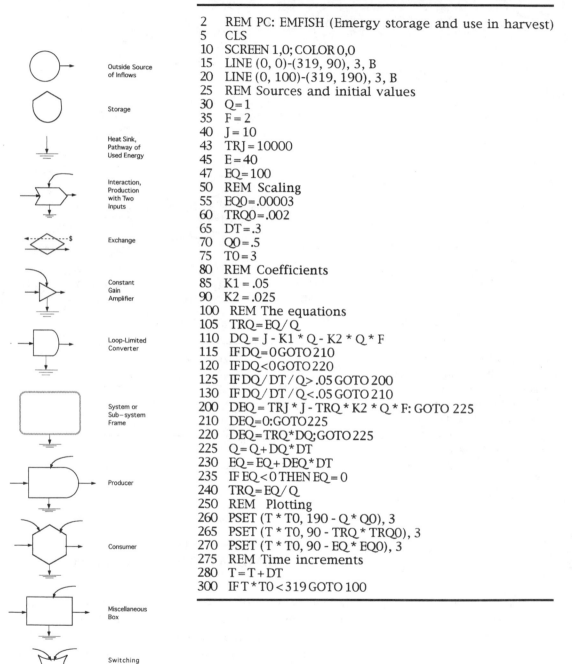

	Outside Source of Inflows
	Storage
	Heat Sink, Pathway of Used Energy
	Interaction, Production with Two Inputs
	Exchange
	Constant Gain Amplifier
	Loop-Limited Converter
	System or Sub–system Frame
	Producer
	Consumer
	Miscellaneous Box
	Switching Actions, On–Off Processes

```
2    REM PC: EMFISH (Emergy storage and use in harvest)
5    CLS
10   SCREEN 1,0; COLOR 0,0
15   LINE (0, 0)-(319, 90), 3, B
20   LINE (0, 100)-(319, 190), 3, B
25   REM Sources and initial values
30   Q = 1
35   F = 2
40   J = 10
43   TRJ = 10000
45   E = 40
47   EQ = 100
50   REM Scaling
55   EQ0 = .00003
60   TRQ0 = .002
65   DT = .3
70   Q0 = .5
75   T0 = 3
80   REM Coefficients
85   K1 = .05
90   K2 = .025
100  REM The equations
105  TRQ = EQ / Q
110  DQ = J - K1 * Q - K2 * Q * F
115  IF DQ = 0 GOTO 210
120  IF DQ < 0 GOTO 220
125  IF DQ / DT / Q > .05 GOTO 200
130  IF DQ / DT / Q < .05 GOTO 210
200  DEQ = TRJ * J - TRQ * K2 * Q * F: GOTO 225
210  DEQ = 0:GOTO 225
220  DEQ = TRQ*DQ;GOTO 225
225  Q = Q + DQ * DT
230  EQ = EQ + DEQ * DT
235  IF EQ < 0 THEN EQ = 0
240  TRQ = EQ / Q
250  REM  Plotting
260  PSET (T * T0, 190 - Q * Q0), 3
265  PSET (T * T0, 90 - TRQ * TRQ0), 3
270  PSET (T * T0, 90 - EQ * EQ0), 3
275  REM Time increments
280  T = T + DT
300  IF T * T0 < 319 GOTO 100
```

of a tank and graphs the change in the quantity stored (Figure 11.1c).

EQUATIONS FOR EMERGY OF STORAGE

Shown below the regular equations in Figure 11.1b are equations for the emergy of a storage. Heat sinks are omitted because their emergy flow is zero. Three emergy equations are required to cover three stages of storage change: increasing, constant, and decreasing. These are used in the computer program to calculate emergy and transformity.

The energy dispersed in heat sink depreciation is a necessary part of the process of storing emergy. Therefore, emergy storage is what is required to build the storage in spite of the concurrent dispersal of some energy and matter through the depreciation pathway. Because the energy in the heat sink has no availability (no ability to do work), its emergy flow is zero. Therefore, the summing of what goes into emergy storage does not include a term for the dispersal pathway. There is no depreciation term in emergy storage equations, whereas there is in the regular equations.

The program chooses one of three emergy equations to tally the emergy, depending on whether storage is growing, constant, or decreasing, as discussed next.

If Storage is Increasing (dQ/dT > 0). When storage is increasing, then the equation for its emergy storage is the sum of the inflows of emergy minus the outflows of usable emergy, but the concurrent necessary depreciation is not subtracted. The energy lost from availability in depreciation is a necessary part of the process of storing emergy. By definition, emergy only includes a tally of the available energies contributing positively to its formation and all expressed in joules of one kind of energy. The computer equations are:

$$\text{IF DQ} > 0 \text{ THEN DEQ} = \text{EJ} - \text{EH}$$

where: $\quad \text{EJ} = \text{TrJ*J}$

and $\quad\quad \text{EH} = \text{TrQ*K2*Q*F}$

If Storage is Constant (dQ/dT = 0). When there is no change in storage or change in its nature, there is no change in emergy storage. The computer equation is:

$$IF\ DQ = 0\ THEN\ DEQ = 0$$

If Storage is Decreasing (dQ/dT < 0). When the change in emergy storage is negative, the loss in emergy is the loss of the energy times the transformity of the storage, whether it is due to depreciation loss or whether it is due to the transfer of useful emergy out. The computer equation is:

$$DEQ = TrQ*DQ$$
$$EQ = EQ + DEQ*DT$$

The transformity of the stored quantity TrQ is calculated on each iteration by dividing the emergy tally for that storage EQ by its energy storage Q. By definition, the transformity of the storage at any time is the emergy of the storage divided by the energy stored:

$$TrQ = EQ/Q$$

"What If" Experimental Problems for the EMFISH.bas Program in Table 11.1

11.3–1 If more of the yield were sent to users, what happens to values of natural capital? Increase K2 and observe the emergy level.

11.3–2 What indicates the quality of the fish stock? How does it change in a year when there is less input? Reduce input by half and rerun.

◆

11.4   EMERGY STORAGE MODEL EMGTANK WITH DEPRECIATION OUTFLOW ONLY

The simulation of a storage model (EMGTANK), with one inflow and only depreciation outflow in Figure 11.2, shows what happens when there is a growth of a quantity to a steady state (balance between inflow and outflow). The emergy and the transformity also increase. The storage reaches a value within 5% of the steady state quickly, but a longer time is required to reach the steady state as it approaches a constant level asymptotically.

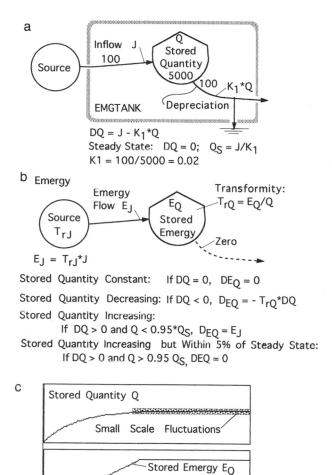

FIGURE 11.2.
Simulation of emergy and transformity in storage with the convention of ending emergy growth at 95% of asymptotic steady state. (a) Systems diagram and equation; (b) notation for simulation of emergy flows and storage; (c) simulation with BASIC (program EMGTANK.bas in Table 11.2).

During the last 5% of the growth in real-world examples, the emergy input is not really adding new storage value because fluctuations of this magnitude are present from the smaller scale, sometimes called noise. It is within the range of fluctuations caused by statistical fluctuations (Chapter 22). Therefore, programs are given a statement that stops adding new emergy when the growth is within 5% of the steady state.

EQUATIONS

Emergy is calculated for the growth to 95% of the asymptotic level only. Qs (the steady state) equals J/K1 (found by setting

DQ = 0 where DQ = J − K1*Q). Equations used to tally the emergy depend on whether storage is growing, within 5% of the asymptote, or decreasing.

When	Q < Qs
The equation is	IF Q < 0.95*Qs THEN DEQ = EJ
Where:	EJ = TrJ*J
When	Q = Qs, DQ = 0 and DEQ = 0
When	DQ < 0
	DEQ = TrQ*DQ
And	TrQ = EQ/Q
	EQ = EQ + DEQ*DT

The simulation of EMGTANK (Figure 11.2c) shows the emergy storage and its transformity growing like a ramp, stopping at the 95% level.

"What If" Experimental Problems for the Program EMGTANK.bas in Table 11.2

11.4–1 If depreciation were faster, how would the emergy of the storage change? Increase K1.

11.4–2 If the storage and its inputs were higher in the scale of size and energy hierarchy, how would its emdollar value change? Make a change and rerun the simulation. For the United States in the late 1990s, the emergy/money ratio is 1.3×10^{12} solar emjoules per dollar.

11.5   EMERGY IN INFLOW-LIMITED AUTOCATALYTIC MODEL RENEMGY

Simulation equations are given for RENEW, a model of autocatalytic growth on a source limited flow, shown in Chapter 13, in Figure 13.1. The same model with emergy calculations is RENEMGY (Figure 11.3), which simulates emergy and transformity with the program in Table 11.3. Whereas Figure 11.3a has the regular rate of change equations, Figure 11.3b also has the emergy equations obtained by multiplying the terms in each pathway by the transformity from the source of flow at the left. There is one input source, the sun (solar transformity: $T_{rS} = 1$).

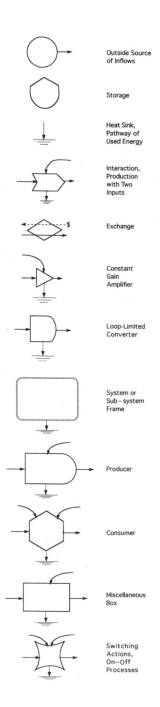

Outside Source of Inflows

Storage

Heat Sink, Pathway of Used Energy

Interaction, Production with Two Inputs

Exchange

Constant Gain Amplifier

Loop-Limited Converter

System or Sub–system Frame

Producer

Consumer

Miscellaneous Box

Switching Actions, On–Off Processes

TABLE 11.2
Program for Emergy in Simple Storage (EMGTANK.bas)
(Figure 11.2)

```
5    REM PC
10   REM EMGTANK.bas (Emergy of a simple storage)
15   SCREEN 1,0: Color 0,0
20   LINE (0, 0)-(319, 90), 3, B
25   LINE (0, 100)-(319, 190), 3, B
30   REM Sources and initial values
35   Q=10
45   J=100
50   TrJ=10
55   EQ=10
60   TRQ=EQ/Q
100  REM Scaling
110  EQ0=.0005
120  TrQ0=1
130  DT=1
140  Q0=.01
150  T0=1
160  REM Coefficients
170  K1=.02
180  S = J/K1:  REM Steady State Q
200  REM Equations
210  DQ=J-K1*Q
220  IF DQ=0 THEN DEQ=0:GOTO 270
230  IF DQ<0 THEN DEQ=TRQ*DQ:GOTO 270
240  IF Q<.95*S THEN GOTO 260
250  IF Q>.95*S THEN DEQ=0: GOTO 270
260  DEQ= TrJ*J:GOTO 270
270  Q=Q+DQ*DT
280  EQ=EQ+DEQ*DT
290  IF EQ<0 THEN EQ=0
300  TRQ=EQ/Q
305  REM Time increments
310  T=T+DT
315  REM Plotting
320  PSET (T * T0, 190 -EQ * EQ0), 3
340  PSET (T * T0, 90 - Q * Q0), 3
350  PSET (T*T0, 190 - TRQ*TrQ0),3
400  IF T * T0 < 319 GOTO 200
```

Emergy inflow is $E_J = T_{rJ}*K_0*R*Q$. The emergy within the storage is the net balance between the input emergy flow minus the yield outflow $E_Y = T_{rQ}*K_3*Q$. Because depreciation dispersals are not included in equations for stored emergy, equations that cover the growth period (while Q is increasing) omit depreciation.

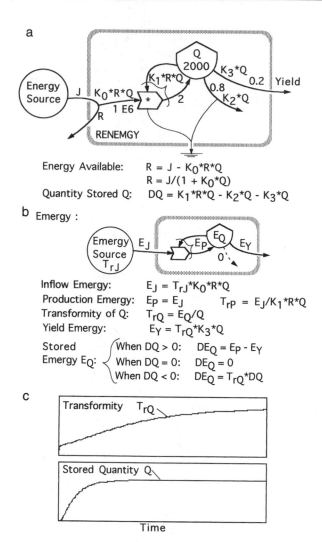

Energy Available: $R = J - K_0*R*Q$
                       $R = J/(1 + K_0*Q)$
Quantity Stored Q: $DQ = K_1*R*Q - K_2*Q - K_3*Q$

b Emergy :

Inflow Emergy: $E_J = T_{rJ}*K_0*R*Q$
Production Emergy: $E_P = E_J$ $T_{rP} = E_J/K_1*R*Q$
Transformity of Q: $T_{rQ} = E_Q/Q$
Yield Emergy: $E_Y = T_{rQ}*K_3*Q$

Stored   When DQ > 0: $DE_Q = E_P - E_Y$
Emergy E_Q: When DQ = 0: $DE_Q = 0$
                  When DQ < 0: $DE_Q = T_{rQ}*DQ$

FIGURE 11.3.
Autocatalytic model RENEMGY
with equations for simulating
emergy. (a) Energy systems
diagram, equations, and
numbers for calibration;
(b) notation for simulating
transformity and emergy;
(c) simulation using BASIC
(program RENEMGY.bas in
Table 11.3).

The computer equations are as follows:

$$\text{IF DEQ} < 0 \text{ THEN DEQ} = DQ*TrQ$$
$$EQ = EQ + DEQ*DT$$

Transformities of storage increase with growth (Figure 11.2c),
representing the higher quality of stored energy concentrated
over its inflow. Its emergy is transmitted forward with a higher
transformity because of the dispersal and degrading necessary to
the storing process.

TABLE 11.3
Program for Autocatalytic Emergy Storage Using Renewable Energy (RENEMGY.bas) (Figure 11.3)

```
10   REM PC: RENEMGY.bas (Emergy with autocatalytic growth on
     renewable source)
20   CLS
30   SCREEN 1,0: COLOR 0,0
50   LINE (0, 0)-(319, 90), 3, B
60   LINE (0, 100)-(319, 190), 3,B
70   REM Scaling
80   DT=20
85   T0=.02
90   Q0=.03
95   TRQ0=.00003
100  REM Sources and Starting Conditions
105  J=1100000!
110  TRJ = 1: REM Transformity of Energy Source
115  Q=1
120  REM Constant Coefficients
130  K0=.01
150  K1=2E-08
160  K2=.0008000001#
165  K3=.0002
170  REM Plotting
180  PSET (T * T0, 190 - Q * Q0), 3
190  PSET (T * T0, 90 - TRQ * TRQ0), 3
195  R = J / (1 + K0 * Q)
197  EJ = TRJ * (K0 * R * Q):REM Emergy of Inflow
200  TRP = EJ / (K1 * R * Q):REM Transformity of Production
230  EP = TRP * (K1 * R * Q): REM Emergy of Production Flow
240  EY = TRQ * (K3 * Q): REM Emergy of Yield
260  DQ = K1 * R * Q - K2 * Q - K3 * Q
270  DEQ = EP - EY:REM Change in Emergy Stored
280  Q = Q + DQ * DT
290  EQ = EQ + DEQ * DT:REM Emergy of Storage
295  TRQ = EQ / Q:REM Transformity of Stored Emergy
300  T = T + DT
310  IF T * T0 < 319 GOTO 170
```

"What If" Experimental Problems for the RENEMGY.bas Program in Table 11.3

11.5–1 How does the output transformity of the storage and its output flow compare with that of the input? Compare the dialog box of the input with that of the simulation.

11.5–2 What happens to the quality of the storage if the inputs stop? Referring to Figure 11.3, identify and change the input source in the program and rerun.

◆

11.6   SIMULATING INTERNATIONAL EMERGY EXCHANGE

Emergy is useful in evaluating the net benefit of foreign trade and other financial exchanges where emergy per unit money differs by country and where the emergy value of products and services does not correspond to their prices. In Part III, Chapter 19, a model of international trade (FREEMARK) is simulated. Then emergy equations are included (program EMEXCHNG. bas) to simulate the benefits as evaluated on a common basis (Figure 19.2e).

◆

11.7   EMERGY SIMULATION WITH EXTEND

The general systems blocks for EXTEND (Chapters 4 and 5) automatically calculate the emergy and transformity of storages and outputs based on the transformity of the inputs. The equations used in the scripts of icon blocks for EXTEND programs are similar to those in Figures 11.1, 11.2, and 11.3.

To use them, add and connect the special plotters for emergy storage, empower (emergy flow), and transformity from the GNSYSLIB library to the models assembled on screen. The simulation plots graphs of these properties along with the usual plots of storage and flow.

PART THREE

FUNDAMENTAL MINIMODELS

The minimodels in Part III have configurations and energy relationships like those found in many kinds of systems. Chapters 12 through 16 contain general systems models. Chapters 17 and 18 contain fundamental microeconomic and macroeconomic models with flows of money. Chapter 19 covers international exchange, and Chapter 20 aggregates features of global perspective. Each model is accompanied by a brief explanation, a systems diagram, change equations, explanation of equations, BASIC program, examples from more than one field, and a plot of a typical simulation. Appendix C provides answers to the "What If" questions. The programs are included in the book disk (Appendix E).

These minimodels include mathematical relationships found in many fields, often introduced as fundamental concepts in textbooks on biochemistry, biophysics, biology, populations, geology, ecology, and economics. Learning to recognize basic relationships in network form enhances the ability to use them in more complex systems. The study of and simulation of these minimodels develop the ability to relate and think interchangeably about systems in five languages: verbal description, network diagrams, change equations, BASIC programs, and graphs of functions. By example, different models illustrate different ways of introducing data and displaying output.

In this book each of the model sections stands alone. They can be used in any order. The chapters in Part III are arranged in order of the chapters in the book *Ecological and General Systems,* available from the University Press of Colorado, where more complete discussions of their mathematics are given (Odum, 1994; a reprint of *Systems Ecology,* 1983). Other models and simulations of energy webs and material cycles are given by DeAngelis (1992).

Starting in this Part III, coefficients are shown on pathways of the systems diagram, but not the whole mathematical term found in the equation below the diagram.

SUGGESTIONS FOR SIMULATION

To study a minimodel, read the text, peruse the diagram noting the way the equations follow from the network, and run the program. Using a PC or Macintosh, load this book's CD-ROM onto the computer. To run a minimodel, load some form of BASIC and use its menus to open a program file (example: TANK.bas). Because the programs are text files, they can also be used by other kinds of programs and computers. The programs can be modified for other kinds of BASIC by changing a few statements; for example, the statements for plotting may differ. Rerun the programs with the suggested "What If" changes. Compare your results with answers given in Appendix C. To disable a line without erasing it, type REM just after the number. The program ignores anything on the line after the REMARK command.

Directions for PC

Load Qbasic.exe (Quickbasic) from PC Windows. Then load the book CD-ROM. Use the Qbasic File Menu to load a program from the folder called Basicprg in the Windows part of the CD. The program listing is displayed in the upper part of a split screen. In the small lower screen (command window) you can type instructions, such as requests to print values. Toggle between screens with the F6 key. Toggle from program screen to output screen with the F5 key. To "screen dump" the output graph to the clipboard, use the Print Screen key. From the clipboard you can paste the graph into a drawing or word processing program. Toggle between the graph and the desktop with Alt-Tab.

To print the numerical values of quantities, rates, flows, etc, at the end of a run, use the command window to type in ? T, K2*N*Q (or whatever items you wish), each separated by a comma. The question mark means print. The numbers appear at the top of the screen. More details are provided in Appendix B including use of the old BASICA.

Directions for Macintosh

Load the book CD-Rom, find the Macintosh section, and click on the folder called Basicprg. Open CHIPMUNK BASIC and use menus to open a simulation program. Type Run or use the control menu to RUN the program. The graph plot appears in the middle of the screen. Type LIST or use the control menu to list the program. You can drag the graph window to the right so that the list is visible. Below the list you can type the next command or use the menu.

To compare one run with another after making a change, you can plot the new graph with the first graph. If you make a change after the first run type at the bottom of the program list Run 60 (or whatever the line number is below the graphics statements at the tope of the program). To save a modified program, use SAVE AS and assign a different name. To print the graph, pull down PRINT from the FILE menu. More details on use of the programs are provided in Appendix B including use of the older Macintosh QUICKBASIC.

Chapter Twelve

◆

MODELS OF PRODUCTION AND RECYCLE

$\mathbf{T}$his chapter considers the production process and material cycles found in systems on all scales. NETPROD shows how net production results from the balance of production and consumption. The minimodel FACTORS simulates an external limiting material, whereas INTLIMIT simulates a limiting internal cycle. In DAYP&C a production–consumption–recycle system responds to a day–night sunlight regime. OPENAQ relates production and changes of dissolved oxygen in open aquatic ecosystems in exchange with the air. AUTOCYCL is a logistic model of autocatalytic production limited by an internal cycle. These models illustrate systems principles on materials and energy. The designs are common building blocks in more complex systems.

◆

12.1 MODEL OF NET PRODUCTION (NETPROD)

The model NETPROD (Figure 12.1a) illustrates *net production,* the difference between gross production and consumption. For example, plant photosynthesis produces organic matter (gross production) that flows into storage Q. At the same time the organic matter is being utilized from that storage by plant respiration and animal consumption. The difference is net production, which is added to or subtracted from the stored organic matter Q. In this example gross production is proportional to the incoming sunlight, which is changed in each of four seasons, and the same sequence repeated in the years following. Consumption is proportional to the stored quantity Q.

The net production rate can be measured in the field as the observed changes in organic matter storage with time. Net production is less than the gross production rate. Large errors result when people try to use net production as a measure of gross production or vice versa.

The simulation with the computer program (Table 12.1) in Figure 12.1b starts with a low initial storage that increases over

170

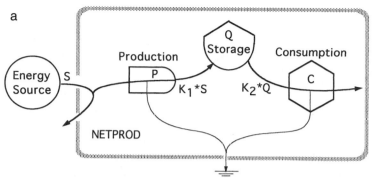

a

Production Rate: $P = K_1*S$ Consumption Rate: $C = K_2*Q$
Change of Storage, time step 3 months: $DQ = P - C$

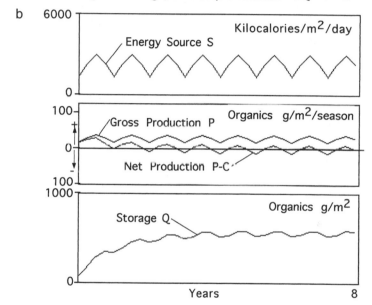

FIGURE 12.1.
Model of net production (program NETPROD.bas). (a) Systems diagram; (b) simulation of energy input, production rates, and storage with one point plotted every 3 months.

an 8-year period. The graphs plot the seasonal inputs of light energy, the gross production rate, the net production rate by difference, and the accumulated storage of product. Whereas gross production is always positive, net production is negative when consumption exceeds gross production. In Figure 12.1b net production is positive in the summer and negative in the winter. Net production is greatest during early growth, declining as the system gets close to steady state. (A steady state is one in which storages are constant due to the balance of inflows and outflows.) The state in Figure 12.1b is slightly fluctuating around a steady average.

TABLE 12.1
Program for Net Production (NETPROD.bas) (Figure 12.1)

```
20 REM PC: NETPROD.bas (Net Production, P minus C)
30 CLS
40 SCREEN 12, 0: REM COLOR 0, 0
90   LINE (0, 0)-(310, 60), , B
110  LINE (0, 70)-(310, 150), , B
130  LINE (0, 160)-(310, 260), , B
140  LINE (0, 120)-(315, 120)
150 T0 = 10
160 T1 = .001
170 D0 = .3
175 S0 = .01
180 Q0 = .1
190 DQ1 = .001
200 Q = 1
210 K1 = .0225
220 K2 = .09
230 N = 1
240  IF N = 1 THEN S = 2000!
250  IF N = 2 THEN S = 3500!
260  IF N = 3 THEN S = 4500!
270  IF N = 4 THEN S = 3500!
280 N = N + 1
290 IF N = 5 THEN N = 1
310 P = K1 * S
320 C = K2 * Q
330 DQ = P - C
340 Q = Q + DQ
345 LINE (T1 * T0, 60 - S1 * S0)-(T * T0, 60 - S * S0)
350 LINE (T1 * T0, 260 - Q1 * Q0)-(T * T0, 260 - Q * Q0)
360 LINE (T1 * T0, 120 - DQ1 * D0)-(T * T0, 120 - DQ * D0)
365 LINE (T1 * T0, 120 - P1 * D0)-(T * T0, 120 - P * D0)
370 Q1 = Q
380 DQ1 = DQ
390 S1 = S
395 P1 = P
400 T1 = T: REM One season (3 months)
410 T = T + 1
500  IF T * T0 < 315 GOTO 240
```

EXPLANATION OF EQUATIONS

Production ($P = K_1 * S$) is proportional to the incident light S. Consumption ($C = K_2 * Q$) is proportional to the storage of product Q. The net production during each time interval is the difference: $DQ = P - C$. In the program (Table 12.1) each time unit is 3 months. Instead of plotting isolated points for each season, the program plots short lines between the last point and the one for the previous season (lines 340–365).

SIMULATION WITH CALCULATION TABLES

With only two pathways, this model can also be simulated by hand easily (without a computer) by creating a calculation table like that shown earlier in Figure 9.2. Suggested headings for the columns are Iteration Number, Season (Winter, Spring, Summer, or Autumn), Sunlight S, Gross production $P = 0.001*S$, Consumption Rate $C = 0.2*Q$, Change in Storage $DQ = P - C$; and Q = previous $Q + DQ$. These headings can also be entered into a spreadsheet for automatic calculation (see Chapter 8).

EXAMPLES

Net production and gross production are the basis for all the ecological, agricultural, and forestry systems of the world. These include ponds, lakes, rivers, oceans, fields, forests, swamps, etc. In agriculture and forestry where emphasis is on yields, consumption by competing insects and microbes is reduced in order to make net production as high as possible. However, wild ecosystems prevail by putting their production into a diversity of other organisms that contribute to soils and improve gross production in this and other ways.

The principles also apply to economic production such as the work of an industry. For example, the total output of shoes is gross production, whereas the net production is the balance between the shoes made and those consumed and dispersed.

The term *net production* has been used in one way in this chapter. However, net production can be used to refer to the difference between any two flows and thus be misunderstood. Diagrams can be drawn for clarity.

"What If" Experimental Problems for the Program in Table 12.1

12.1–1 What is the effect of increasing the rate of consumption by increasing K_2?

12.1–2 What coefficient in Figure 12.1a represents the efficiency with which the sunlight S is converted into gross production? What is the effect of increasing this coefficient?

12.1–3 Predict what kinds of graphs will result if the product storage starts off at a higher value than the steady state in the first simulation in Figure 12.1b. Set Q = 1000.

12.2   EXTERNAL LIMITING FACTORS (FACTORS)

The model FACTORS (Figure 12.2a) relates production to the availability of necessary materials coming from outside. Whenever two or more different items are required for a process, the relationship is shown with the interaction symbol. The output is a new product, and the process is called *production*. Increasing the amount of any one of the required inputs increases the output, unless the rate of supply of another factor is limited. In Figure

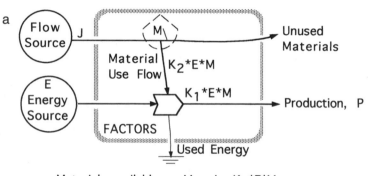

Materials available: $M = J - K_2*E*M$
Therefore: $M = J/(1 + K_2*E)$
Production: $P = K_1*E*M$

FIGURE 12.2.
Model of production as a function of two interacting inputs (program FACTORS.bas). (a) Energy systems diagram and equations; (b) typical simulation graphs of production as it varies with changing availability of energy E for several levels of materials inflow J.

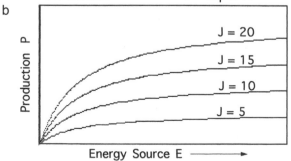

12.2a, the two required factors are indicated by the letters E and M. The amount of material M available to the production process depends on the supply provided by flow J. If you increase the source concentration (force) E, the production increases rapidly at first, but as E is increased further, the other factor M is used up faster than it can be supplied by inflow J. In this situation, M is limiting because of the fixed rate of inflow J. The flow of J is the external limiting factor keeping production from increasing further. The limitation of an external factor to production is sometimes called a Monod model, named after its author who showed the appropriate equation for an example of external nutrient limitation.

Whereas most of our simulations graph quantities with time, Figure 12.2b shows production as the factor E is increased. Increasing E has a large effect at first, which becomes less and less as the other factor M begins to limit more because of limited flow J.

An example is the photosynthetic production of organic matter by grass in a field. Term E refers to light intensity (insolation), J is inflow of nutrients from rain, and M is the concentration of nutrients in the soil next to the plant roots. When light first increases, more grass is produced. But soon production is limited by nutrients: Even with more and more sunlight, there can be little increase in production. At high light intensities the amount of production depends on the supply of available nutrients, in this case from rain.

The simulation (Figure 12.2b) plots a point representing the calculated production P as a function of the energy concentration E with J held constant. Then E is increased, P is recalculated, and another point plotted. The program (Table 12.2) repeats the process in an iteration (lines 120–160). After the curve is complete the program increases J, resets E to 0 (lines 170–190), and returns to the first iteration, generating another curve. Figure 12.2b shows a family of four curves.

EXPLANATION OF EQUATIONS

The FACTORS model uses the "flow use" design (Chapter 6, Section 6.4) for representing what is still available in a flow J as some of the flow K_2*E*M is drawn into a production process (interaction symbol). The resulting production P by that interac-

TABLE 12.2
Program of External Limiting Factors
(FACTORS.bas)
(Figure 12.2)

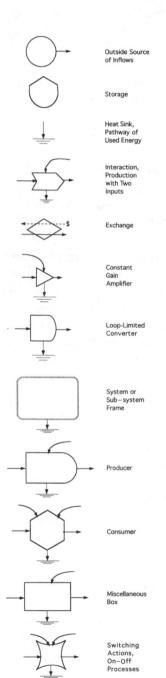

Outside Source
of Inflows

Storage

Heat Sink,
Pathway of
Used Energy

Interaction,
Production
with Two
Inputs

Exchange

Constant
Gain
Amplifier

Loop-Limited
Converter

System or
Sub–system
Frame

Producer

Consumer

Miscellaneous
Box

Switching
Actions,
On–Off
Processes

```
10 REM PC: FACTORS (External limiting factors)
20 SCREEN 1, 0: COLOR 7, 0
40 LINE (0, 0)-(240, 180), 3, B
50 J = 5
60 E = 1
80 K1 = .08
90 K2 = .01
110 DE = 2
115 E0 = 2
118 P0 = 1
120 M = J / (1 + K2 * E)
130 P = K1 * E *M
140 E = E + DE
150 PSET (E / E0, 180 - P / P0), 1
160 IF (E / E0) < 240 GOTO 120
170 J = J + 5
180 E = 0
190 IF J < 25 GOTO 120
```

tion unit is proportional to the available concentrations of both E and M and thus a product of the two K_1*E*M. $J/(1 + K_2*E)$ is the form of the equation that provides a continuous value of available material M to go into the interaction process. The model has no storages and thus no time delay.

EXAMPLES

Another example is the production of cars on an island using a flow of metals from another country that is unchanging. Increasing the energy to the process has a diminishing effect as the available stock of metals locally available gets lower and lower. Limiting factor curves like this are found in all fields of science and are called *diminishing returns* in economics. There are thousands of experimental studies with graphs similarly shaped. Think of a different example. What are J, E, and P in your example?

"What If" Experimental Problems for the Program in Table 12.2

12.2–1 In the external limiting model what is the effect of increasing the inflow of limiting factor J?

12.2–2 How and why would the graph be different if the production process were more efficient? Increase K1 and run the program.

12.2–3 Efficiency of energy use is the ratio of P to E (P/E). Explain when efficiency is greatest by examining the output graphs (Figure 12.2b). How is efficiency related to the energy level used, E?

12.3 INTERNAL LIMITING FACTORS (INTLIMIT)

The model INTLIMIT (Figure 12.3a) shows how the materials recycling within a system can limit its production. For example, an ecosystem in an aquarium receiving only light energy has a photosynthetic production rate that is dependent on the nutrient materials available to recycle.

As shown in the diagram, the model has an unlimited energy source E, a production process requiring materials M, storing of the product Q, and a consumption process that recycles the materials. The total quantity of materials is held constant (conservation of materials) so that materials are either in the available pool M or bound within the product storage Q. Whereas limitations in Figure 12.2a were controlled from outside the system boundary, in this model similar mathematics results from relating production to the total materials available inside the system for production and recycle. If the recycle process is too slow, it can limit the production process so that an increase in factor E has a diminishing return. This model is sometimes called a Michaelis-Menten relationship, named after the authors, which recognizes the appropriate equations for a biochemical example of internal cycle limitation.

When the quantity of product storage is simulated starting at a low value (Figure 12.3b), it grows rapidly at first and then levels off as the availability of materials M becomes limited by the rate of recycle. The rate of production P is highest at the

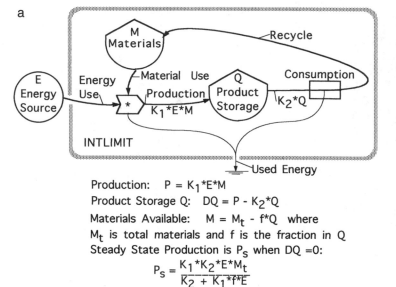

Production: $P = K_1*E*M$

Product Storage Q: $DQ = P - K_2*Q$

Materials Available: $M = M_t - f*Q$ where
M_t is total materials and f is the fraction in Q
Steady State Production is P_s when $DQ = 0$:

$$P_s = \frac{K_1*K_2*E*M_t}{K_2 + K_1*f*E}$$

FIGURE 12.3.

Model of production and consumption which becomes limited by the rate of recycle of materials M when input energy E is increased (program INTLIMIT.bas). (a) Systems diagram and equations; (b) simulation of product storage over time; (c) simulation graph of steady-state production P_s as a function of increasing energy availability E for several levels of total materials M_t.

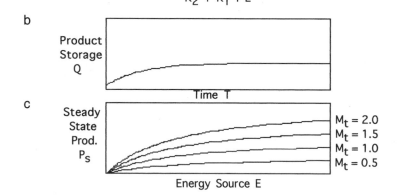

start. The slope of the growth curve of Q is steepest at the start. The growth curve is convex (not S shaped).

To show the diminishing returns of increasing the energy where there are internal material limits, the steady-state level P_s is plotted as a function of energy E. Level P_s was calculated by setting DQ to zero and substituting the expression for production P (Figure 12.3a).

After the program (Table 12.3) plots one graph of Q with time in the upper panel, it calculates the steady-state production P_s and plots it as a function of the energy E (Figure 12.3c). The rate of production achieved at steady-state P_s increases with increased E but with diminishing returns. Materials M become

TABLE 12.3
Program of Internal Limiting Factors (INTLIMIT.bas)
(Figure 12.3)

```
2 REM PC
3 REM INTLIMIT (Internal limiting factor)
5 SCREEN 1, 0: COLOR 0, 0
6 LINE (0, 0)-(319, 90), 3, B
10 LINE (0, 100)-(319, 190), 3, B
50 MT = .5
60 E = 15
70 Q = 1
80 K1 = 2
90 K2 = .5
100 F = .05
120 E0 = 20
130 P0 = 10
140 Q0 = 5
150 DT = .01
160 T0 = 100
200 REM Simulation of changes over Time
210 M = MT - F * Q
220 P = K1 * E * M
230 DQ = P - K2 * Q
240 Q = Q + DQ * DT
250 T = T + DT
260 PSET (T * T0, 90 - Q * Q0)
270 IF T * T0 < 319 GOTO 200
300 REM  Simulation of Steady State production v.s. Energy E
310 E = .1
315 P0 = 5
317 DE = .05
320 Q = (K1 * MT * E) / (K2 + K1 * F * E): REM Steady state is where DQ = 0
325 M = MT - F * Q
330 DE = .05
340 P = K1 * E * M
350 E = E + DE
355 PSET (E * E0, 190 - P * P0)
365 IF (E * E0) < 319 GOTO 320
370 Q = 1
375 MT = MT + .5
380 E = .1
400 IF MT < 2.5 GOTO 300
```

less and less available to production because there are not enough total materials M_t in the cycle. After plotting one graph, the program increases the total material M_t and plots another graph, making a family of curves. As shown by Figure 12.3c, increasing the total materials in the system M_t increases the productivity. Higher productivity means energy use is more efficient.

EXPLANATION OF EQUATIONS

The production process K_1*E*M is a product of the available energy concentration E and the available materials M. Available materials M are the total materials M_t minus those bound in the products f^*Q, where f is the material fraction in the product storage Q. The storage consumed by the consumption process K_2^*Q returns its material to the pool. For example, consumers in an aquarium ecosystem release the nutrient materials that were in the food to the water.

The value of production at steady state is labeled P_s, which is the value after the materials have become optimally distributed between storages M and Q. For steady state, set DQ equal to zero and algebraically solve for Q so that $Q = P/K_2$. Then substitute Q in the materials equation: $M = M_t - f^*Q$, and substitute the new expression for M in $P = K_1^*E^*M$. The result, $P_s = (K_1^*K_2^*E^*M_t)/(K_2 + K_1^*f^*E)$, is the equation for the curves that the program generates in Figure 12.2b.

EXAMPLES

An example of internal limitation is the recycle of chlorophyll from a state ready to receive light, after which it becomes electrically charged, to the dark process where it is recycled for use again. One of the first examples ever studied as a simulation model was the process of using an enzyme in biochemistry. In Figure 12.3a the enzyme is M and the input "substrate" (energy source) is E. In economic production, the recycle of aluminum cans may be an internal limiting factor M with a diminishing returns curve (Figure 12.3b) when other inputs E are increased.

"What If" Experimental Problems for the Program in Table 12.3

12.3–1 From the family of curves generated by the simulation what is the effect of increasing the total materials MT available for recycle?

12.3–2 What is the effect of increasing the coefficient K2 (the fraction of the storage Q recycled per unit time)? Explain.

12.3–3 What happens to the organisms in a system with a small nutrient content but large energy availability like the open tropical sea or the plants growing on granite rock outcrops? Run the simulation with MT at a very low value. Considering the quantity stored and its turnover, comment on the size of organisms.

◆

12.4 MODEL OF PRODUCTION AND CONSUMPTION (DAYP&C)

The model DAYP&C (Figure 12.4a) examines production, storage, and recycle of systems driven by a limited flow source supplying energy that varies with night and day. Examples are ecosystems of ponds or grassland based on sunlight. The sunlight J increases and then decreases each day, with none coming in at night. The available nutrient materials M are used by the photosynthesis in the daytime and released by respiration (by plants and animals) day and night. The quantity of organic matter Q produced goes up with photosynthesis in the daytime and down as it is consumed during both day and night. Quantity Q also represents oxygen, which is produced by photosynthesis and

a

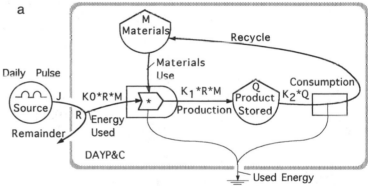

Energy Available: $R = J - K*R*M$ and $R = J/(1 + K*M)$

Materials Available: $M = M_t - f*Q$

Product Stored Q: $DQ = K_1*R*M - K_2*Q$

b

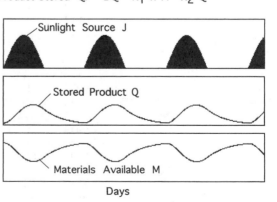

FIGURE 12.4.
Model of production and consumption simulated to show diurnal effects (program DAYP&C.bas). (a) Systems diagram and equations; (b) simulation of accumulated storages Q and available materials M for a several day period also plotting the sunlight input J typical of clear days (half sine wave).

used by respiration. In the daytime the quantities of organic matter and oxygen produced by photosynthesis are greater than those used in respiration; at night there is only respiration using both organic matter and oxygen.

In Figure 12.4b, the top section shows the daily variation of the sun, the second the organic matter and the bottom the nutrients. The simulation program (Table 12.4) plots a half sine wave in the upper panel of the graph, representing the sunlight input J for each day. The inflow J rises and falls in a pattern very similar to sunlight intensity on a clear day. Inflow J is turned off to represent the night. For details on use of sine waves to represent sunlight, see Appendix B. The organic storage Q rises fastest in the morning when nutrients pool M is greater. As storage Q rises, so does the consumption rate. Nutrients are least when the

⃝→	Outside Source of Inflows
⬡	Storage
⊥	Heat Sink, Pathway of Used Energy
▷	Interaction, Production with Two Inputs
◇--$	Exchange
▷	Constant Gain Amplifier
⬭	Loop-Limited Converter
▭	System or Sub-system Frame
◗	Producer
⬡	Consumer
▭	Miscellaneous Box
⋈	Switching Actions, On-Off Processes

TABLE 12.4
Program of Diurnal Production and Consumption (DAYP&C.bas) (Figure 12.4)

```
10 REM PC: DAYP&C (Daily production and consumption)
15 CLS
20 SCREEN 1, 0
30 COLOR 0, 0
35 LINE (0, 0)-(319, 60), 3, B
40 LINE (0, 70)-(319, 130), 3, B
50 LINE (0, 140)-(319, 190), 3, B
60 Q = 5
80 MT = 1
90 K = .9
100 K1 = .2
110 K2 = .04
120 K3 = .01
130 Q0 = .5
140 M0 = 50: REM .02
150 T0 = 1
160 DT = 1
170 J = 40 * SIN(T / 15.9)
180 IF J < 0 THEN J = 0
190 R = J / (1 + K * M)
200 M = MT - K3 * Q
210 DQ = K1 * R * M - K2 * Q
220 Q = Q + DQ * DT
230 PSET (T * T0, 130 - Q * Q0), 1
240 PSET (T * T0, 190 - M * M0), 3
250 LINE (T, 60)-(T, (60 - J)), 3
260 T = T + DT
300 IF T * T0 < 319 GOTO 170
```

organic storage is greatest. The organic storage decreases most rapidly as the light stops at sunset. This program was calibrated for an aquatic system with small storages and rapid turnover time. The model can be use for the labile storages of organic matter in leaves and soils by not including the large storages of wood and soil organic matter that are not affected much by diurnal processes.

EXPLANATION OF EQUATIONS

Available light energy R is that remaining after the uses K_0*R*M in production. Shown in Figure 12.4a, the labile organic matter storage Q is increased by photosynthetic production K_1*R*M and decreased by consumption (respiration and dispersal of plants, animals, and microbes) K_2*Q. Production is proportional to the available sunlight R and the soil nutrients M. Respiration K_2*Q is proportional to the organic content Q that is labile enough to be involved in the diurnal cycle.

The pool of materials (nutrients) M required for production is calculated as the difference between the total materials M_t and that bound in products stored $f*Q$, where f is the fraction of organic storage Q that is nutrient material.

EXAMPLES

Many biological systems show diurnal responses that fit this model including forest leaves, algae in ponds, whole microcosms, plankton in the sea, and coral reefs. A similar model can be used for the seasonal variations in solar-based resources. The model can represent economic processes where goods are produced, used, and then the worn-out parts and wastes are recycled into making new goods. Many parts of our economy operate on diurnal pulses associated with the daily cycle of activity and sleep of humans.

"What If" Experimental Problems for the Program in Table 12.4

12.4–1 Suppose half of the nutrient materials blow away in a fire. How is the diurnal pattern affected? How does this situation compare with the experiment reducing materials with the INTLIMIT model (Figure 12.3)?

12.4–2 What happens if the daily input energy is increased as in summer? In line 170, change the energy flow: J = 60 * SIN (T/15.9).

12.4–3 Suppose severe environmental conditions interfere with consumption (for example Arctic cold, briny waters). After your original run, use the command line to print the last value for the production rate by typing a question mark (which means print) followed by K1*R*M. Reduce K2 to 0.01 and run again. How does the productivity with reduced consumption compare when consumption is more normal?

12.4–4 Suppose a biotic system that fits this model is put in a greenhouse with continuous light equivalent to daylight. Disable line 170 by typing REM after the line number (thus not erasing the original statement). To make light constant, insert a line 175   J = 40 and rerun. Is the production rate affected? Print out the last value of production as explained in What If 12.4–3.

◆

12.5   MODEL OF PRODUCTION AND OXYGEN IN AQUATIC ECOSYSTEMS (OPENAQ)

The model OPENAQ (Figure 12.5a) relates production, consumption, and recycle including dissolved oxygen and its exchange with the air. Examples are ponds and aquarium ecosystems open to the air. When light is shining, production of oxygen and organic matter occurs. These are then used by consumers, which in turn regenerate the nutrient raw materials N (including carbon dioxide) that are recycled back to plant use. Oxygen diffuses back and forth between the gas phase in the air and the dissolved oxygen state within the water. When oxygen is supersaturated it diffuses or bubbles into the gas phase. When respiratory consumption uses up dissolved oxygen at night, it is somewhat replenished by diffusion in from the air.

The simulation in Figure 12.5 shows diurnal variations over a 5-day period. At the top is sunlight (without clouds). The

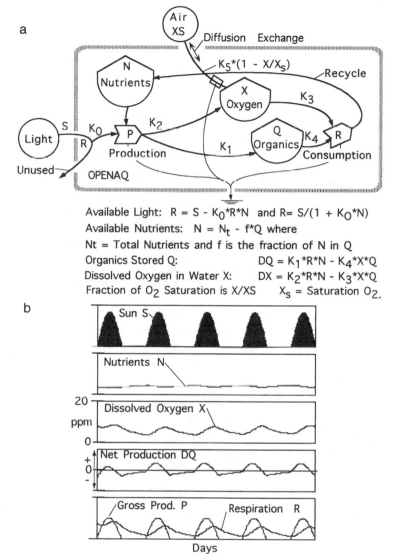

Available Light: $R = S - K_0*R*N$ and $R = S/(1 + K_0*N)$
Available Nutrients: $N = N_t - f*Q$ where
N_t = Total Nutrients and f is the fraction of N in Q
Organics Stored Q: $\qquad DQ = K_1*R*N - K_4*X*Q$
Dissolved Oxygen in Water X: $\quad DX = K_2*R*N - K_3*X*Q$
Fraction of O_2 Saturation is X/XS $\quad X_s$ = Saturation O_2.

FIGURE 12.5.
Model of an open aquatic
ecosystem with atmospheric
oxygen exchanging with
dissolved oxygen in the water
(program OPENAQ.bas).
(a) Systems diagram and
equations; (b) typical simulation
for 5 days showing metabolic
rates, levels of dissolved
oxygen, and nutrients.

dissolved oxygen rises and falls with the maximum level occurring
late in the day and the minimum at sunrise. Plotted below is rate
of production P, rate of respiration R, and the net production
DQ, which is the difference between the two. Net production is
positive in the daytime and negative at night. This is a good
model to use with measurements of day–night oxygen change
for purposes of evaluating aquatic productivity.

EXPLANATION OF EQUATIONS

As shown in Figure 12.5a, available light energy is R remaining after light use K_0*R*N is drawn from the inflowing sunlight S. The production of oxygen K_2*R*N and the concurrent production of organic matter K_1*R*N are both dependent on the light energy R and the nutrients available N (in parts per billion, abbreviated ppb). The available nutrients N are those not bound up in organic storages $f*Q$, where f is the fraction of the nutrients that is part of the organic biomass Q. The organic biomass is a balance between inflowing production K_1*R*N and the consumption K_4*X*Q, which is dependent on the dissolved oxygen X and the metabolically active biomass Q.

The dissolved oxygen is a balance between the oxygen production K_2*R*N, the use of oxygen by consumption K_3*X*Q, and the diffusion exchange with the atmosphere. The oxygen at saturation, X_s, is the parts per million (ppm) oxygen in the water in equilibrium with oxygen in the air. Oxygen passes from air into the water when the water is undersaturated in proportion to the difference in the pressure exerted by the gas molecules. For a diffusion equation, oxygen pressure in the air is taken as 1 and that by the dissolved oxygen in the water as the fraction of saturation X/X_s. When the dissolved oxygen is less than saturation, X/X_s is less than 1. In the middle of the day the oxygen may be much greater than saturation, with X/X_s larger than 1. The diffusion pathway in Figure 12.5 is $K_7*(1 - X/X_s)$, where K_7 is the rate of diffusion, mainly dependent on the stirring of the water by currents, winds, waves, etc. Rate K_7 is very low in a quiet aquarium.

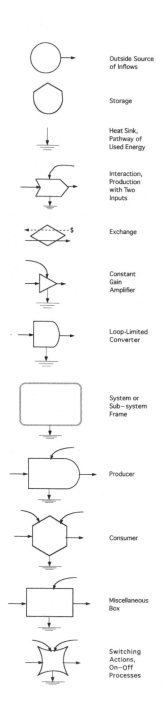

Outside Source of Inflows

Storage

Heat Sink, Pathway of Used Energy

Interaction, Production with Two Inputs

Exchange

Constant Gain Amplifier

Loop-Limited Converter

System or Sub-system Frame

Producer

Consumer

Miscellaneous Box

Switching Actions, On-Off Processes

EXAMPLES

Ponds, pools, lakes, and shallow seas often fit this model. Aquaria are found in the school rooms of the world by the millions. Some grow in bright light; some have animals; some contain saltwater ecosystems. Some have been observed working for decades. Because these small contained ecosystems have many of the features of larger ecosystems, they are sometimes called *microcosms*.

"What If" Experiments for the Program in Table 12.5

12.5–1 What would be the effect of adding a larger quantity of organic matter? Set initial value of Q = 40. Explain.

12.5–2 What would happen if light were turned on continuously? Change line 405   S = 1. What happens to dissolved oxygen? To net production? Why?

12.5–3 What would be the effect of reducing the inorganic nutrients? Set NT = 0.3.

12.5–4 What is the effect of increasing the diffusion coefficient K5 to 1.0? What is analogous in a pond and in an aquarium? Suppose the coefficient were reduced to 0.1 and many fish added, or suppose a large amount of organic matter were added (excess fish food in the aquarium and excess wastes in a pond).

◆

12.6 MODEL OF LOGISTIC AUTOCATALYTIC PRODUCTION AND RECYCLE (AUTOCYCL)

The model AUTOCYCL (Figure 12.6a) is an autocatalytic consumer unit with materials (example: nutrients) cycling between an available pool N and bound state in the consumer assets C. The energy source E is constant and thus unlimited, an example of autocatalytic growth on a constant energy source being exponential (Figure 9.4). Growth in this model is limited by the availability of nutrient materials T_n. The role of the cycling materials is dramatized in the diagram by the heavily shaded recycle loop. The production process is affected both by the energy level E and by the available materials N. Where the nutrients are free, they are darkly shaded; where nutrients are incorporated in the products they are stippled. The model may apply to phytoplankton in tropical seas where the nutrients are much more limiting than the light.

The typical simulation in Figure 12.6b shows an S-shaped growth of assets Q and a matching down curve of available nutrients as they are used in the production process. With the autocatalytic feedback to production, growth accelerates at first and then levels off as the nutrients required for new production are tied up in the storage C.

TABLE 12.5
Program of Production and Oxygen in Aquatic Ecosystems (OPENAQ.bas) (Figure 12.5)

```
3 REM PC OPENAQ.BAS (Production & dissolved oxygen in aquatic systems)
20 SCREEN 12, 0
30 LINE (O, O)-(320, 60), , B
35 LINE (0, 70)-(320, 130), , B
40 LINE (0, 140)-(320, 200), , B
45 LINE (0, 210)-(320, 270), , B
50 LINE (0, 280)-(320, 340), , B
55 LINE (0, 240)-(325, 240)
60 S0 = .04
65 Z = 1: REM .5 depth
70 DT = .1
75 T0 = 3
80 N0 = 10
85 Q0 = 15
90 S0 = 50
95 X0 = 3
100 P0 = 30
105 R0 = 30
110 REM STARTING VALUES:
115 XS = 8
120 X = 8
125 NT = 1.5
130 F = .05: REM fraction of nutrients in Q
140 Q = 10
200 REM COEFFICIENTS:
210 K0 = 9!
220 K1 = 10
230 K2 = 10
240 K3 = .01
250 K4 = .01
260 K5 = .1
300 REM PLOTTING TIME CURVES:
325 LINE (T * T0, 60)-(T * T0, (60 - S * S0))
335 PSET (T * T0, 130 - N * N0)
345 PSET (T * T0, 200 - X * X0)
355 PSET (T * T0, 240 - DQ * Q0): REM Net production rate
365 PSET (T * T0, 340 - P * P0): REM Gross production rate
375 PSET (T * T0, 340 - R * R0): REM Respiration rate
400 REM EQUATIONS
405  S = SIN(T / 3.78)
410 IF S < 0 THEN S = 0
415 N = ((NT - F * Q) / Z)
420 IF N < .000001 THEN N = .000001
425 R = S / (1 + K0 * N)
430 DQ = K2 * R * N - K4 * X * Q
435 DX = K1 * R * N - K3 * X * Q + K5 * (1 - X / XS)
440 Q = Q + DQ * DT
445 X = X + DX * DT / Z
450 P = K1 * R * N
460 R = K3 * X * Q
470 T = T + DT
500 IF T * T0 < 320 GOTO 300
```

EXPLANATION OF EQUATIONS

As shown in Figure 12.6b, autocatalytic production $K_1*E*N*C$ is proportional to the energy concentration E, the

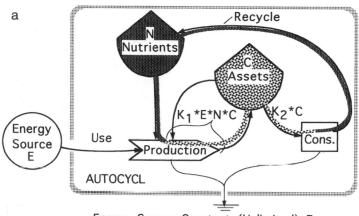

Energy Source Constant (Unlimited) E
Nutrients available: $N = T_n - f*C$ where
Total Nutrients is T_n and f is the fraction in C
Assets C: $DC = K_1*E*N*C - K_2*C$
The combined equation is Logistic:
$$DC = (K_1*E*T_n - K_2)*C - (K_1*E*f)*C^2$$

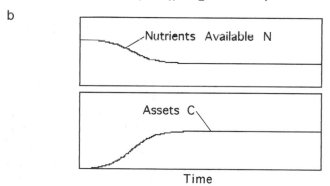

FIGURE 12.6.
Model of autocatalytic growth with a limiting material cycle, which is logistic (program AUTOCYCL.bas). (a) Systems diagram and equations. Nutrient materials when free are shown by dark shading and by stippling when they are combined in stored products and assets; (b) simulation of stored assets C and available nutrients N.

nutrient availability N, and the feedback use of the stored structure C. The stored assets C is a balance between the production inflow and the various losses K_2*Q in proportion to the quantity stored. The nutrients bound up in the stored assets are $f*Q$, where f is the fraction of the stored quantity that is nutrient material. For example, the phosphorus in biomass may be 0.01 or less. With total nutrients T_n held constant, the nutrients N available to production are the difference between the total in the system T_n and that bound in assets.

When the equation for consumer growth ($DC = K_1*E*N*C$ $1 - K_2*Q$) is algebraically combined with the expression for conservation of materials ($N = T_n - f*C$), a combined equation for change per unit time step results (Figure 12.6a). It has a positive growth term proportional to storage C and a negative drain that is quadratic (C^2). E and T_n are included and can increase the growth rate and the steady-state level.

Such equations with the general form $DQ = aQ - bQ^2$ (with a and b constants) are said to be logistic. Units with logistic growth equations have a very stable steady state, since the negative square term increases faster than the positive term as the quantity increases.

EXAMPLES

Many ecosystem models have constant coefficients of growth (unlimited energy), but are stabilized by including limited nutrient cycles. Some authors suggest that civilization could ultimately be limited by difficulties in recycling critical materials. Bernard cells are hexagonal patterns in viscous heated liquids (structural assets C). As the heated water in the pattern cools, it recirculates to the bottom of the container N, where it is available to be heated again.

TABLE 12.6
Program of Logistic Autocatalytic Production and Recycle (AUTOCYCLE.bas) (Figure 12.6)

```
10 REM PC
20 REM AUTOCYCL (Autocatalytic production & recycle logistic on flow source)
30 CLS
40 SCREEN 1, 0: COLOR 7, 0
50 LINE (0, 0)-(320, 90), 3, B
55 LINE (0, 100)-(320, 190), , B
60 E = 1
70 TN = 200
75 F = .1
80 C = 10
86 DT = 1
90 T0 = .3
100 N0 = .3
105 C0 = .05
120 K1 = .0002
130 K2 = .02
140 PSET (T * T0, 190 - C * C0), 3
150 PSET (T * T0, 90 - N * N0), 3
170 N = TN - F * C
175 IF N < .001 THEN N = .001
180 DC = K1 * E * N * C - K2 * C
190 C = C + DC * DT
200 T = T + DT
210 IF T * T0 < 320 GOTO 140
```

"What If" Problems for the Program in Table 12.6

12.6–1 Can you increase the growth and steady-state level of assets by increasing the material nutrient pool TN? What happens with more nutrients added (TN = 300)?

12.6–2 Can you increase the growth and steady-state level of assets by increasing energy level E?

12.6–3 Compare the shapes of the growth curve in the simulation of AUTOCYCL with that in the simulation of INT-LIMIT, which is also limited by its internal material cycle. Explain.

Chapter Thirteen

◆

MODELS OF GROWTH

Introduced in this chapter are minimodels for growth of a single storage that apply to many kinds of systems and scales. In introductory chapters, the models RAMP, TANK, and DRAIN were used to simulate the behavior of a storage with simple inflows and outflows. EXPO and the models in this chapter are autocatalytic with feedback of storage to accelerate input production—designs that self-organize for maximum power. RENEW grows on a source-limited inflow. NONREN grows on a stored reserve without other sources. 2SOURCE grows on both renewable and nonrenewable sources. SLOWREN grows on storage reserves that are slowly renewed. LOGISTIC has exponential growth limited by quadratic losses. These are different ways of connecting energy to a storage. These models are often used to overview complex systems simply. These designs are also the building blocks of more complex designs.

13.1 MODEL OF AUTOCATALYTIC GROWTH ON A RENEWABLE SOURCE (RENEW)

The model RENEW (Figure 13.1a) has an autocatalytic unit based on an inflow of energy limited from outside. For example, a forest grows biomass (leaves, trunks, roots, animals, bacteria, etc.) using the regular inflows of sunlight energy. This kind of energy source is renewable but externally limited. How the sunlight is used cannot affect the amount that is flowing in. A forest that uses the sunlight grows, increasing biomass until it is using almost all the sunlight that falls each day. When the amount of biomass growth equals the amount that is decomposed, the quantity of stored biomass Q becomes constant, and the system is in a steady state.

In the simulation (Figure 13.1b) the growth in biomass Q is almost exponential at first, while there is more sunlight than the organisms can use. Then Q levels off at a steady state as the

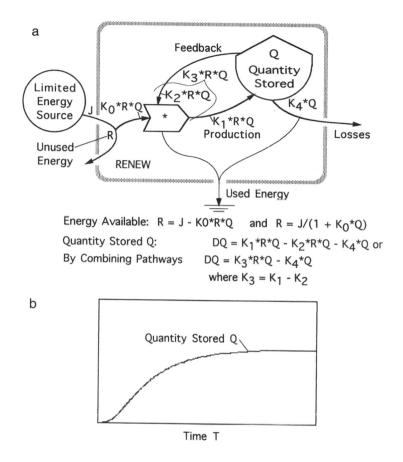

Energy Available: $R = J - K0*R*Q$ and $R = J/(1 + K_0*Q)$

Quantity Stored Q: $DQ = K_1*R*Q - K_2*R*Q - K_4*Q$ or

By Combining Pathways $DQ = K_3*R*Q - K_4*Q$

where $K_3 = K_1 - K_2$

FIGURE 13.1.
Model of growth on a renewable source (program RENEW.bas). (a) Systems diagram and equations; (b) typical simulation graph.

light becomes limiting, and production balances the losses due to depreciation, dispersal, etc.

EXPLANATION OF EQUATIONS

As shown in Figure 13.1a, J is the steady inflow of energy, (sunlight in the example discussed). The energy used by the production process is K_0*R*Q. Energy R is the energy remaining that is available for further use: $R = J - K_0*R*Q$.

Because the flows of light energy are almost instantaneous, the equation has to be changed into a form that represents R at any time without requiring iteration. Algebraically solving for R, the expression to use in the program is:

$$R = J/(1 + K_0*Q).$$

The quantity stored Q is a balance between the positive contribution by autocatalytic production flow K_1*R*Q and the drains by linear loss pathway K_4*Q and by the feedback from the storage to aid production K_2*R*Q. In the forest example, the production of biomass K_1*R*Q is proportional to the sunlight available R and the amount of biomass Q already growing. As in many other autocatalytic growth models (Figure 13.1a), the production and feedback loops are combined as a net production flow K_3*R*Q, where coefficient K_3 is a difference between coefficients K_1 and K_2.

The death and decomposition of biomass K_4*Q is a proportion K_4 of the stored biomass of forest organisms Q. Therefore, the equation for the change in forest biomass on each iteration DQ is:

$$DQ = K_3*R*Q - K_4*Q.$$

The amount of biomass at the end of each calculation by the program is the biomass at the beginning Q plus the change DQ during the iteration interval DT:

$$Q = Q + DQ*DT.$$

The changes are multiplied by DT (the time change) so that the amount of change added is appropriate for the time interval of each iteration.

EXAMPLES

This model is appropriate for natural systems (forests, fields, marshes, rivers, lakes, oceans) growing on inputs that are renewable source limited (sun, rain, wind, tides, waves). An example is successional growth of vegetation on a bare field, from herbs to mature forest trees. At first the weeds grow quickly, then shrubs, tree seedlings, and finally a forest using all the inflowing energy of the sun and rain as it becomes available.

An economic example is a business that has a steady inflow of a raw material such as leather. The business uses the leather to produce belts and gain capital to buy more. The number of belts in stock increases quickly. Then the stock levels off when the number of belts is limited by the rate of supply of leather from the source. Human civilizations that are based on limited renewable energies may also follow this model.

"What If" Experiments for the Program in Table 13.1

13.1–1 Compare and contrast systems that grow exponentially on unlimited energy (Figure 9.7) with those that grow up and then level off because of source-limited energy (Figure 13.1). How do the equations differ?

13.1–2 How would an increase in environmental sun and rain affect forest growth represented with this minimodel? Run the simulation with increased J. What is the effect of reducing energy inflow by half?

13.1–3 Let's consider a successional forest that is already at the shrub stage. Increase the starting value of biomass on line 50: set $Q = 25$. Does the new graph support a higher biomass level? Explain why.

13.1–4 Compare the growth rate and sustainable level of this forest to one that has higher decomposition rates. What do you change to simulate this? Why?

TABLE 13.1
Program of Growth on a Renewable
Source (RENEW.bas) (Figure 13.1)

```
10 REM PC: RENEW (Renewable source)
15 CLS
20 SCREEN 1, 0
30 COLOR 0, 0
35 LINE (0, 0)-(319, 180), 3, B
40 J = 45
50 Q = .1
60 K = .1
70 K3 = .008
80 K4 = .03
85 DT = 1
90 T0 = 1
95 Q0 = 1
100 PSET (T * T0, 180 - Q * Q0), 3
110 R = J / (1 + K * Q)
120 DQ = K3 * R * Q - K4 * Q
130 Q = Q + DQ * DT
140 T = T + DT
150 IF T * T0 < 319 GOTO 100
```

◆

13.2 SLOWLY RENEWABLE GROWTH (SLOWREN)

The model SLOWREN (Figure 13.2a) has two storages in series. Inflow J, from an outside source into the system, accumulates in the first storage E. The storage E becomes a reserve that supplies resource for growth of an autocatalytic consumer unit, storing assets Q. Without the consumer unit, a large reserve storage E develops because outflows are small. Then, if a consumer unit with feedback actively increasing consumption is connected, it has a surge of growth of its assets A, which reduces E to a low level. The growth based on feedback from the consumer storage to pump more input is *autocatalytic*. Thus, it starts its growth like the minimodel EXPO of Figure 9.4.

The simulation plotted in Figure 13.2b starts with a large accumulated reserve E before the consuming unit starts its use. The storage of consumer unit Q grows rapidly, drawing more and more energy, reducing the reserve E to a low level. With less

a

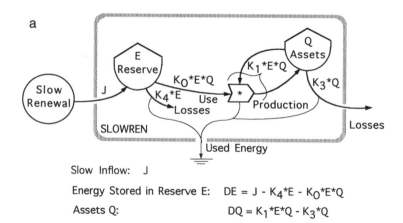

Slow Inflow: J

Energy Stored in Reserve E: $DE = J - K_4*E - K_0*E*Q$

Assets Q: $DQ = K_1*E*Q - K_3*Q$

b

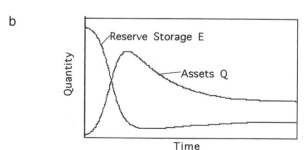

FIGURE 13.2.
Model of growth on a slowly renewable source (program SLOWREN.bas). (a) Energy systems diagram and equations; (b) typical simulation graph.

energy available, assets Q decline again, and the reserve recovers a bit. Although the reserve storage still receives its slow, steady inflow J from outside, its storage is used by the consumer unit as fast as it added. A new balance develops between the inflow and outflows, with the consumer unit thereafter sustained at a lower level.

This simulation graph resembles that generated by the 2SOURCE model, discussed in Section 13.4. In both there is a surge of growth based on using a previous reserve, growth that cannot be supported after the reserve is used.

EXPLANATION OF EQUATIONS

The reserve energy storage E is a balance between inflow J and two outflows. As shown in Figure 13.2a, losses K_4*E are in proportion to storage E. The use K_0*E*Q by assets development is autocatalytic. Changes in assets are a balance between autocatalytic production K_1*E*Q and losses K_3*Q, representing depreciation, consumption, and dispersal.

EXAMPLES

The arrangement of a tank and an "autocatalytic" consumer unit is found in many kinds of geological, chemical, biological, and economic systems. This minimodel is not a bad representation of the way resources are supplying our high energy consuming civilization. The reserve tank represents the great storages of coal, oil, natural gas, soil, wood, and minerals available a hundred years ago. We have been in steep accelerating growth, using up these reserves, and are now slowing in growth. If our economic system follows this oversimplified model, our civilization will have to be reduced in quantity to be supported because the generation of organic matter (biomass and fuels) is slower than our present rate of use of them.

Another example is the population of fishes and other aquatic animals in a reservoir behind a new dam that drowns a forest. All the dead organic matter from the decomposing trees supports a very high level of fish for a few years, but later the populations are reduced, living only on the regular inflow of organic matter from the tributary stream and local photosynthesis.

Another example is a town that develops around the cutting of a virgin forest, but later has to live on the regular, renewable growth of replanted trees, cutting them about as fast as they grow.

"What If" Experimental Problems for the Program in Table 13.2

13.2–1 What happens if the reserve tank is initially zero? In statement 60, set $E = 0$. In this run how does the quantity of consumers Q compare with the run that starts with a large initial reserve? Explain.

13.2–2 The level attained by the consumer in the long run is its *carrying capacity*. What would happen to the carrying capacity if the regular inflow were doubled? Make $J = 4$ in statement 50.

13.2–3 What would happen if there were no inflow? Set $J = 0$ in line 50.

TABLE 13.2
Program of Growth on a Slowly Renewable Source (SLOWREN.bas) (Figure 13.2)

```
10 REM PC: SLOWREN (Slowly renewable)
15 CLS
20 SCREEN 1, 0
30 COLOR 0, 0
35 LINE (0, 0)-(319, 180), 3, B
40 Q = 3
50 J = 2
60 E = 159
70 K0 = .001
80 K1 = .001
90 K3 = .03
100 K4 = .01
110 Q0 = 1
120 E0 = 1.2
130 DT = 1
140 T0 = 1
200  PSET (T * T0, 180 - Q * Q0), 1
210  PSET (T * T0, 180 - E * E0), 3
220 DQ = K1 * E * Q - K3 * Q
230 DE = J - K0 * E * Q - K4 * E
240 E = E + DE * DT
250 Q = Q + DQ * DT
260 T = T + DT
270 IF T * T0 < 319 GOTO 200
```

◆

13.3 GROWTH ON A NONRENEWABLE SOURCE (NONRENEW)

The model NONRENEW (Figure 13.3a) represents the pulse of consumption when autocatalytic consumers use a stored resource that is not being replaced. Because there are no inflows, such a resource is said to be *nonrenewable*. For example, a dead log downed by a storm is a large storage of wood available to be eaten by beetles. The population of beetles will increase almost exponentially as they eat the log, and the quantity of log decreases

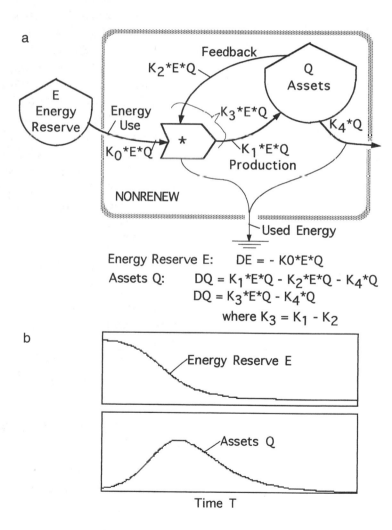

Energy Reserve E: $DE = -K0*E*Q$

Assets Q: $DQ = K_1*E*Q - K_2*E*Q - K_4*Q$

$DQ = K_3*E*Q - K_4*Q$

where $K_3 = K_1 - K_2$

FIGURE 13.3.
Model of growth on a nonrenewable source (program NONRENEW.bas). (a) Energy systems diagram and equations; (b) typical simulation graphs.

rapidly. However, after the log is used up, the population of beetles disappears also.

In the simulation (Figure 13.3b), the assets Q grow in a pulse and return to zero. The energy reserve E is pulled down to a low value, but in this particular calibration is not entirely used. There is a minimum quantity of log necessary to support beetles.

EXPLANATION OF EQUATIONS

In Figure 13.3a, E is the nonrenewable resource being used by the autocatalytic consumers Q at a rate $K_0{*}E{*}Q$ that depends on both the resource concentration E (the log) and the quantity of consumer assets Q (beetles). The production of consumers (beetles) $K_1{*}E{*}Q$ is also a function of the resource quantity E and the consumer assets Q. The term $K_2{*}E{*}Q$ represents the feedback effort by the feeding consumers. The change in the consumer assets Q is a balance between the production, feedback, and the other losses $K_4{*}Q$ (depreciation, dispersal, and mortality). In the program, the production and feedback pathways in the autocatalytic loop are combined into one term, $K_3{*}E{*}Q$, where $K_3 = K_1 - K_2$ (shown with a bracket in Figure 13.3a). At the end of each iteration by the program, the quantity of consumer assets Q (beetles) is the number at the beginning Q plus the change DQ multiplied by the time interval of iteration: Q = Q + DQ*DT.

EXAMPLES

This model represents systems using storages that are not being renewed. An economic example is a mining town based on a limited source of gold ore. As the ore is mined, the town grows on its revenues from the gold. When the mine runs out, the town economy decreases until everyone leaves and it becomes a "ghost town."

"What If" Experimental Problems for the Program in Table 13.3

13.3–1 What would happen to the beetle population if the log that fell in the storm were larger? Would the population be greater, live longer, or both? Change E to 250 in

TABLE 13.3
Program of Growth on a Nonrenewable Source
(NONRENEW.bas) (Figure 13.3)

```
10 REM PC: NONRENEW (Nonrenewable source)
15 CLS
20 SCREEN 1, 0
30 COLOR 0, 0
35 LINE (0, 0)-(319, 180), 3, B
40 E = 160
50 Q = .1
60 K = .001
70 K1 = .001
75 K4 = .03
80 DT = 1
85 T0 = 1
90 Q0 = 1
95 E0 = 1
100 PSET (T * T0, 180 - Q * Q0), 3
120 PSET (T * T0, 180 - E * E0), 1
130 DQ = E * K1 * Q - K4 * Q
140 DE = -K * E * Q
150 Q = Q + DQ * DT
160 E = E + DE * DT
170 T = T + DT
180 IF T * T0 < 319 GOTO 100
```

line 40. To show both graphs on the screen on some computers, type RUN 40 in the command panel.

If this model applies, what would you predict our world economy would do if large new deposits of fossil fuels were found? Would the system use up the fuels faster or would the fuels sustain the economy longer?

13.3–2 What would happen to the log E and beetles Q if you started with 100 times more beetles? Change Q from 0.1 to 10 in line 50. Explain the result using the beetle example and then the mining town example.

13.3–3 What effect does a different species of beetles have if its growth is more efficient? Change K1 to 0.0015. What happens to Q? To E? Then assume you have another beetle species that is less efficient; change K1 to 0.0004.

13.3–4 What would you change to increase the death rate of the beetles? How would E and Q change? Simulate and explain your results.

◆

13.4 GROWTH ON TWO SOURCES: RENEWABLE AND NONRENEWABLE (2SOURCE)

In the model 2SOURCE (Figure 13.4a), energy for the growth of assets by an autocatalytic consumer comes from two sources, one a steadily flowing renewable source J. The other source is a storage that is not being replenished, a nonrenewable storage E. The model combines two other models (RENEW and NONRENEW), and the equations are a synthesis. This mini-model represents one view of our own global society. The world economy grew up on both fossil fuel reserves and renewable sources. If the model is correct, the economy will have to adjust to a lower level as the nonrenewable fossil fuels get scarce.

The simulation in Figure 13.4b is typical, with the nonrenewable storage E being reduced as the growth of assets Q surges, decreasing again as the energy reserve disappears.

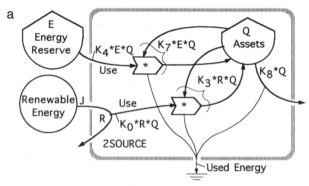

Renewable Energy Available: $R = J - K_0*R*Q$ and $R = J/(1 + K_0*Q)$
Energy Reserve E: $DE = - K_4*E*Q$
Assets Q: $DQ = K_7*E*Q + K_3*R*Q - K_8*Q$

FIGURE 13.4.
Model of growth on two sources, one renewable, one nonrenewable (program 2SOURCE.bas). (a) Energy systems diagram and equations; (b) typical simulation graph.

EXPLANATION OF EQUATIONS

Referring to the diagram (Figure 13.4a), nonrenewable energy storage E outside the system is decreased by autocatalytic use K_4*E*Q. At the interaction symbol, part of the energy is transformed into production flow K_7*E*Q. Energy flow K_0*R*Q is drawn into use from the inflow of renewable energy J, and at the second interaction symbol is used to generate production flow K_3*R*Q. The available renewable energy R is the difference between the inflow J and the use. The quantity of assets Q is a balance between the two production inflows and the losses K_8*Q that are in proportion to the storage.

As indicated by brackets and explained in previous models, the gross production and feedback loops of autocatalytic production are combined, with one coefficient each (K_7 and K_3) representing the contributions of net production.

EXAMPLES

An example is the growth of a population of microbes that decomposes leaves raked into a pile. The microbes grow up quickly, but when the pile of leaves is used up, their population decreases to the number that can live on the fewer leaves that fall regularly from evergreen trees.

"What If" Experimental Problems for the Program in Table 13.4

13.4-1 If more fuels are found for the reserve E, will the assets peak higher? Last longer? Set E = 200. On the command line on some computers, type RUN 40 to display both graphs on the screen together. Explain.

13.4-2 Suppose pollution reduces renewable resource inflow J. How are assets Q affected? Decrease J to 30. Explain.

13.4-3 If the depreciation rate of society increases as structures get older, how will this affect assets? Add line 195: IF T > 80 THEN K8 = .04. Explain the results.

13.4-4 If the simulation is started with Q high, will the graph differ from the first run? Describe what you changed in the program and explain the results using an example.

TABLE 13.4
Program of Growth on Renewable and Nonrenewable
Sources (2SOURCE.bas) (Figure 13.4)

```
10 REM PC: 2SOURCE (Renewable and nonrenewable)
15 CLS
20 SCREEN 1, 0
30 COLOR 0, 0
35 LINE (0, 0)-(319, 180), 3, B
40 J = 80
50 N = 150
60 Q = 1
70 K = .1
80 K3 = .002
90 K4 = .0007
100 K7 = .0008
110 K8 = .03
120 DT = 1
125 T0 = 1
130 Q0 = 1
135 N0 = 1
140 PSET (T * T0, 180 - Q * Q0), 3
150 PSET (T * T0, 180 - N * N0), 1
160 R = J / (1 + K * Q)
170 DQ = K7 * N * Q + K3 * R * Q - K8 * Q
180 DN = -K4 * N * Q
182 Q = Q + DQ * DT
185 N = N + DN * DT
190 T = T + DT
200 IF T * T0 < 319 GOTO 140
```

◆

13.5   MODEL OF LOGISTIC GROWTH (LOGISTIC)

The model LOGISTIC (Figure 13.5) is a combination of auto-catalytic production using a constant (unlimited) energy source and a quadratic drain representing the effect of crowding. The equation that results has been widely used to represent growth and leveling of populations. An example is a population of microbes that grows with a food supply that is in excess. At first the population increases exponentially, as in Figure 9.4. However, the death rate increases even more rapidly than the production because of interactions and crowding, which cause stress and toxicity. Because of the quadratic term representing these negative interactions, the population levels off. Even though there

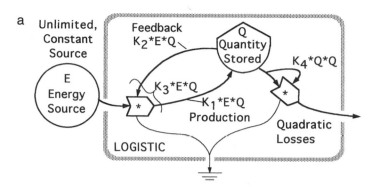

Quantity D: $DQ = K_3*E*Q - K_4*Q*Q$ where $K_3 = K_1 - K_2$

Quantity at Steady State: $DQ = 0$ and $Q_s = (K_3/K_4)*E$

FIGURE 13.5.
Model of autocatalytic growth
and quadratic loss that is
mathematically logistic
(program LOGISTIC.bas).
(a) Systems diagram and
equations; (b) a family of
simulation graphs of storage
with time. Each curve was
generated by increasing the
energy source E between runs.

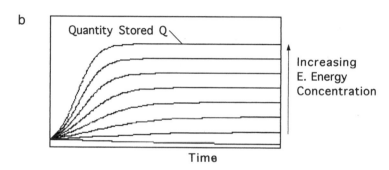

is no lack of food to limit growth, crowding rapidly increases
the death rate until the deaths equal the births and the population
reaches a steady state.

A family of simulation curves is shown in Figure 13.5b. After
each simulation curve is plotted, the lines at the bottom of the
program (lines 160–200 in Table 13.5) increase the outside energy available E and restart the simulation, plotting another
curve. In the first simulation graph, the production does not
exceed the drains and the assets decrease with time. In the next
runs with more energy, growth starts off nearly exponential and
then levels, making an S-shaped growth curve. Seven such curves
are shown evenly spaced, each growing faster and leveling with
a greater storage.

This model is only one of many system designs that generate logistic equations for different reasons. For another, see
AUTOCYCL in Section 12.6.

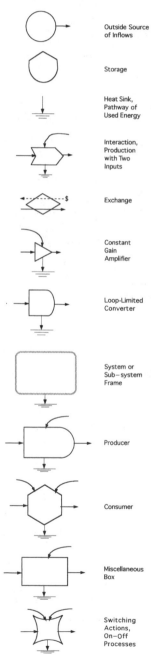

Outside Source
of Inflows

Storage

Heat Sink,
Pathway of
Used Energy

Interaction,
Production
with Two
Inputs

Exchange

Constant
Gain
Amplifier

Loop-Limited
Converter

System or
Sub-system
Frame

Producer

Consumer

Miscellaneous
Box

Switching
Actions,
On-Off
Processes

TABLE 13.5
**Program of Logistic Growth with
Quadratic Losses (LOGISTIC.bas)
(Figure 13.5)**

```
10 REM PC: LOGISTIC (Logistic growth)
15 CLS
20 SCREEN 1, 0
30 COLOR 0, 0
35 LINE (0, 0)-(319, 180), 3, B
40 E = .2
50 Q = 10
60 K3 = .005
70 K4 = .0005
80 Q0 = 1
90 T0 = 1
95 DT = .5
100 PSET (T * T0, 180 - Q * Q0), 1
120 DQ = K3 * E * Q - K4 * Q * Q
130 Q = Q + DQ * DT
140 T = T + DT
150 IF T * T0 < 319 GOTO 100
160 REM Energy E increaseed and rerun
170 T = 0
180 Q = 10
190 E = E + 2
200 IF E < 15 GOTO 100
```

EXPLANATION OF EQUATIONS

The form of the logistic equation in Figure 13.5a was derived from the arrangement of pathways in the systems diagram. Stored quantity Q is increased by the flow of net production K_3*E*Q (the gross production K_1*E*Q minus the feedback K_2*E*Q). The storage is decreased by the outflow, which is proportional to the square of Q (K_4*Q*Q), a self-interactive, quadratic drain. At steady state DQ is zero, and by algebraic rearrangement of the change equation, steady-state Q_s is found to be proportional to the energy level E, with the relationship: $Q_s = K_3*E/K_4$.

However, a different form of the logistic equation is often found in ecological textbooks: $DQ = r*Q*(K - Q)/K$. It is often verbalized by saying that the change in population DQ for each time step is the product of the population Q times a constant intrinsic reproduction rate r but diminished by the fraction that the difference $(K - Q)$ is of the steady-state level K. The logistic derived in this way can be related to the one in Figure 13.5 as follows: Because r is constant, an unlimited energy source is implied, and r can be replaced with $K_3 * E$. Then if K is the steady-state quantity Q_s, it can be replaced by $K_3 * E/K_4$. The result of algebraic substitutions is the equation in Figure 13.5:

$$DQ = K_3 * E * Q - K_4 * Q * Q.$$

To make the model more realistic in following the second energy law, there should be a linear depreciation pathway on the storage (i.e., $-K_5 * Q$). To keep this model in its simplest form, the term was not included.

OTHER EXAMPLES

When mice are crowded into cages, their frequent interactions cause physiological stress that interferes with breeding and stops population growth.

"What If" Experimental Problems for the Program in Table 13.5

13.5–1 Suppose a city growth is following the logistic model. What is the effect of increasing the availability of fuels and electric power? Examine the series of plots in Figure 13.5b, each with a higher value of E.

13.5–2 Consider a population of mice in a cage that has a steady inflow of food pellets each day. What happens to the size of the population if you start with a larger number of mice? Set Q = 160 in line 50 and in line 180 and rerun.

13.5–3 If you tried to substitute growth of stressed mice with another species that was less sensitive to crowding, which variable would you change? How does that affect the final population level? Explain.

13.4–4 Not all of the losses from a population are quadratic. What is the effect of adding a linear depreciation pathway $-K_5 * Q$ to the DQ equation in line 120 and adding a new line 75 $K_5 = 0.005$?

Chapter Fourteen

◆

MODELS OF COMPETITION AND COOPERATION

Chapter 14 contains pairs of autocatalytic units in parallel for the study of populations and other fields. Depending on their relationship to energy sources and their interactions with each other, units in parallel may be independent, competing, or cooperating. In some models, competitive exclusion by one population drives the other population to low levels. In others, coexistence occurs. The model EXCLUS has two similar units competing for a limited energy source. The model TWOPOP compares the exponential growths of two populations in the same place with unlimited energy and without interactions. The model INTERACT has two logistic populations with negative interactions. The model CO-OP has two mutually beneficial populations using the same limited energy source.

Words like *cooperation* and *competition* have many meanings that can cause confusion. Relationships are clarified when given as minimodels and simulated. More parallel designs are included in models in Chapter 19 on international relationships. Readers are encouraged to generate other models.

The models in this chapter are labeled and discussed as populations measured as numbers of individuals. Production flows are reproduction, and losses are mortality and emigration. However, the models also apply to parallel relationships in other systems where the storages may be in mass or energy units.

Minimodels have been well used to understand laboratory experiments where two populations have been isolated. Rarely are two populations isolated the way they are in the models of this chapter, but the behavior of these designs helps us understand the larger systems in which parallel pairs may be included.

14.1 MODEL OF COMPETITION FOR A LIMITED SOURCE (EXCLUS)

If two species are using a common source of food that is in short supply, the growth of one may deprive the other of its source of food (see Figure 14.1a). Under some conditions the species that grows faster causes the other population to die out, as in Figure 14.1b. This process is called *competitive exclusion*. With other values of their growth coefficients, the two species may coexist without competitive exclusion.

The simulation generates curves of population number with time in Figure 14.1b, and plots the same points on a graph of one population as a function of the other (phase plane) in Figure 14.1c. At first both populations increase. Then as population Q_1 grows more, population Q_2 decreases to nearly zero.

EXPLANATION OF EQUATIONS

The available energy R is that left unused as the two populations draw energy on pathways K_1*R*Q_1 and K_2*R*Q_2. These

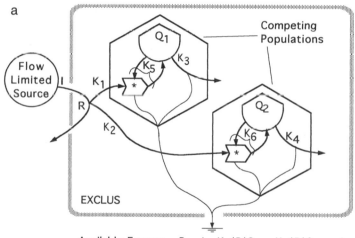

Available Energy: $R = I - K_1*R*Q_1 - K_2*R*Q_2$ and
$R = I/(1 + K_1*Q_1 + K_2*Q_2)$
Population Q_1: $DQ_1 = K_5*R*Q_1 - K_3*Q_1$
Population Q_2: $DQ_2 = K_6*R*Q_2 - K_4*Q_2$

FIGURE 14.1.
Model of two populations competing for a limited inflow of energy (program EXCLUS.bas). (a) Energy systems diagram and equations; (b) graph of population with time; (c) phase plane, graph of one population as a function of another.

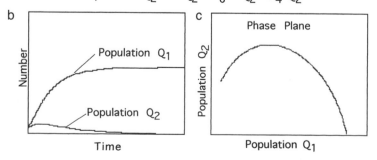

pathways feed back action of the storages to aid production for itself, a design we call *autocatalytic*. Both populations Q_1 and Q_2 are a balance between autocatalytic production (reproduction) K_5*R*Q_1 and K_6*R*Q_2 and losses. The mortality, consumption, dispersal, and other losses are represented by the linear pathways K_3*Q_1 and K_4*Q_2.

EXAMPLES

An ecological example of competitive exclusion is competition between two species of water fleas in a jar, which are supplied some yeast cells to eat each day. If a few of either species are put into the jar alone, the number of animals increases up to a steady level, as in the logistic example of Chapter 13 (Figure 13.5). If put together, the more efficient species consumes more of the food, and the other species dies out.

An example from business is competition between two timber companies developing a limited area of growing trees. The one that is more efficient with more production may drive the other out of business.

"What If" Experimental Problems for the Program in Table 14.1

14.1–1 What is the effect on the growth of the populations if you start with large populations? Set Q1 and Q2 = 100.

14.1–2 Suppose one population Q1 has cooperative self-interactions (like helping each other), so that the growth rate is quadratic (proportional to $Q*Q = Q^2$, introduced in Figure 6.3d). To compare growth of a species with quadratic growth with a species with ordinary autocatalytic growth, revise two lines in the program EXCLUS. bas to read:

$$230 \ R = I \ / \ (1 + K8*Q1*Q1 + K2*Q2)$$
$$250 \ DQ1 = K7*R*Q1*Q1 - K_3*Q1$$

where K7 and K8 are the coefficients for the quadratic growth already included in the program (Table 14.1). Describe and explain the result.

14.1–3 By varying the growth rate and decay rate coefficients, find a set of values of constants that allows for coexistence.

TABLE 14.1
Program of Competitive Exclusion (EXCLUS.bas)
(Figure 14.1)

```
2 REM PC EXCLUS.bas (Competitive exclusion)
10 CLS
20 SCREEN 1, 0
25 COLOR 0, 0
30 LINE (0, 0)-(319, 100), , B
40 LINE (50, 110)-(270, 190), 3, B
50 I = 5
60 DT = .1
70 T0 = 1
80 Q0 = 1
90 G1 = 2
95 G2 = 5
120 Q1 = 8
130 Q2 = 8
140 K1 = .08
150 K2 = .04
160 K3 = .05
170 K4 = .05
180 K5 = 9.000001E-02
190 K6 = .05
195 K7 = .003
197 K8 = .003
200 PSET (G1 * Q1 + 50, 190 - G2 * Q2), 3
205 PSET (T * T0, 100 - Q1 * Q0), 1
210 IF Q1 > 180 THEN Q1 = 180
220 PSET (T * T0, 100 - Q2 * Q0), 2
230 R = I / (1 + K1 * Q1 + K2 * Q2)
240 IF R < 0 THEN R = 0
250 DQ1 = K5 * R * Q1 - K3 * Q1
260 DQ2 = K6 * R * Q2 - K4 * Q2
270 Q1 = Q1 + DQ1 * DT
280 Q2 = Q2 + DQ2 * DT
290 T = T + DT
300 IF T * T0 < 230 GOTO 200
```

◆

14.2 MODEL OF TWO POPULATIONS IN EXPONENTIAL GROWTH (TWOPOP)

There is a natural tendency for populations with unlimited food supplies to grow increasingly rapidly with exponential growth, as was shown in Figure 9.4. As storage grows, the auto-catalytic feedback also increases, accelerating growth. In Figure

14.2a two populations are shown, both growing from the same unlimited source.

The simulation in Figure 14.2b shows the steep upward curve of exponential growth for both populations, one faster than the other because its growth coefficient K_1 is larger. When growths are both exponential, the difference expands. Even though resources are not in short supply, and even though there are no negative interactions, one population becomes increasingly dominant. The business that borrows wins competition by starting sooner.

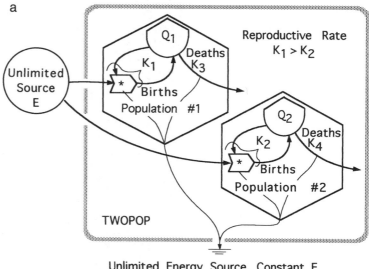

Unlimited Energy Source, Constant E

Population Q_1: $DQ_1 = K_1*E*Q_1 - K_3*Q_1$

Population Q_2: $DQ_2 = K_2*E*Q_2 - K_4*Q_2$

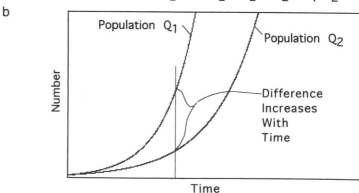

FIGURE 14.2.
Model of two populations in the same place, both using an unlimited energy source (source with constant force) (program TWOPOP.bas). (a) Energy system diagram and equations; (b) simulation.

EXPLANATION OF EQUATIONS

In Figure 14.2 reproduction of new individuals of population Q_1 and Q_2 is proportional to the population numbers and also to the concentration of available energy E. Reproduction of new individuals in the systems diagram (Figure 14.2) is drawn as a feedback loop from the storage to the interaction symbol interacting with the energy source. A loop reinforcing an inflow back to its own storage is called *autocatalytic,* named after chemical reactions where stored quantities help in their own duplication. Where a flow is proportional to the interaction of two factors both necessary to the process, the flow is a mathematical product of the two. The birth flows are K_1*E*Q_1 and K_2*E*Q_2. Loss of population by mortality flows is linear in this model. This means that deaths are in proportion to the populations, and the flows are K_3*Q_1 and K_4*Q_2.

Some populations have individuals that work cooperatively in their production processes. That kind of population growth is proportional to the interaction among its individuals, which makes growth proportional to the square of the population number. It is said to be *quadratically autocatalytic.* If one of the populations has a quadratic feedback and is not energy limited, it grows faster. (For its simulation, see "What If" problem 14.2–3.)

EXAMPLES

Examples of the rapid growth of populations on unlimited food resources are the early stages of growth of microorganisms in food, the growth of new plants when a field is abandoned, and the growth of new industries during human colonization of a new area.

Examples of quadratic growth are the growth of fishes that school and birds that nest in colonies. The growth of the economy of the United States prior to 1973 was autocatalytically quadratic.

"What If" Experimental Problems for the Program in Table 14.2

14.2–1 If the two populations have the same initial number of individuals, but the efficiency of growth is greater in one than the other, the first outdistances the other very

TABLE 14.2
**Program Comparing Two Populations in Exponential Growth
(TWOPOP.bas) (Figure 14.2)**

```
10 REM PC: TWOPOP.bas (Comparing exponential growth)
15 CLS
20 SCREEN 1, 0
30 COLOR 0, 0
35 LINE (0, 0)-(319, 180), 3, B
40 E = 1
50 Q1 = 3
60 Q2 = 3
70 K1 = .07
80 K2 = .08
90 K3 = .05
100 K4 = .05
105 K5 = .018
110 Q10 = 1
120 Q20 = 1
130 T0 = 1
140 DT = 1
150 PSET (T * T0, 180 - Q1 * Q10), 1: REM GREEN
155 IF Q1 > 180 THEN Q1 = 180
160 PSET (T * T0, 180 - Q2 * Q20), 2: REM RED
165 IF Q2 > 180 THEN Q2 = 180
170 DQ1 = K1 * E * Q1 - K3 * Q1
180 DQ2 = K2 * E * Q2 - K4 * Q2
190 Q1 = Q1 + DQ1 * DT
210 Q2 = Q2 + DQ2 * DT
230 T = T + DT
240 IF T * T0 < 319 GOTO 150
```

rapidly, as shown in Figure 14.2b. Can you generate this result by leaving the initial conditions for Q1 and Q2 the same in lines 50 and 60 of the program, making the production coefficients the same, and changing the death rates K3 and K4? Did one population begin to outdistance the other?

14.2–2 What happens if the growth characteristics are the same, but one starts ahead of the other? Change the starting numbers for each population so that one population starts out with more, but with the same growth rates (K1 = K2). How does the graph compare with that in Figure 14.2b?

14.2–3 To see the effect of having one population grow with quadratic feedback, substitute the following line:

$$250 \ DQ1 = K5*E*Q1*Q1 - K3*Q1.$$

Which population has the faster growth?

◆

14.3 MODEL OF TWO POPULATIONS IN COMPETITIVE INTERACTIVE GROWTH (INTERACT)

In the model INTERACT (Figure 14.3a), two populations compete by negative effects of their interaction. In this model

a

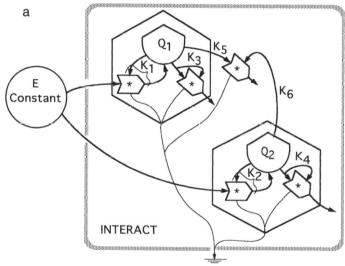

Population Q_1: $DQ_1 = K_1*E*Q_1 - K_3*Q_1*Q_1 - K_5*Q_1*Q_2$

Population Q_2: $DQ_2 = K_2*E*Q_2 - K_4*Q_2*Q_2 - K_6*Q_1*Q_2$

b

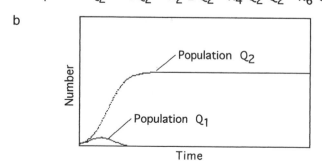

FIGURE 14.3.
Model of two populations, both using an unlimited energy source but affecting each other through an energy draining interaction (program INTERACT.bas). (a) Energy systems diagram and equations; (b) graphs with time.

each of the populations has autocatalytic production and quadratic losses, which make their growth equations logistic (Chapter 13). Both populations have unlimited food concentrations, because E is constant. Without interactions each population grows and levels off as in Figure 13.5. With negative interactions, neither can maintain as much growth nor as high a population at steady state. With sufficient negative effects, one population may drive out the other, an example of competitive exclusion. (See competitive exclusion that occurs via a different mechanism in Figure 14.1b.)

In the simulation (Figure 14.3b), population Q_2 predominates, displacing population Q_1 because of a more efficient growth rate, lower mortality rate, and fewer losses from interactions.

This minimodel has unlimited energy sources and thus is unrealistic in only being limited by the quadratic crowding effects or interunit interactions. The model is included here because it is the one used most often in introductory textbooks to consider competition. Usually the production term (growth production) is given as r^*Q_1, where r is a constant "intrinsic rate of natural increase," a specific reproduction rate (reproduction rate per individual):

$$DQ_1 = r^*Q_1 - (r/K)^*Q_1^2 - K_5^*Q_1^*Q_2.$$

The equation says that the change on each iteration is the growth at the intrinsic rate, with a negative effect proportional to self-interaction, minus a negative effect due to interaction with the competing population. When rearranged algebraically the equation is

$$DQ_1 = r^*Q_1^*(K - Q_1)/K - K_5^*Q_1^*Q_2,$$

which says that the rate of change is the growth according to the intrinsic rate, but decreased by the difference $(K - Q_1)$ relative to the steady-state level K.

EXPLANATION OF EQUATIONS

In our equations (Figure 14.3a), we break r into an external energy-driven force and a growth constant $r = K_1^*E$, which recognizes the energy basis. For the first population the autocatalytic production (reproduction) is K_1*E*Q_1, and the mortality drain is $K_3*Q_1^2$, written in the program as Q_1*Q_1. The death rate *per individual* K_3*Q_1 (calculated by dividing $K_3*Q_1^2$ by Q_1) is proportional to Q_1 and is described as *density dependent*. The interaction with the other population is proportional to the product of the numbers of each $K_5*Q_1*Q_2$.

To find the population steady state, set $DQ_1 = 0$ and solve for Q_1. Without interaction, the steady-state population level is $Q_1 = K_1*E/K_3$, increasing with energy concentration. With the interaction between populations, the steady-state population level is $Q_1 = (K_1*E - K_5*Q_2)/K_3$, less because of the second population Q_2.

EXAMPLES

An example of competitive exclusion is competition between two species of grain beetles in a jar supplied with abundant flour on which they feed. If only one species is present, the population grows and levels off. With two species together, one species displaces the other. This model is appropriate if the actions of one species negatively affect the other through waste products, crowding, disturbing of egg production, or other means. Another example is the competition between two species of weeds, where each secretes chemical substances that inhibit the root growth of the other.

"What If" Experimental Problems for the Program in Table 14.3

Study the way the two logistic-growing populations interact by making changes and running the simulation. After each simulation run, type ? Q1, Q2 (the question mark used in this way means PRINT), and the last values of populations Q_1 and Q_2 will appear on the screen. In this way you can see which population curve is Q_1 and which is Q_2.

14.3–1 First, set the two interaction coefficients (K5 and K6) to equal zero. Then run. What happens to each population? Is there competitive exclusion? How does your result compare with Figure 14.3b?

14.3–2 Next, with interactions K5 and K6 still set at zero, de-

TABLE 14.3
Program of Competitive Interactive Growth (INTERACT.bas)
(Figure 14.3)

```
3 REM PC: INTERACT (2 populations in iterative competition)
4 SCREEN 1, 0: COLOR 0, 1
6 LINE (0, 0)-(319, 180), 3, B
10 E = 1
20 Q1 = 3
30 Q2 = 3
40 K1 = .07
50 K2 = .08
60 K3 = .002
70 K4 = .001
80 K5 = .002
83 K6 = .001
85 Q10 = 1
88 Q20 = 1
90 T0 = 1
95 DT = 1
100 PSET (T * T0, 180 - Q1 * Q10), 1
120 PSET (T * T0, 180 - Q2 * Q20), 2
150 DQ1 = K1 * E * Q1 - K3 * Q1 * Q1 - K5 * Q1 * Q2
160 DQ2 = K2 * E * Q2 - K4 * Q2 * Q2 - K6 * Q1 * Q2
162 Q1 = Q1 + DQ1 * DT
165 Q2 = Q2 + DQ2 * DT
170 T = T + DT
180 IF T * T0 < 319 GOTO 100
```

crease the available food supply (energy concentration maintained E) to half of what it was in simulation problem 14.3–1. How were growth and steady-state levels changed?

14.3–3 Next decrease the interaction coefficient. Set K5 = 0.0005 and run again. What happens to populations now? Is there competitive exclusion? Explain.

14.3–4 Next set the second interaction coefficient K6 = 0.002 and run again. How is competition affected now? Can you increase K6 to a high enough value to displace, the other species? (Change the coefficient and run again.)

◆

14.4 MODEL OF TWO POPULATIONS THAT COOPERATE (CO-OP)

In the model CO-OP (Figure 14.4a) two populations grow on the same renewable, limited-flow energy source as in the

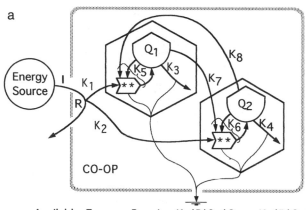

Available Energy: $R = I - K_1*R*Q_1*Q_2 - K_2*R*Q_1*Q_2$

$R = I/(1 + K_1*Q_1*Q_2 + K_2*Q_1*Q_2)$

Population Q_1: $DQ_1 = K_5*R*Q_1*Q_2 - K_3*Q_1 - K_7*R*Q_1*Q_2$

Population Q_2: $DQ_2 = K_6*R*Q_1*Q_2 - K_4*Q_2 - K_8*R*Q_1*Q_2$

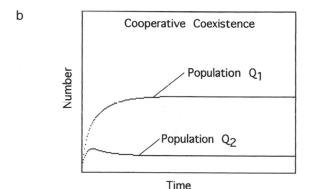

FIGURE 14.4.
Model of two populations using the same limited inflow source but with an interaction of mutual benefit (program CO-OP.bas). (a) Energy systems diagram and equations; (b) graphs with time.

competition model EXCLUS (Figure 14.1a). However, in this system some of the assets of each population are used to aid the growth of the other population, resulting in *mutual cooperation*. In the simulation (Figure 14.4b) both populations grow and level off. Instead of one displacing the other, they both coexist. The growth of each population is dependent on the other.

EXPLANATION OF EQUATIONS

The production (reproduction) of each is autocatalytic and also a function of the other $K_5*R*Q_1*Q_2$ and $K_6*R*Q_1*Q_2$. The available energy R is that remaining from uses by both popula-

TABLE 14.4
Program for Cooperative Coexistence (CO-OP.bas)
(Figure 14.4)

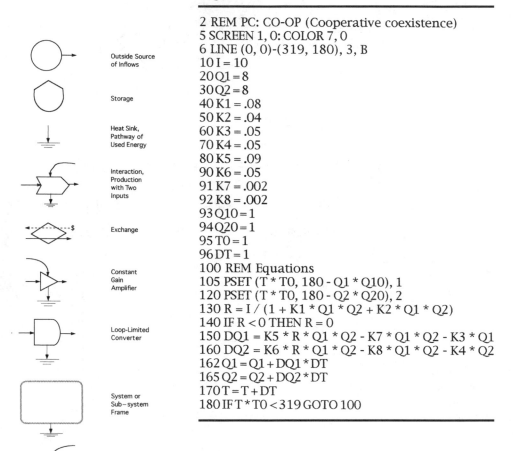

Outside Source of Inflows

Storage

Heat Sink, Pathway of Used Energy

Interaction, Production with Two Inputs

Exchange

Constant Gain Amplifier

Loop-Limited Converter

System or Sub–system Frame

Producer

Consumer

Miscellaneous Box

Switching Actions, On–Off Processes

```
2 REM PC: CO-OP (Cooperative coexistence)
5 SCREEN 1, 0: COLOR 7, 0
6 LINE (0, 0)-(319, 180), 3, B
10 I = 10
20 Q1 = 8
30 Q2 = 8
40 K1 = .08
50 K2 = .04
60 K3 = .05
70 K4 = .05
80 K5 = .09
90 K6 = .05
91 K7 = .002
92 K8 = .002
93 Q10 = 1
94 Q20 = 1
95 T0 = 1
96 DT = 1
100 REM Equations
105 PSET (T * T0, 180 - Q1 * Q10), 1
120 PSET (T * T0, 180 - Q2 * Q20), 2
130 R = I / (1 + K1 * Q1 * Q2 + K2 * Q1 * Q2)
140 IF R < 0 THEN R = 0
150 DQ1 = K5 * R * Q1 * Q2 - K7 * Q1 * Q2 - K3 * Q1
160 DQ2 = K6 * R * Q1 * Q2 - K8 * Q1 * Q2 - K4 * Q2
162 Q1 = Q1 + DQ1 * DT
165 Q2 = Q2 + DQ2 * DT
170 T = T + DT
180 IF T * T0 < 319 GOTO 100
```

tions $K_1*R*Q_1*Q_2$ and $K_2*R*Q_1*Q_2$. Losses include the linear pathways K_3*Q_1 and K_4*Q_2 and the losses from the interaction $K_7*R*Q_1*Q_2$ and $K_8*R*Q_1*Q_2$.

EXAMPLES OF COOPERATION
BETWEEN POPULATIONS

Many examples of cooperation are seen, both in the natural world and between human populations. In ecology this is called *symbiosis*. Insects pollinate flowers, which supply pollen as insect food. Squirrels plant acorns, which grow into trees, which produce acorns to feed the next generations of squirrels. Trade is an example of cooperation between countries that is considered further with models in Chapter 19.

"What If" Experimental Problems for the Program in Table 14.4

14.4–1 Is cooperation necessary for the survival of both populations? If you make Q1 = 0 and run the program, what happens to Q2? Explain.

14.4–2 If you start with a much greater population in country Q1 than in Q2, with the same rate of exchange, how do the final populations differ from the original graph? Explain.

14.4–3 What changes in the program would you make to have both populations grow to the same level?

Chapter Fifteen

◆

MODELS OF SERIES AND OSCILLATION

Sometimes called a *pulsing paradigm,* is a general systems concept that says self-organizing systems on all scales pulse. Storages in one unit accumulate, followed by pulsed consumption and momentary surge of temporary structure in another unit. Then the sequence repeats its oscillation. There is no steady state in the short run, but the repeating pattern has average levels when viewed over a longer range. This chapter contains models with units in series that pulse. For example, one species of mammals builds up a population that is consumed by a surge of growth of a consuming population of carnivores. These models are standard for introducing population studies, in which case the storages are in numbers of individuals, growths are reproductive (birth) rates, and losses are death rates. However, the same models are used in other fields, calibrated in mass or energy units. Similar phenomena are observed from the molecules to the stars. Apparently pulsing designs prevail during self-organization because pulsing designs maximize performance and are a reinforcing pattern.

PREYPRED is a simple model with two autocatalytic units in series, often representing prey–predator relationships in introductory ecological textbooks. OSCILLAT connects the prey–predator design to a more realistic source of energy and production. The FIRE model pulses with an on–off switch mechanism. In the very general model PULSE, a typical consumer unit pulses its system because of the combination of linear and autocatalytic pathways. In the model DESTRUCT, a pulse from outside aids processes of recycle and reset. When pulsing flows are large relative to the turnover times of storages, the oscillations vary in a complex pattern called *mathematical chaos,* illustrated in the model CHAOS.

15.1 MODEL OF PREY–PREDATOR OSCILLATION (PREYPRED)

The model PREYPRED (Figure 15.1a) is a diagram of the standard textbook model relating two autocatalytic populations

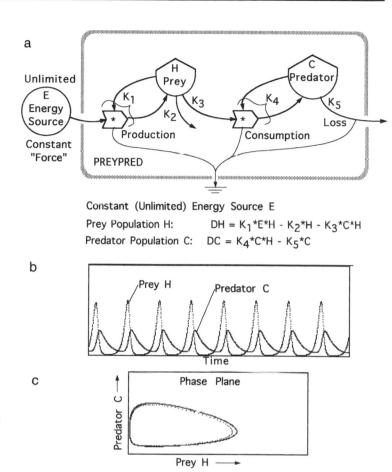

Constant (Unlimited) Energy Source E

Prey Population H: $\quad DH = K_1{*}E{*}H - K_2{*}H - K_3{*}C{*}H$

Predator Population C: $\quad DC = K_4{*}C{*}H - K_5{*}C$

FIGURE 15.1.
Textbook model showing the idea of prey–predator oscillation with unlimited energy source (program PREYPRED.bas). (a) Energy systems diagram and equations; (b) simulation plot with time; (c) phase plane plot relating the quantities in the two units.

in series. Since the equations assume a constant intrinsic rate of increase, a constant food concentration is implied. The diagram indicates constant availability E by the way the line connects directly with the interactive production symbol. The populations oscillate. The prey population starts to grow exponentially. Then the population of predators grows so fast that the numbers of prey are reduced again. With less to eat, the carnivore population declines again. The turnover time of the predators is larger than that of the prey, and consequently their pulses last longer.

The simulation generates Figure 15.1b, the traditional graph of a prey–predator model system. Shown in Figure 15.1c is the concurrent plot of one population as a function of the other (a phase plane plot).

EXPLANATION OF EQUATIONS

In the diagram, E is a source with constant concentration of food for the prey; H is the quantity of prey and C the quantity of predators. The production of prey is K_1*E*H. The production of predators is K_4*H*C, which draws energy from the prey K_3*H*C. Loss of prey due to other causes (death or emigration) is term K_2*H; loss of predators is pathway K_5*C.

OTHER SYSTEMS

Similar oscillations are observed in host–parasite relationships and in other carnivore–herbivore relationships. Examples are spruce trees and spruce budworms in Canada, and week-to-week pulses of phytoplankton and zooplankton in the sea. Supply and demand processes in economics show these oscillations. When a product Q is made, its stock on the market increases and prices decline until consumers H buy it, then its supply goes down until more is produced in another cycle.

"What If" Experimental Problems for the Program in Table 15.1

15.1–1 If the efficiency of growth of the carnivores doubles, how does the graph change? Increase K4. Explain.

15.1–2 Decrease the turnover time of predators by reducing the death rate of the predators: K5 = 0.1. What happens to the quantities and frequency of oscillations as a result?

15.1–3 What is the effect of reducing the energy source in half? What would this mean in an ecosystem with wildcats and rabbits?

♦

15.2   MODEL OF OSCILLATION AT THREE LEVELS (OSCILLAT)

A more realistic model for "prey–predator" type oscillations is OSCILLAT (Figure 15.2a), a system that uses a storage P, being maintained within the system by a steady source-limited inflow energy J (e.g., sun and rain). The design has linear produc-

TABLE 15.1
Program of Prey–Predator with Constant Force Source
(PREYPRED.bas) (Figure 15.1)

```
20 REM PC: PREYPRED  (Prey-predator, constant force source)
30 CLS
40 SCREEN 1!: COLOR 0, 0
50 LINE (0, 0)-(319, 100), 3, B
60 LINE (50, 110)-(270, 190), 3, B
70 E = 700
80 H = 10
90 C = 5
120 K1 = .002
130 K2 = .05
140 K3 = .1
160 K4 = .02
170 K5 = .3
180 C0 = 1
190 C1 = .5
200 H0 = 1
202 H1 = .5
207 E0 = 12.5
208 T0 = .3
210 DT = .05
220 PSET (T / T0, 100 - H / H0), 1
230 PSET (T / T0, 100 - C / C0), 3
240 PSET (50 + H / H1, 190 - C / C1), 1
260 DH = K1 * E * H - K2 * H - K3 * C * H
270 DC = K4 * C * H - K5 * C
280 C = C + DC * DT
290 IF C < 1 THEN C = 1
300 H = H + DH * DT
305 IF H < 1 THEN H = 1
330 T = T + DT
340 IF T / T0 < 319 GOTO 220
```

tion, which aggregates the details of plants. The herbivores H and the carnivores C are an oscillating pair as in the model PREYPRED (Figure 15.1a). The growth of herbivores stimulates a surge of growth of carnivores, which consume most of the herbivores. This causes the carnivores to decrease or emigrate. Then the cycle repeats. The OSCILLAT model is dependent on an energy storage being maintained by the outside source. If the prey–predator chain is connected to a limited-flow source without storage P, the oscillation is damped out, becoming a steady state.

The simulation in Figure 15.2b shows the plants oscillating as the consumption varies with ups and downs of the herbivores. In many ecosystems like this, oscillating pairs cause the rest of the system to oscillate.

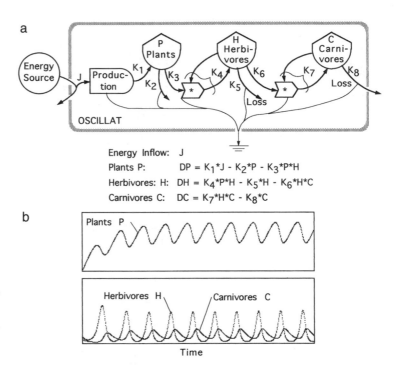

Energy Inflow: J

Plants P: $DP = K_1*J - K_2*P - K_3*P*H$

Herbivores: H: $DH = K_4*P*H - K_5*H - K_6*H*C$

Carnivores C: $DC = K_7*H*C - K_8*C$

FIGURE 15.2.
Model of prey–predator
oscillation with an energy
source from plant productivity
(program OSCILLAT.bas).
(a) Energy systems diagram and
equations; (b) simulation
results.

EXPLANATION OF EQUATIONS

The plant storage P is a balance between the linear production
inflow K_1*J, the plants consumed by the herbivores K_3*P*H,
and the other losses K_2*P (includes respiration, depreciation,
mortality, emigration). Because the plant production flow is linear, it is entirely dependent on the external energy inflow J.
With a linear production, a constant J does not make the source
unlimited. The storage of herbivores H is a balance between
their autocatalytic production K_4*P*H, the consumption by the
carnivores C, and other losses and outflow K_5*H. The storage
of carnivores C is a balance between the autocatalytic production
K_7*H*C and losses and outflows K_8*C. Inflow J is constant.

OTHER EXAMPLES

Oscillations are observed in many Arctic populations. The
oscillations of lemmings and carnivores are often accompanied
by sharp variations in stored vegetation biomass. Since there is

a storage of plant matter on which herbivores depend and which oscillates with the animal populations, the model in Figure 15.2 better represents Arctic ecosystems than Figure 15.1. The predators of the lemmings, foxes and owls, oscillate with the lemming population. Thus, the populations of producers and two levels of consumers go up and down, following each other.

"What If" Experimental Problems for the Program in Table 15.2

15.2–1 In the model in Figure 15.2, change to a system for a lower latitude where there is more sun: Increase J to

TABLE 15.2
Program for Prey–Predator with Limited Source (OSCILLAT.bas)
(Figure 15.2)

```
20 REM PC: OSCILLAT  (Prey-predator oscillation, renewable source)
30 CLS
35 SCREEN 1, 0: COLOR 7, 0
40 LINE (0, 0)-(319, 80), 3, B
50 LINE (0, 90)-(319, 180), 3, B
70 J = 10000: 'Kilocal. per m2/summer
80 C = 5
90 H = 10
100 K1 = .01
110 K2 = .005
120 K3 = .01
130 K4 = .002
140 K5 = .05
150 K6 = .1
155 K7 = .02
160 K8 = .3
180 C0 = 1
190 P0 = .08
200 H0 = 1
205 T0 = 3
210 DT = .1
220 PSET (T * T0, 180 - C * C0), 1
230 PSET (T * T0, 180 - H * H0), 2
240 PSET (T * T0, 90 - P * P0), 3
250 DP = K1 * J - K2 * P - K3 * P * H
260 DH = K4 * P * H - K6 * C * H - K5 * H
270 DC = K7 * C * H - K8 * C
280 C = C + DC * DT
290 IF C < 1 THEN C = 1
300 H = H + DH * DT
305 IF H < 1 THEN H = 1
310 P = P + DP * DT
320 IF P < .001 THEN P = .001
330 T = T + DT
340  IF T * T0 < 319 GOTO 220
```

40,000 kilocalories per square meter per year. What happens to the oscillations? Explain why. Remember the original oscillations in this model are about every 10 years. Now move to a higher latitude where the sun's energy is less: Decrease J to 2,000. What happens to the oscillations?

15.2–2 Do the animal populations oscillate if the plants are constant? Insert line 325   P = 500.

15.2–3 What happens if the carnivores are absent? Insert line 295   C = 0. To see the herbivores' graph, make H0 = 10. Suppose the turnover time of the carnivores is increased, which might happens if parasites were substituted for the carnivores. Increase K6 to 0.3, K7 to 0.15, and K8 to 6.

◆

15.3 MODEL OF SWITCHING OSCILLATION (FIRE)

FIRE (Figure 15.3a) is a pulsing model that includes a switch symbol (labeled "fire"). Plants grow using available sunlight energy R and pool of nutrients N while some biomass is used in consumption (respiration, animals, microbes). When the vegetation biomass Q has reached a critical mass G_1, then the fire switches on. When the grass has burned down to a low concentration G_2, there is no longer enough to burn and the fire switches off. While the fire burns, it releases nutrients to the storage of available nutrients N, which stimulate growth, and nutrients are bound up in the grass again. This model has a "conservation of nutrients" provision with no inflows or losses of nutrients. Nutrients are either free in the environmental pool N or bound in the biomass Q.

In the simulation (Figure 15.3b) with the switch off, growth of biomass B occurs accompanied by decrease of the nutrient pool N. When the fire switch turns on, vertical lines mark the reset of conditions to high nutrients and low biomass again.

EXPLANATION OF EQUATIONS

Plant production K_1*R*N requires both energy (sun, wind, rain) and nutrient, and thus is a product of these resources avail-

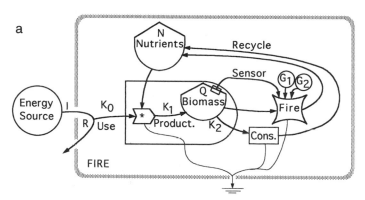

Energy Available:　$R = I - K_0*N*R$　and　$R = I/(1 + K_0*N)$

Nutrients N:　$N = N_t - f*Q$　　　N_t = Total Nutrients

Biomass Q:　　$DQ = K_1*N*R - K_2*Q$　(Without Fire)

Fire Switch:　IF $Q > G1$ THEN $Q = G2$

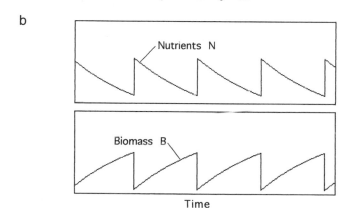

FIGURE 15.3.
Model of oscillation based on threshold switching (program FIRE.bas). "Cons." is abbreviation for animal and microbial consumers. (a) Energy systems diagram and equations; (b) simulation.

able (R and N). Available energy R is the difference between the inflow I and that used by plant production K_0*R*N. The total nutrients in the system N_t include the nutrients in the soil N plus those bound up in the grass ($f*Q$), where f is the fraction of the biomass that is nutrients.

In the program (Table 15.3), K_2*Q is the respiration of the biomass without fire. The fire switch is turned on when $Q > G_1$ (line 200). Value G_2 is the quantity of grass left after the fire; G_1 and G_2 are thresholds and are shown at the top of the switch symbol in the figure. The program statements 180 and 190 draw lines from the values Q and N just before the fire to their values right after the fire. When these are plotted in red, they illustrate the fire dramatically.

TABLE 15.3
Program for Fire with Recycled Nutrients (FIRE.bas) (Figure 15.3)

```
10 REM PC: FIRE  (Fire with recycled nutrients)
30 CLS
35 SCREEN 1, 0: COLOR 0, 1
40 LINE (0, 0)-(319, 180), 3, B
50 LINE (O, 90)-(319, 90), 3
60 I = 10
70 Q = 1000
80 NT = 100
90 G1 = 5000
100 G2 = 2000
110 K = .9
120 K1 = 8
130 K2 = .01
140 F = .01
150 Q0 = .01
160 N0 = .9
163 T0 = 1
165 DT = 1
170 R = I / (1 + K * N)
180 IF Q > G1 THEN LINE (T / T0, 180 - G1 * Q0)-(T / T0, 180 - G2 * Q0)
190 IF Q > G1 THEN LINE (T / T0, 90 - (NT - F * G1) * N0)-(T / T0, 90 - (NT - F * G2)
       * N0)
200 IF Q > G1 THEN Q = G2
210 PSET (T * T0, 180 - Q * Q0), 1
220 PSET (T * T0, 90 - N * N0), 2
230 DQ = K1 * N * R - K2 * Q
240 Q = Q + DQ * DT
250 N = NT - F * Q
260 T = T + DT
270 IF T * T0 < 319 GOTO 170
```

EXAMPLES OF SWITCH MODELS

Many natural systems have fire as part of their normal pattern. Pine forests in southeastern United States, chaparral in southern California and the Mediterranean lands, and grasslands around the world have regular fires. Sometimes fire comes every 4 or 5 years as in the Southeast; sometimes it's every 100 years or so, as in the lodge pole pine forest in Yellowstone National Park. A spurt of growth follows the fire as the nutrients are recycled into the soil.

An example from business is a buildup of inventories in a store to surplus, followed by a special sale to sell most of them. The money from the sales is used to restock the shelves.

"What If" Experimental Problems for the Program in Table 15.3

15.3–1 What is the effect of adding more solar energy? Increase
        I to 15. Why? With the original solar energy, increase

nutrients. How does a change due to nutrients compare with the change due to solar energy? Why?

15.3–2 What happens to the fire regime if the total nutrients are reduced by half? Why?

15.3–3 What happens to a fire-cycle system if fire is kept out completely? Bypass the fire statements by adding a line: 175   GOTO 210.

15.3–4 Suppose a type of vegetation grows that is more fire resistant. Set a higher threshold for burning: G1 = 10000. What happens?

15.4   MODEL OF PULSE AND RECYCLE (PULSE)

PULSE (Figure 15.4a) is a production–consumption–recycle model that pulses without switches because the autocatalytic consumer unit has both linear and quadratic-autocatalytic intake pathways. A process accumulates a storage Q, which is being consumed by the consumer unit C.

There are two modes of consumption, the slow gradual one and an epidemic destructive one. As the consumer unit grows, it reaches a point where the autocatalytic pathway accelerates in a frenzy, using up the storage Q while giving a brief surge to the consumer storage C. Recycled outputs of consumption (wastes) are used by production again. Because pulsing designs prevail in self-organization by generating more production in the long run, we can expect to find features of this model in all kinds of systems.

PULSE is a more realistic model for producer–consumer systems than the prey–predator models (Figures 15.1a and 15.2a). For example, arid grasslands have grasshoppers (locusts) that have two modes of behavior: a slow consumption mode and a frenzied consumption mode. For several seasons the quantity of grasshoppers increases slowly, eating only a small portion of accumulating grasses. When there is enough accumulated vegetation for the grasshoppers to have a surge of reproduction and rapid consumption, they change behavior, developing hordes that consume vegetation to a low level (Figure 15.4b).

The simulation (Figure 15.4b) shows the buildup of product storage Q followed by the pulse of consumers C, reduction of products, and increase of materials. The available free materials

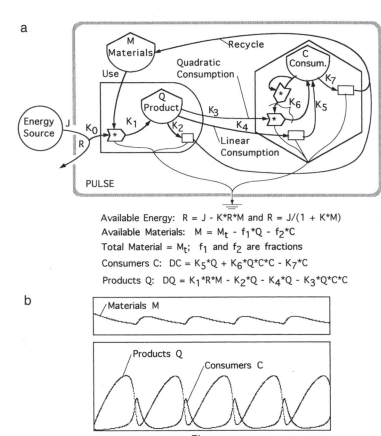

a

Available Energy: $R = J - K*R*M$ and $R = J/(1 + K*M)$
Available Materials: $M = M_t - f_1*Q - f_2*C$
Total Material = M_t; f_1 and f_2 are fractions
Consumers C: $DC = K_5*Q + K_6*Q*C*C - K_7*C$
Products Q: $DQ = K_1*R*M - K_2*Q - K_4*Q - K_3*Q*C*C$

b

FIGURE 15.4.
A pulsing model of production, consumption, and recycle based on linear and autocatalytic consumption (program PULSE.bas). (a) Energy systems diagram and equations; (b) simulation.

M used up during the product storing phase are released during the consumer pulse.

EXPLANATION OF EQUATIONS

Primary production K_1*R*M requires both energy and material, and thus is a product of these resources available (R and M). Available energy R is the difference between the inflow I and the use K_0*R*N by the primary production process. Materials are conserved in this model: Total materials M_t include those available for production M, those incorporated in the producers (f_1*Q) and those in the consumers (f_2*C). Storage of accumulating products Q is a balance between the primary production K_1*R*M, the steady losses K_2*Q (depreciation and consumption), and two pathways of consumption K_4*Q and $K_3*Q*C*C$. The linear pathway of slow consumption is proportional to the quantity of producers K_4*Q. The rapid consumption pathway is caused by a superaccelerated growth of consumers. Mathematically, this is quadratic autocatalytic ($K_3*Q*C*C$), which represents processes that go faster due to self-interactions. (See the discussion on quadratic feedback in Section 9.7.) The surge of consumer growth C is initiated automatically without switches when the quadratic-autocatalytic term $K_6*Q*C*C$ for producing new consumers exceeds the term for consumer recycling and depreciation K_7*C.

OTHER EXAMPLES

In rainforest areas like the Amazon River basin, tribal people practiced "slash and burn" shifting agriculture. Trees were cut and burned, the area was planted and harvested for several years, and then it was abandoned to regrow when the people moved to a fresh site. In 30 years or so, after some trees had regrown, the process was repeated.

This model applies to geological cycles where a process gradually builds up a storage of energy that later sets up a frenzied consumption process, such as an earthquake or volcanic eruption that disperses energy and materials.

The model fits some theories for the rise and decline of human civilizations such as the Mayan developments of Mexico. For example, productive processes of the landscape build up storages of soil and forests until there is enough to support a pulse of human development including new architecture, knowledge, and armies. Perhaps our own civilization is a pulse of consumption based on the previous long accumulation of fossil fuels and mineral deposits.

"What If" Experimental Problems for the Program in Table 15.4

15.4–1 What happens if the total materials are reduced a tenth or increased by 10 times?

15.4–2 What happens if the energy available is reduced a tenth or increased by 10 times? Consider at what level energy is well used.

TABLE 15.4
Program in BASIC for Pulse and Recycle (PULSE)
(Figure 15.4)

```
5 REM PC PULSE.bas (Pulsing and recycle)
10 CLS
20 SCREEN, 1, O
30 COLOR 0, 0
40 LINE (0, 0)-(319, 50), , B
45 LINE (0, 60)-(319, 180), , B
50 J = 10:
60 Q = 2
70 C = 2
80 MT = 100  '
100 K0 = .1
105 K1 = .03
110 K2 = .001
115 K3 = .0004
120 K4 = .01
125 K5 = .005
130 K6 = .0003
132 K7 = .2
135 F1 = .5
140 F2 = 1
145 Q0 = .8
150 C0 = 1
165 M0 = .3
170 T0 = 1
175 DT = .2
200 R = J / (1 + K0 * M)
205 M = MT - F1 * Q - F2 * C
207 IF M < .1 THEN M = .1
210 DQ = K1 * R * M - K2 * Q - K4 * Q - K3 * C * C * Q
220 DC = K5 * Q + K6 * Q * C * C - K7 * C
230 Q = Q + DQ * DT
235 IF Q < .001 THEN Q = .001
240 C = C + DC * DT
245 IF C < .001 THEN C = .001
250 PSET (T * T0, 180 - Q * Q0), 1
260 PSET (T * T0, 180 - C * C0), 2
265 PSET (T * T0, 50 - M * M0), 1
270 T = T + DT
280 IF T * T0 < 319 GOTO 200
```

◆

15.5   MODEL OF ENERGY APPLIED TO
RECYCLE (DESTRUCT)

The model DESTRUCT (Figure 15.5a) illustrates an important feature of many, if not all, systems that develop assets A in excess of that necessary to maximize production. The model applies

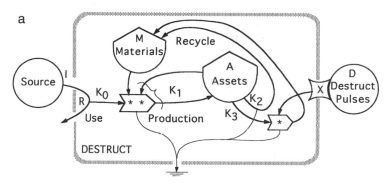

Energy Available: $R = I - K_0*R*M$ and $R = I/(1 + K_0*M)$
Materials Available: $M = M_t - f*A$
Total Materials = M_t;   f is the fraction in Assets A
Production Rate: $P = K_1*R*M*A$
Stored Quantity Q: $DQ = P - K_2*A - X*K_3*A*D$
Destruct Pulse flows when $X = 1$.

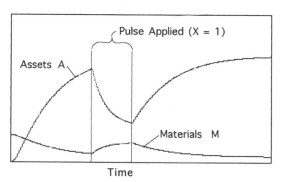

FIGURE 15.5.
Model of energy-assisted destruction and recycle (program DESTRUCT.bas). (a) Energy systems diagram and equations; (b) simulation.

pulses of available energy D to destroy structures A and make materials M available again.

Normally energy transformations use considerable energy of one quality R to make products of higher quality, assets A. Some processes of dispersal and deterioration reduce the assets and recycle the materials (disorder). We sometimes call this an *order–disorder cycle*. As in photosynthesis, most of the energy is required on the left of the cycle, building up the products (order). It takes less energy to complete the down cycle because it is downhill thermodynamically.

However, when the normal depreciation processes are not fast enough, energy can be applied to the down-recycle side of the system. Since energy is being applied in the direction of depreciation (downhill), a smaller amount of energy is required for the destruction than for the production. The model DESTRUCT directs downhill energy D to a simple design of production and recycle. This model has provision for conservation of materials, since total materials are the sum of those available M and those bound within assets f*A.

In the simulation (Figure 15.5b) the model starts with available materials and few assets. After growth has occurred for 100 time units, the pulse D is turned on for 50 days and then turned off again (lines 200 and 210 in the program in Table 15.5). During the pulse, assets decrease and materials increase. Afterward growth continues.

Although not included in this model, the downhill effects may release energy that causes temporary storms of destruction. Excess energy may self-organize temporary turbulent autocatalytic systems in the environment, which may become disasters (Alexander, 1987). In Figure 15.5 this can be drawn by adding

TABLE 15.5
Program Illustrating a Destructive Energy Pulse
(DESTRUCT.bas) (Figure 15.5)

```
10 REM PC
20 REM DESTRUCT (Destruction pulse after autocatalytic construction)
30 SCREEN 1, 0
40 COLOR 0, 0
50 LINE (0, 0)-(319, 180), , B
60 I = 4
65 F = .2
70 D = 1
80 MT = 100
90 A = 1
95 DT = 1
100 T0 = 2
110 M0 = 3
120 A0 = 3
130 K0 = .0009
140 K1 = .001
150 K2 = .01
160 K3 = .02
170 REM equations
180 PSET (T / T0, 180 - A / A0)
190 PSET (T / T0, 180 - M / M0)
200 IF T / T0 > 100 THEN X = 1
210 IF T / T0 > 150 THEN X = 0
220 R = I / (1 + K0 * M * A)
230 DA = K1 * R * M * A - K2 * A - X * K3 * A * D
270 A = A + DA * DT
280 M = MT - F * A
290 T = T + DT
300 IF T / T0 < 319 GOTO 170
```

another autocatalytic consumer unit drawing energy from the interaction of the energy from D and assets A.

EXPLANATION OF EQUATIONS

In DESTRUCT, production is an autocatalytic process $K_1*R*M*A$ requiring externally limited inflow energy R, available materials M, and the structure of the assets A. The steady depreciation–consumption–dispersal process K_2*A is in proportion to assets A. The pulse starts when the program (Table 15.5, line 200) makes $X = 1$, which turns on the pulse pathway $X*K_3*A*D$. Materials available M are those not bound in assets $f*A$, where f is the fraction of the assets that are materials.

EXAMPLES

Although human and forest systems are breaking down all the time, hurricanes and fires speed up the destruction and recycle of materials. War adds destructive energy to breaking down a society and, in this model, causes it to rebuild. In briny waters where organic matter has accumulated, photorespiration uses the sun to speed the processes of respiration and recycle by the pink bacterial populations of *Halobacteria*. Organelles and genes in living cells are organized to destroy structures as they become nonfunctional. In ecosystems diseases destroy accumulated populations that become ill adapted.

"What If" Experimental Problems for the Program in Table 15.5

15.5–1 In the example of a hurricane impacting a forest, describe the process from Figure 15.5a: What are D, A, and M?

15.5–2 What does increasing the energy source I do to the quantities of A and M? Explain.

15.5–3 What is the effect of increasing the energy of the pulse D 10 times?

◆

15.6 MODEL OF THE CHAOS OF ONE OSCILLATING PAIR DRIVING ANOTHER (CHAOS)

The CHAOS model (Figure 15.6a) illustrates the phenomenon of *mathematical chaos,* which is common where complex systems oscillate. With mathematical chaos, pulsing causes variation in patterns of oscillation. The word *chaos* was selected because the values on graphs appear erratic. However, choice of the word *chaos* was unfortunate because it implies disorder, whereas the patterns of mathematical chaos are determined exactly by the equations just like other dynamic models. Chaotic pulsing is usually increased with increased energy level.

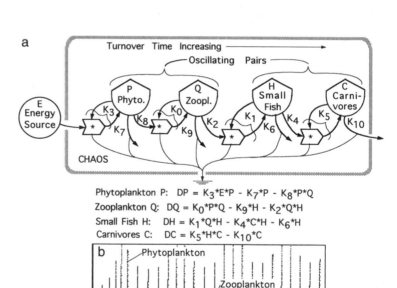

Phytoplankton P: $DP = K_3*E*P - K_7*P - K_8*P*Q$
Zooplankton Q: $DQ = K_0*P*Q - K_9*H - K_2*Q*H$
Small Fish H: $DH = K_1*Q*H - K_4*C*H - K_6*H$
Carnivores C: $DC = K_5*H*C - K_{10}*C$

FIGURE 15.6.
Model of chaos generated by one oscillating pair connected to another (program CHAOS.bas). (a) Energy systems diagram and equations; (b) oscillations by the pair on the left; (c) oscillations by the pair on the right; (d) phase plane plot of the pair on the left; (e) phase plane plot of the pair on the right.

One type of chaos occurs when pulsing of part of a system affects oscillations elsewhere within the network. The CHAOS model (Figure 15.6a) shows how one oscillating pair of units on one time scale interacting with another oscillating pair on a faster time scale causes both to change their patterns of pulsing. In other words, there are oscillations in oscillation. Putting the model in terms of an example can help the reader visualize the way units of different size have different scales of oscillation. Although this model does not have other features necessary to represent ecosystems well, the four storages are labeled phytoplankton P, zooplankton Q, small fish H, and larger carnivores C.

The simulations in Figures 15.6b and 15.6c contain the graphs with time. All the units are oscillating, but the magnitudes are changing as they affect each other. Turnover times increase up the scale of units from left to right in order of size and position in the energy chain (unit order: P, Q, H, C). The duration of the pulses increases in the same order as turnover times. The same oscillations are plotted on phase plane graphs (one storage as a function of another) in Figures 15.6d and 15.6e. Without chaos, the loops formed by an oscillation at steady state are simple rings on a phase plane as shown in the PREYPRED model (Figure 15.1). However, in this CHAOS model the rings shift back and forth. The outer border of the family of curves is called the *attractor*.

EXPLANATION OF EQUATIONS

The equations for the four autocatalytic units are similar. For example, the rate of change DP of stored phytoplankton P is a balance between the production K_3*E*P minus losses K_7*P minus the autocatalytic consumption by the next unit K_8*P*Q. The source E is constant (unlimited). In this model the only loss pathway from carnivores is the linear pathway $K_{10}*C$, since no higher level carnivore is included.

OTHER SYSTEMS

Mathematical chaos has been widely found in almost every kind of system where searches were conducted. Chaos may be a

TABLE 15.6
**Program for Oscillator Pair at One Scale Driving Another
(CHAOS.bas) (Figure 15.6)**

```
10 REM PC
20 REM CHAOS
25 REM (Chaos generated by prey–predator oscillator through fast
       phytoplankton)
30 CLS
40 SCREEN 12, 0: REM  COLOR 7, 0
50 LINE (0, 0)-(319, 90), 3, B
60 LINE (0, 100)-(319, 190), 3, B
65 LINE (0, 210)-(150, 300), 3, B
70 LINE (170, 210)-(319, 300), 3, B
75 E = 500
80 C = 5
90 Q = 1
100 H = 1
110 P = 1
120 K0 = .2
130 K1 = .002
140 K2 = .1
150 K3 = .02
160 K4 = .1
170 K5 = .02
180 K6 = .05
190 K7 = .5
200 K8 = 1
210 K9 = 3
220 K10 = .3
240 C0 = 1
245 C1 = 3
250 Q0 = .2
255 Q1 = .5
260 H0 = .9
265 H1 = 1.4
270 P0 = .2
275 P1 = .2
280 T0 = 5
290 DT = .01
292 E = 500
300 REM Plotting Phase Planes
310 PSET (P1 * P, 300 - Q1 * Q), 3
315 PSET (H1 * H + 170, 300 - C1 * C), 1
320 REM Plotting Graphs with Time
330 PSET (T0 * T, 190 - C0 * C), 3
340 PSET (T0 * T, 190 - H0 * H), 2
345 PSET (T0 * T, 90 - P0 * P), 1
350 PSET (T0 * T, 90 - Q0 * Q), 3
360 DP = K3 * E * P - K7 * P - K8 * Q * P
365 DQ = K0 * P * Q - K9 * Q - K2 * Q * H
370 DH = 50 * K1 * Q * H - K4 * C * H - K6 * H
380 DC = K5 * C * H - K10 * C
390 C = C + DC * DT
```

(continues)

TABLE 15.6 (*Continued*)

```
400 IF C < .001 THEN C = .001
410 H = H + DH * DT
420 IF H < .001 THEN H = .001
430 Q = Q + DQ * DT
440 IF Q < .001 THEN Q = .001
450 P = P + DP * DT
460 IF P < .00001 THEN P = .00001
470 T = T + DT
480  IF T0 * T < 320 GOTO 300
```

complex way in which designs formed by self-organization use excess energy to increase productivity.

Regular oscillations of populations of the snowshoe hare, its predator, the lynx, and other animals were recorded by pelt counts in Canada by the Hudson Bay Company from 1845 to 1935. The oscillations shift somewhat over time, suggesting chaotic mechanisms.

In Chapter 6 we explained how chaotic high–low jumping of values can occur in digital simulation when the flows of each iteration are large compared to the tanks into which they flow. The storages fill to high values, which causes their outflows on the next iteration to become excessive and the tank storages drop to low values again. Where continuous processes are being simulated with the discrete iterations of digital simulation, such chaos is an artifact of using too large a time step (DT) for Euler integration. To make sure that this artifact is not present in a simulation, it can be tested by running with smaller and smaller DT values until no change is observed.

High–low jumping may occur in the real world where a storage quantity receives frequent high-energy pulses.

"What If" Experimental Problems for the Program in Table 15.6

15.6–1 What happens with less available energy? Reset E to 50.

15.6–2 What happens if the two oscillating pairs are disconnected. Set K2 to zero and apply available energy to the equation for small fish H in place of zooplankton. Disable line 370 by inserting REM and a space after the line number. Insert a line 372 DH = K1*E*H − K4*C*H − K10*C

15.6–3 Increase the iteration time DT from 0.01 to 0.1. What happens? Why?

Chapter Sixteen

◆

MINIMODELS OF SUCCESSION AND EVOLUTION

T he sequence of stages in self-organization of systems is called *succession* on the time scale of ecosystems and *evolution* on longer scales. This chapter discusses minimodels for the growth and development of ecosystems, but the principles of growth and organization of species diversity may apply to other systems also. WETLAND is a simple model of an ecosystem accumulating peat. CLIMAX relates growth of biomass to species diversity. NUTRSPEC shows an inverse relationship of nutrient excess and diversity. The model SPECAREA relates species diversity to the square root of the support area. SPECIES includes migration and evolution in development of diversity on islands. After growth and organization these models develop steady states. However, steady states are normally reset and restarted later by pulses on a larger scale (Chapter 15).

16.1 MODEL OF WET GRASSLAND (WETLAND)

The model WETLAND (Figure 16.1a) contains the main features of a wetland ecosystem. In a wet grassland nutrients flow in with rain and runoff. Grasses photosynthesize using the nutrients rapidly and building up standing vegetation biomass Q_2. As the grass dies, some is decomposed and recycled back to nutrients in the soil and waters Q_1. The rest of this dead grass accumulates as peat Q_3. Consumption processes of animals and fire are in proportion to the organic accumulations. When excess nutrients flow in, they are readily filtered and bound into the vegetation and peat.

The simulation (Figure 16.1b) shows growth of plants and peat over a 10-year period. The program inputs monthly averages for solar energy I from Florida with seasonal maximum in early summer and minimum in winter. The same set of data is repeated each year. Grasses grow in the summer and decrease in the winter. During the 10 years there is an increasing buildup of peat until

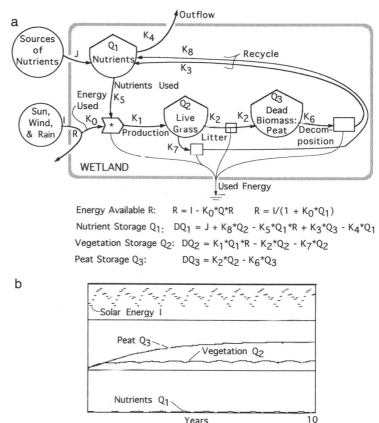

FIGURE 16.1.
Model of wet grassland
(program WETLAND.bas).
(a) Energy systems diagram and
equations; (b) simulation.

Energy Available R:　$R = I - K_0 * Q * R$　　$R = I/(1 + K_0 * Q_1)$

Nutrient Storage Q_1:　$DQ_1 = J + K_8 * Q_2 - K_5 * Q_1 * R + K_3 * Q_3 - K_4 * Q_1$

Vegetation Storage Q_2:　$DQ_2 = K_1 * Q_1 * R - K_2 * Q_2 - K_7 * Q_2$

Peat Storage Q_3:　　　$DQ_3 = K_2 * Q_2 - K_6 * Q_3$

its decomposition equals its production, and the nutrient outflow
equals the inflow. The figures used to calibrate the program
(Table 16.1) were obtained from a class study of Paynes Prairie
in northcentral Florida.

EXPLANATION OF EQUATIONS

Nutrients are increased by outside inflow J and by recycle of
nutrients released from consumption $K_3 * Q_3$ and $K_8 * Q_2$. Nutrients are decreased by water outflows $K_4 * Q_1$ and uptake by plants
$K_5 * R * Q_1$ as part of production. Photosynthetic production is
represented as a product $K_1 * R * Q_1$ of the unused light R and
the nutrients in the waters and soil Q_1. By not including an
autocatalytic action of the vegetation Q_2, the production function
does not have short-term growth acceleration that might be of

TABLE 16.1
Program for the Wet Grassland Model (WETLAND.bas)
(Figure 16.1)

```
1 REM PC
2 REM WETLAND.bas (Wet grassland)
3 SCREEN 1, 0: COLOR 7, 0
4 LINE (0, 0)-(319, 180), 3, B
5 LINE (0, 50)-(319, 50), 3
6 LINE (0, 120)-(319, 120), 3
10  DIM A(12)
11  DATA  2300,2000,2800,3400,4000,4200,4400,3600,3700,3300,3000,3200
12 FOR M = 0 TO 11
13  READ A(M)
14  NEXT
20 J = .005
30 Q1 = .05
40 Q2 = 100
50 Q3 = 200
60 K0 = 113
65 K1 = .268
70 K2 = .005
75 K3 = .0000016
80 K4 = .03
85 K5 = .000268
90 K6 = .0016
95 K7 = .005
100 K8 = 1.34E-06
110 N0 = 20
120 Q0 = .02
125 P0 = .02
130 I0 = .01
135 TX = 3560
140 T0 = 320 / TX
145 DT = 4
150 I = A(M)
200 R = I / (1 + K0 * Q1)
205 D1 = J + K8 * Q2 - K4 * Q1 + K3 * Q3 - K5 * Q1 * R
210 D2 = K1 * Q1 * R - K2 * Q2 - K7 * Q2
220 D3 = K2 * Q2 - K6 * Q3
230 Q1 = Q1 + D1 * DT
235 IF Q1 < .0000001 THEN Q1 = .0000001
240 Q2 = Q2 + D2 * DT
300 Q3 = Q3 + D3 * DT
310 T = T + DT
315 Y = INT(T / 365)
318 W = 12 * Y
320 M = INT(T / 30.4) - W
400  PSET (T0 * T, 180 - Q1 * N0), 1
410  PSET (T0 * T, 120 - Q2 * Q0), 2
420  PSET (T0 * T, 120 - Q3 * P0), 1
430  PSET (T0 * T, 50 - I * I0), 1
440  IF T * T0 < 319 GOTO 150
445 KEY OFF
```

interest on a smaller time scale. Losses of vegetation Q_2 to respiration, consumption, and fire are represented by K_7*Q_2, and losses of peat by K_6*Q_3.

The changes in sunlight are programmed as an array with the monthly averages listed as DATA in line 11 of the Table 16.1 program. How to program an array is explained in Appendix B. In this program TX is the number of days; the time spaces are divided by TX to get the scale for time T0 (lines 135 and 140).

EXAMPLES

All around the world are wetlands that accumulate peat in freshwaters and saltwaters. This model is appropriate for tropical, temperate, and arctic areas. The model may apply to other kinds of systems that use a material to build products and later pass remainders to a waste storage before being recycled.

"What If" Experimental Problems for the Program in Table 16.1

16.1–1 What would change if a city used this wetland to dump its treated sewage? Increase the inflow of nutrients J to 0.025. Would it change the looks of the wetland?

16.1–2 If this area were partially drained so that the peat decomposed twice as fast, how would you change the program to indicate this? How would the graph change?

16.1–3 What happens if this model is adapted to some arctic conditions? Reduce the consumption rates K6 and K7 to one-tenth of the original, representing lower temperature and shorter season for animals. Also change the solar regime by subtracting 2000 kcal/m^2/day in line 150: I = A(M) − 2000.

♦

16.2 MODEL OF SUCCESSION (CLIMAX)

Figure 16.2a shows a model of succession (CLIMAX) for the development of complex systems. Succession starts with the production of organized material followed by the acquisition and organization of information. Labeled and calibrated for a tropical

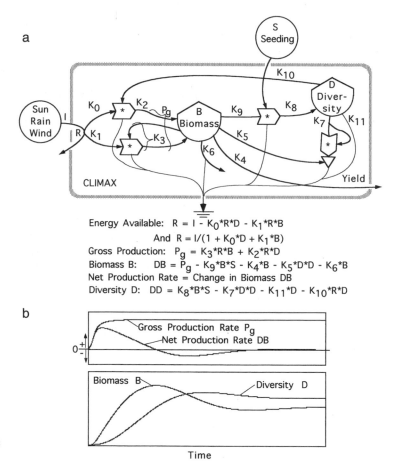

Energy Available: $R = I - K_0*R*D - K_1*R*B$
And $R = I/(1 + K_0*D + K_1*B)$
Gross Production: $P_g = K_3*R*B + K_2*R*D$
Biomass B: $DB = P_g - K_9*B*S - K_4*B - K_5*D*D - K_6*B$
Net Production Rate = Change in Biomass DB
Diversity D: $DD = K_8*B*S - K_7*D*D - K_{11}*D - K_{10}*R*D$

FIGURE 16.2.
Model of succession (program
CLIMAX.bas). (a) Energy
systems diagram and equations;
(b) simulation.

rain forest ecosystem in Puerto Rico, the model develops biomass
B first, which provides the habitat and energy basis for acquisition
of species (information) D from seeding sources S outside. In
typical succession starting with a bare field, a few grasses and
herbs grow, first producing a small amount of biomass (B in
Figure 16.2a). In later seasons there are taller grasses and shrubs,
tree seedlings, and eventually large trees. The growth of this
biomass depends on renewable outside energy sources I, but also
on species diversity D, which represents the number of different
kinds of plants and animals.

As the quantity of biomass increases, it provides niches for
seeding from outside to increase the diversity. In this model,
diversity is the number of species per 1000 individual organisms
counted. The number of species D is a balance between new

species seeded and those lost by local extinction. Losses on one extinction pathway K_7 are caused by competition of species.

The more biomass and the more diversity in a system, the more productivity required to maintain them. In the model the extra energy needed to support the diversity is indicated by the quadratic drain from the biomass storage.

For example, in the tropical rain forest there are many species of frogs. It has been necessary for them to use extra energy to develop different colors, habitats, croaks, and mating behaviors to enable them to occupy different niches and survive together. However, diversity increases division of labor and improves efficiency. Net production is the increase in biomass in excess of that used by respiration, consumption, and support of diversity.

The simulation graph in Figure 16.2b shows changes in the production rate Pg, biomass B, and diversity D. Gross production Pg, which is proportional to the outside energy sources I, increases with biomass and diversity and levels quickly. Net production DB is the growth of biomass minus its respiration and use for diversity. At first most of gross production is net gain going into biomass. When gross production is no longer increasing, net production decreases as more of the production goes for maintenance.

In Figure 16.2b the rate of production and biomass increase first. Then as the species diversity increases, more of the organic matter is required for supporting the complexity, and the biomass decreases somewhat. The production rate is sustained with the help of increased efficiency aided by the diversity. At first most of the production is net production contributing the growth of biomass. Later more and more of the production goes into consumption and diversity, and the net production decreases. Some biomass is yielded to humans who selectively harvest wood or wildlife without clear cutting. At one stage the net production is temporarily negative and biomass declines briefly. The simulation develops a steady state, which in ecology has been called a *climax*. Not included in this model is the restart necessary when the system is pulsed by large-scale phenomena such as hurricanes.

EXPLANATION OF EQUATIONS

Renewable resources remaining for use R are the inflow I minus that used for production, including K_1*R*B, the autocatalytic use in producing green biomass B, and additional use K_0*R*D for diversity-aided production. Gross production is the sum of the two production processes K_2*R*D and K_3*R*B. Depreciation and respiration of biomass are through pathway K_6*B, and the yield from the system is pathway K_4*B. The drain on biomass due to number of species is quadratic K_5*D*D because the number of species interactions increases as the square of the number. These interactions drain energy either because of negative effects or because of the energy used to make separate niches and prevent negative effects. New species are added K_8*B*S in proportion to biomass B and seeding that is available S. Some biomass is used K_9*B*S by the introductions (e.g., soil providing initial habitat for animals and microbes). Diversity D, the numbers of different species of organisms, is a balance between the additions and extinctions. Linear extinction $K_{11}*D$ is proportional to the number of species; K_7*D*D is quadratic extinction due to species interactions and $K_{10}*D*R$ is stress losses of species not adapted to the energy conditions.

EXAMPLES OF SUCCESSION AND CLIMAX SYSTEMS

Ecosystems with similar species of plants and animals are found in similar climate zones. The major types of ecosystems with their typical organisms are called *biomes*. The climax species in the grasslands biome are grasses; in coral reefs they are coral; in the tundra special grasses and herbs develop that can live in a short growing season and permafrost soil. Different species occupy the same niche in different areas. An example is the large herbivores in the tundra: The caribou live in the tundra in North America and the reindeer in the same niche of the tundra in Northern Europe.

Human civilizations also go through succession and climax. They start with a few people in an area; the population grows as the economic assets grow. The culture may reach a climax with high diversity of occupations (comparable to species in ecosystems). Businesses may go through stages of growth and diversification.

"What If" Experimental Problems for the Program in Table 16.2

16.2–1 What happens if there is no seeding from outside? Change statement 80 to S = 0. Run the original program,

TABLE 16.2
Program for Model of Succession (CLIMAX.bas) (Figure 16.2)

```
10 REM PC CLIMAX.BAS (Succession with growth and diversity)
20 SCREEN 1,0:COLOR 0,0
40 LINE (0,0)-(320,70),,B
45 LINE (0,80)-(320,180),,B
55 LINE (0,50)-(325,50)
60 REM COEFFICIENTS
70 I = 1
80 S = .1
90 B = 100
100 D = 1
110 K0 = .01
120 K1 = .0002
130 K2 = .5
140 K3 = .007
150 K4 = .000025
160 K5 = .01
170 K6 = .000175
180 K7 = .000001
190 K8 = .000001
200 K9 = .0001
203 K10 = .0001
205 K11 = .0001
207 TX = 30000!
210 T0 = 360/TX
220 B0 = .0008
230 G0 = 1.2
240 P0 = 1.2
250 D0 = 1.5
260 DT = 20
300 PSET (T * T0,180 - D * D0)
305 PSET (T * T0,180 - B * B0)
310 PSET (T * T0,50 - PG*P0)
315 PSET (T*T0, 50-DB*G0)
320 R = I/ (1 + K0 * D + K1 * B)
325 DD = K8 * B * S - K11 * D - K7 * D * D -K10*R*D
330 PG = K2 *R*D + K3*R*B: REM gross production
340 DB = PG   - K6 * B - K4 * B - K5 *  D * D -K9*B*S
350 B = B + DB * DT
355 IF B <.1 THEN B = .1
360 D = D + DD * DT
370 T = T + DT
380 IF T*T0 < 320 GOTO 300
```

make your change, and run it again. Some computers will graph both runs, if you run it again with the statement RUN 70.

16.2–2 If you increase the harvest yield, K_4*B 20 times, what happens to the biomass and diversity? Change K4 to 0.0005.

16.2–3 What would happen to the system if rain was the principal limiting environmental resource, and a climate change reduced the rain in half? Change I to 0.5.

◆

16.3   MODEL OF EXCESS RESOURCE AND DIVERSITY (NUTRSPEC)

According to Lotka's concept of self-organization for maximum power, systems designs prevail that maximize the intake of available unused resources as a first priority. In that case maximum performance occurs when the species most adapted to fast colonization overgrow the others. However, after excess resources have been incorporated into the system, a systems design develops that is more efficient at recycling materials so materials are not limiting. Maximum performance at this stage is provided by high diversity of species that provide specialization and division of labor, getting more out of the energy budgets and efficiently recycling the materials. If there is steady inflow of excess nutrients, the succession may be arrested in a stage with a few weed species (low diversity) depositing the excess nutrients in organic matter. The principle that diversity is inverse to resource excess was suggested by Yount (1956).

The minimodel NUTRSPEC (Figure 16.3a) contains both mechanisms, the one that develops overgrowth by weedy colonization B and the complex M of efficient, mature, high-diversity species. When excess inflow of nutrients J_n is supplied, the weedy overgrowth occurs, organic matter deposits Y develop, and diversity M remains small. When there is no excess of unused resource, high-diversity species displace the weedy components without organic matter deposition.

In the first simulation (Figure 16.3b) the nutrients J_n flowing in are few, and the concentrations in the water are too small to plot. Weeds B do not overgrow, and organic matter Y does not accumulate. Species diversity M and its productivity gradually increase.

In the second simulation shown (Figure 16.3c) the nutrient inflow J_n was much increased, causing high concentrations of

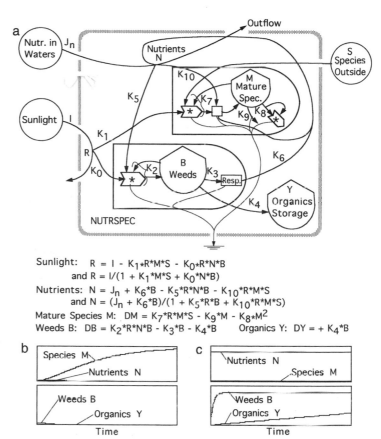

Sunlight: $R = I - K_1*R*M*S - K_0*R*N*B$
and $R = I/(1 + K_1*M*S + K_0*N*B)$

Nutrients: $N = J_n + K_6*B - K_5*R*N*B - K_{10}*R*M*S$
and $N = (J_n + K_6*B)/(1 + K_5*R*B + K_{10}*R*M*S)$

Mature Species M: $DM = K_7*R*M*S - K_9*M - K_8*M^2$

Weeds B: $DB = K_2*R*N*B - K_3*B - K_4*B$ Organics Y: $DY = + K_4*B$

FIGURE 16.3.
Model of inverse relationship of excess nutrients and diversity (program NUTRSPEC.bas).
(a) Energy systems diagram and equations; (b) simulation with little nutrient inflow;
(c) simulation with high nutrient inflow.

nutrients in the system. Weeds B grow rapidly, with net growth that goes into organic storage deposits Y. The complex of high-diversity species M did not develop.

EXPLANATION OF EQUATIONS

The model has two autocatalytic units competing by drawing on the same sunlight I and same nutrients N. Both the solar energy and the nutrients are represented with flow junctions R and N so that they provide a running average without developing artificial chaos, even when values are small. Thus the solar energy remainder equation was calculated by subtracting the uses $K_0*R*N*B$ and $K_1*R*M*S$ from the inflowing light I. The nutrient remainder equation was calculated by subtracting the uses $K_5*R*N*B$ and $K_{10}*R*M*S$ from the sum of the inflow J_n and recycle K_6*B. The coefficient K_{10} is the net use (intake by species M minus their recycle). Nutrients are a main factor in weed production K_2*R*N*B. Because weed production is greater than its respiration K_3*B, there is net flow of organic matter K_4*B into stored organics Y.

The production K_7*R*M*S, which supports species complex M in this model, is not dependent on the nutrients because the mature complex recycles most of what it needs. Loss of species is represented with linear pathway K_9*M and a quadratic pathway K_8*M*M recognizing the interactions of species (explained in Sections 16. 2 and 16.4). The high respiration and consumption of high-diversity species complex are recognized with no deposition of organic matter. Whereas there is a wide range of properties in the species of real ecosystems, this model brings out the consequences of the differences between colonizing and mature vegetation by exaggerating the different factors in the two production functions K_2*R*N*B and K_7*R*M*S.

OTHER SYSTEMS

The concepts of maximum power, materials, and diversity can be applied not only to ecosystems but to self-organization on any scale. The inverse relationship of excess resources and diversity has been observed in the history of human societies. For example, the spread of Europeans colonizing America first produced a society of low occupational diversity with a preponderance of flimsy, temporary housing. In the nineteenth century there was capitalistic expansion by a few corporations that were developing monopolies. Later more complex, diverse, and efficient societies developed.

"What If" Experimental Problems for the Program in Table 16.3

16.3–1 What happens if the nutrient inflow is intermediate between the extremes used in the simulations of Figures 16.3b and 16.3c? In line 170, set JN = 0.8.

16.3–2 For low nutrient condition, what is the effect of more or less seeding of species? After setting JN = 0.1, in line 180 set S = 3; then try S = 0.5.

16.3–3 For high nutrient condition, what is the effect of more

TABLE 16.3
Model of Nutrients and Diversity (NUTRSPEC.bas) (Figure 16.3)

```
5 REM PC
10 REM NUTRSPEC (high nutrient vs high diversity)
15 SCREEN 1, 0: COLOR 0, 0
20 LINE (0, 0)-(319, 90), 3, B
25 LINE (0, 100)-(319, 190), 3, B
30 REM Starting Values
40 B = 200
45 M = 1
47 G = 2
50 N = .1
52 REM Coefficients
55 K0 = .0095
57 K1 = .19
58 K2 = .01
60 K3 = .005
63 K4 = .02
67 K5 = .0001
70 K6 = .00001
73 K7 = .0002
77 K8 = .00001
80 K9 = .0005
84 K10 = .0002
87 REM Scaling
90 DT = 1
100 T0 = 6
110 N0 = .03
120 M0 = 1
130 B0 = 55.5
140 Y0 = 5555
145 REM Sources
150 I = 100
170 JN = .1: REM low nutr run: 0.1; High nutr run 3.
180 S = 1
200 REM Plotting and Equations
219 PSET (T / T0, 90 - N / N0), 1
220 PSET (T / T0, 90 - M / M0), 3
230 PSET (T / T0, 190 - B / B0), 3
240 PSET (T / T0, 190 - Y / Y0), 1
250 R = I / (1 + K0 * N * B + K1 * M * S)
260 N = (JN + K6 * B - K10 * R * M * S) / (1 + K5 * R * B)
270 DY = K4 * B
280 DB = K2 * R * N * B - K3 * B - K4 * B
290 DM = K7 * R * M * S - K9 * M - K8 * M * M
305 IF N < .00001 THEN N = .00001
310 Y = Y + DY * DT
320 B = B + DB * DT
330 M = M + DM * DT
350 T = T + DT
360 IF T / T0 < 319 GOTO 100
```

seeding of species? After setting JN = 3, set S = 3; then try S = 10.

16.4   MODEL OF SPECIES AND SUPPORT AREA (SPECAREA)

The model SPECAREA in Figure 16.4a relates the number of species to the area of energy support. When species are counted in a forest or other ecosystems, a graph can be plotted of species found as more and more area is surveyed. The resulting graphs curve up to the right as in Figure 16.4b. The SPECAREA model generated the curve in Figure 16.4b according to a theory relating species to their resources. The idea is that the energy necessary for each species increases as more species are added. The possible interactions between species increase as the square of their num-

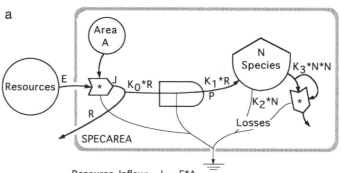

Resource Inflow: $J = E*A$
Resource Available: $R = J - K_0*R$ and $R = (E*A)/(1 + K_0)$
Diversity N: $DN = K_1*R - K_3*N*N - K_2*N$

FIGURE 16.4.
Model of species in relation to support area (program SPECAREA.bas). (a) Energy systems diagram and equations; (b) simulation of species supported as a function of increasing area; (c) same simulation with species plotted as a function of the square root of the support area.

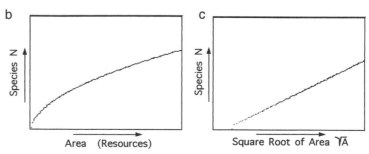

ber. Either energy is lost in competition or energy is used to organize species to avoid interference and competition.

If the model is simulated plotting number of species as a function of time, growth occurs at a decreasing rate until the curve levels. Then if the area of support A is increased, the growth reaches a higher level. By repeating this procedure over and over, the simulation arranged by the program in Table 16.4 plots the species supported as a function of the support area (Figure 16.4b). The program has a time iteration within an area iteration. For each value of area A the program increases time and species numbers N in steps until the species number is near its limit (until the change in species number DN is small). Then the program makes an increase in area (line 310). Then it plots the point,

TABLE 16.4
Model of Species and Support Area (SPECAREA.bas)
(Figure 16.4)

```
5 REM PC: SPECAREA (Species increase with more area resources)
10 CLS
20 REM For square root plot, substitute lines 95,225, and 330
35 SCREEN 1, 0: COLOR 7, 0
40 LINE (0, 0)-(240, 180), 3, B
60 DT = .5
65 DA = 1
70 T0 = 1
80 N0 = 1
89 REM Use line 90 or 91 but not both (disable line by inserting REM)
90 A0 = 2: REM area graph
91 REM A0 = 30: REM for square root graph
110 I0 = 100
120 A = 1
130 N = 1
140 E = 2
150 K0 = 9
160 K1 = 5
170 K2 = .1
190 K3 = .01
200 REM EQUATIONS
210 REM  Use lines 220 or 221 but not both (disable line by inserting REM)
220 PSET (A * A0, 180 - N * N0), 1: REM for area graph
221 REM PSET ((A ^ .5) * A0, 180 - N * N0): REM for area square root graph
230 N = 1
240 J = E * A
250 R = J / (1 + K0)
260 P = K1 * R
280 DN = P - K3 * N * N - K2 * N
290 N = N + DN * DT
300 IF N < 0 THEN N = 0
302 T = T + DT
305 IF ABS(DN) > .1 GOTO 240
310 A = A + DA
315 REM  Use lines 320 or 321 but not both (Disable line by inserting REM).
320  IF A * A0 < 240 GOTO 200: REM For area graph
321 REM IF (A ^ .5) * A0 < 240 GOTO 200: REM For square root area graph
```

increases the area, resets time and N to 0, and builds up N to its carrying capacity, again plots another point, etc.

If the area A necessary to add species increases as the square of the number of species N^2, the number of species increases as the square root of the area, which is written as $A \wedge 0.5$, the one-half power of the area. Figure 16.4c is the graph that is plotted when a print statement is substituted that plots species number N as a function of the square root of area. (See remark in program line 210.)

EXPLANATION OF EQUATIONS

The inflow of resources J is a product E*A of the energy inflow per area E and the area A. The energy available R is that remaining after the uses K_0*R by the production process. The productive support of species depends on that energy $P = K_1$*R. In this model the species number N is a balance between additions dependent on area and energy and two pathways of loss. One is linear K_2*N; the second is quadratic K_3*N*N, representing the increasing energy drain of species interactions or its prevention.

At steady state the change in species number DN is zero, and the diversity equation in Figure 16.4a can be algebraically solved. Omit term K_2*N, which is negligible except when N is small. Then species numbers are proportional to the square root of the energy resource. Where constants are combined, $N = (K*E*A) \wedge 0.5$.

OTHER EXAMPLES

The model may apply to the number of occupations in human societies in relation to the resource base. Curves of number of occupations sampled have a similar shape to the species-area graph. The model may represent information and resources required more generally. The ability to organize, retrieve, and maintain information may go up as the square of the number of units. As you add more resources, the number of books that can be supported and well used in a library may follow curves like that shown in Figure 16.4.

"What If" Experimental Problems for the Program in Table 16.4

16.4–1 What happens if the budget of energy is greater? Double E.

16.4–2 Suppose pathway K_2*N represents the rate of loss of information due to the depreciation of the structures on which information is stored. What is the effect of increasing the depreciation rate? Increase K2.

16.4–3 What would be the shape of the graph of species and area if the horizontal scale were put on a logarithmic scale? Substitute a plot statement with log function; multiplying by 50 spreads the scale as needed:

220 PSET (50*log (A), 180 − N*N0).

When species-area curves are plotted with a semi-log scale (horizontal scale on log paper and vertical scale regular), they curve upward. How does the simulation compare?

◆

16.5 MODEL OF ISLAND SPECIATION AND DIVERSITY (SPECIES)

The model SPECIES (Figure 16.5) relates the processes developing diversity of species and ecosystem productivity on isolated islands over long periods of time. Islands and other isolated locations do not receive as many species by dispersal as in continental areas. Diversity develops slowly, and high levels cannot be sustained. In the famous example of the Galapagos Islands studied by Darwin, some of the species were found transported from the mainland. Other kinds evolved within the islands by the speciation process, apparently by being isolated temporarily on separate islands long enough to develop different characteristics. Although located 800 miles from the mainland of South America, some species became part of the island ecosystem's self-organizational process seeded from the mainland. Birds fly, often carrying seeds; other species may be transported by wind or water.

When there are only a few species available for self-organiza-

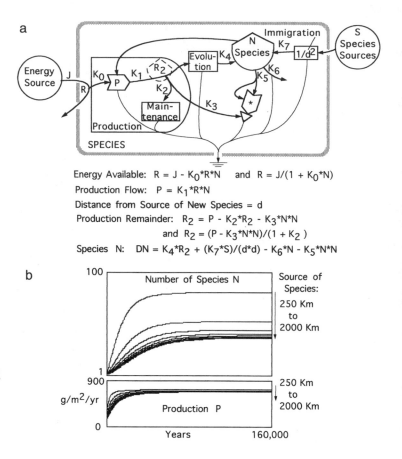

Energy Available: $R = J - K_0*R*N$ and $R = J/(1 + K_0*N)$

Production Flow: $P = K_1*R*N$

Distance from Source of New Species = d

Production Remainder: $R_2 = P - K_2*R_2 - K_3*N*N$

and $R_2 = (P - K_3*N*N)/(1 + K_2)$

Species N: $DN = K_4*R_2 + (K_7*S)/(d*d) - K_6*N - K_5*N*N$

FIGURE 16.5.
Model of species on isolated islands (program SPECIES.bas). (a) Energy systems diagram and equations; (b) family of curves simulated by increasing distances between the island and the source of species.

tion, the community production may be less than its full potential. This is because some of the species may not be as well adapted to perform the various functions as the diversity of species.

On the left in Figure 16.5 the model has species evolving as one of the long-term outputs of the ecosystem production P based on renewable energy flow J. On the right, species are being seeded at a slow rate from distant sources S, but inverse to the square of the distance. Energy required for diversity is increased as the square of the number (explained in Section 16.4).

In the simulation, species increase, causing production to increase. Both level off as the system becomes limited by resources, including the resource of species availability. Each unit of time represents 100 years, and the horizontal scale represents 160,000 years. After the program has plotted one graph, the distance from the mainland, initially 250 kilometers, is increased by 200 and run again. The result is a family of curves, each for a different distance. As shown in Figure 16.5b, increasing distance causes lower diversity and reduces productivity.

EXPLANATION OF EQUATIONS

The model in Figure 16.5 shows the interplay of dispersal, local speciation, and production. As the diagram shows, the more species there are, the more the renewable environmental resources are efficiently used to form products, although production is ultimately limited by the inflow of these resources (sun, wind, rain, nutrients). The production available for use is R_2, which is the difference between the production P and the flows already in use. Respiration and consumption are in proportion to the production K_2*R_2. Production used to support the diversity is in proportion to the square of the number of species K_3*N*N, thus representing the additional energy that goes into competition or into various niche separation mechanisms that prevent that competition. Some energy goes into the speciation process K_4*R_2 at a very slow rate (microevolution). The energy for speciation is the energy to support extra individuals with variations that are necessary to the process of isolation and selection.

Shown in Figure 16.5a the diversity N (expressed as species per 1000 individuals counted) is increased by evolution of species K_4*R_2 and by immigration from a source of species S. Immigration is diminished as the inverse square of the distance $1/d^2$ from the species source. Species are lost by two pathways. One flow is in proportion to the number of species K_6*N; the other is in proportion to the square of the species K_5*N^2, representing the effects of competition, consumption, and other interactions among species in causing extinction.

Lines 410–430 in the program (Table 16.5) demonstrate how to make a simulation run successive curves. Time and initial conditions are reset, the change is made in the variable under study (in this case distance D), and then the program is sent back to its cycle of iteration starting with line 200.

OTHER EXAMPLES

The model may be applied to other situations where seeding of information is limited by distance so that the rate of developing

TABLE 16.5
Model of Island Speciation and Dispersal (SPECIES.bas)
(Figure 16.5)

```
2 REM PC: SPECIES.bas  (Island dispersal, speciation, productivity)
3 CLS
5 SCREEN 1, 0: COLOR 0, 0
7 LINE (0, 0)-(319, 90), 3, B
10 LINE (0, 100)-(319, 190), 3, B
12 DT = 1
14 T0 = .2
16 P0 = .001
18 N0 = 1
20 N = 1
25 J = 1
27 S = 2
30 D = 250
35 K0 = .2666
40 K1 = 20000
45 K2 = 2.5
50 K3 = 13
55 K4 = .000005
65 K5 = .000041
70 K6 = .00067
75 K7 = 10000!
80 Md = 2000
200 REM PLOTTING
230 PSET (T * T0, 190 - P * P0), 1
250 PSET (T * T0, 90 - N * N0), 2
350 R = J / (1 + K0 * N)
360 R2 = (K1 * R * N - K3 * N * N) / (1 + K2)
362 IF R2 < 0 THEN R2 = 0
365 P = K1 * R * N
370 DN = K4 * R2 - K5 * N * N - K6 * N + (K7 * S) / D / D: REM D is distance to
        mainland
380 N = N + DN * DT
390 T = T + DT
400 IF T * T0 < 319 GOTO 200
410  D = D + 200
425 T = 0: N = 1
430 IF D < Md GOTO 200
```

information anew is occurring at the same rate as dispersal. Another biological example is species developing on isolated mountain tops where species that can provide seedings are far away.

This model represents the principles of developing and maintaining information, since the seeding of species is really the seeding of genetic information. The model can also be used for information transfer and maintenance in human systems. Cultural information affecting human productivity was transmitted by travelers to remote areas slowly, before the days of cheap transportation. There was a limit to the information and technology that could be sustained on a small area.

"What If" Experimental Problems

16.5–1 What is the effect on the diversity of increasing the renewable energy? Increase J.

16.5–2 What is the effect of starting with a large diversity? Set N in line 60 to 100.

Chapter Seventeen

◆

MODELS OF MICROECONOMICS

271

T his chapter highlights models of microeconomics, the economic use of resources. These models contain the interface between environment and the economy. With the model SALES, money from selling environmental production is received with varying price. TANKSALE simulates the sales from mining a slowly renewable reserve. BANK simulates a bank receiving deposits and making loans. ECONP&C is a production and consumption model controlled by a countercurrent of money in a closed circulation. ECONUSE models production and sales from a renewable environmental process. RESERVE models the role of biodiversity areas in economic yields requiring land rotation. These models couple the circulation of money to the production and use of real wealth.

◆

17.1 MODEL OF SALES WITH PRICE INVERSE TO SUPPLY (SALES)

The SALES model (Figure 17.1a) shows an environmental production process generating a stock of products sold for money. The price is inverse to the quantity of product, representing the inverse relationship of supply and demand. When the program (Table 17.1) is run (Figure 17.1b), products accumulate and level off, prices go down, and money increases and levels as a steady state is reached. Although prices vary, income to the producer is stable.

EXPLANATION OF EQUATIONS

Products Q are sold in proportion to the stock K_2*Q, and money received in exchange is a product of sales and price $p*K_2*Q$. According to the economic principle of supply and demand, the price p goes up if the supply of products is less and

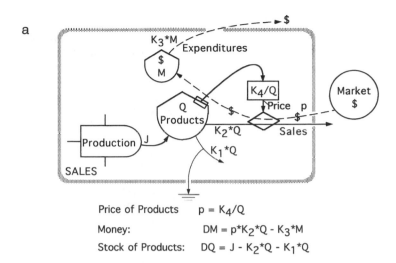

Price of Products   p = K$_4$/Q

Money: DM = p*K$_2$*Q - K$_3$*M

Stock of Products: DQ = J - K$_2$*Q - K$_1$*Q

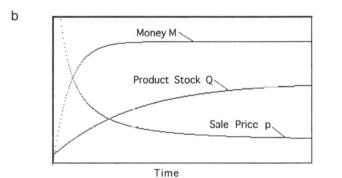

FIGURE 17.1.
Model of sales with price
inverse to supply (program
SALES.bas). (a) Energy systems
diagram and equations;
(b) simulation.

down when the supply Q is greater. Price p is set to be inverse
to the supply K$_4$/Q. People pay more when supply is scarce. The
money received is stored in a bank account M, and then paid
out K$_3$*M in proportion to the amount in the account. The small
rectangle on the Q tank and the line from it to the price box is
a "sensor" expressing the influence of the quantity Q without
reducing it. Depreciation of the stock is K$_1$*Q.

EXAMPLES OF SALES MODELS

Citrus farmers produce oranges, store them, and sell them.
The price they receive fluctuates according to the supply. In this
model the more money the farmer earns, the more he spends.

TABLE 17.1
Model for Price Effect on Sales
(SALES.bas) (Figure 17.1)

```
20 REM PC: SALES (Sales and money)
30 CLS
40 SCREEN 1,0: COLOR 0,1
50 LINE (0,0)-(320,180),3,B
60 Q = 10
70 J = 1
80 K1 = .005
90 K2 = .005
100 K3 = .05
110 K4 = 1500
120 Q0 = 1
130 P0 = 2
140 M0 = 1
150 T0 = 1
160 DT = 1
170  PSET (T * T0,180 - Q * Q0)
180  PSET (T * T0,180 - P * P0)
190  PSET (T * T0,180 - M * M0)
200 P = K4 / Q
210 DQ = J - K1 * Q - K2 * Q
220 DM = (K2 * Q) * P - K3 * M
230 Q = Q + DQ * DT
240 M = M + DM * DT
250 T = T + DT
260 IF T * T0 < 320 GOTO 170
```

***"What If" Experimental Problems for the Program
in Table 17.1***

17.1–1 What happens to the business if the depreciation of the stock increases 10 times?

17.1–2 What effect does doubling production J have on the accumulation of money M and on the amount of product per dollar the consumer can buy? What is the difference between the effect on the business and the effect on consumers?

17.2   MODEL OF ECONOMIC USE OF MINED RESOURCES (TANKSALE)

TANKSALE (Figure 17.2a) is a model of mining and sale of a nonrenewable resource reserve, like coal, oil, or copper. It is

a

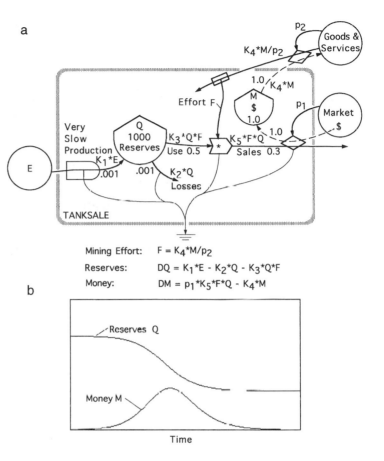

Mining Effort: $F = K_4*M/p_2$

Reserves: $DQ = K_1*E - K_2*Q - K_3*Q*F$

Money: $DM = p_1*K_5*F*Q - K_4*M$

b

FIGURE 17.2.
Model of economic use of
mined resources (program
TANKSALE.bas). (a) Energy
systems diagram and equations;
(b) simulation.

called a *nonrenewable* resource because mining and sales (interaction symbol) occur at a much faster rate than the natural earth process of renewal. Money received from product sales is used to pay for goods and services to do the work of mining. In this model the market prices are constant.

The simulation (Figure 17.2b) starts with large resource reserves. The mining company uses purchased goods and services F to mine, sort, store, and transport the product. At first large quantities are mined with little effort, and money accumulates. Later when reserves are reduced, yield is less, and sales are less. Money required to maintain the mining process is reduced. At the prices used in this simulation, money runs out with some ore left in the ground.

EXPLANATION OF EQUATIONS

The reserve in the ground Q is slowly renewed by a steady geological process of concentration K_1*E. Some of the concentrated ore disperses without use K_2*Q. The mining is a product K_3*Q*F of the quantity Q in the reserve and the availability of necessary inputs F. Most of that mined is sold K_5*F*Q at price p_1. The money on hand M is a balance between the income from sales p_1*K_5*F*Q and the payout for necessary inputs K_4*M. The availability of input force F is generated by money spent for purchases K_4*M divided by the price p_2 of these goods and services K_4*M/p_2.

EXAMPLES OF ECONOMIC USE OF RESERVES

This model represents use of any source where use is much faster than its renewal. Other examples are natural gas, gold, uranium, phosphate for fertilizer, slow growth forests, and soils where agriculture is not based on land rotation.

"What If" Experimental Problems for the Program in Table 17.2

17.2–1 How do money M and storage Q react if the price you receive for the yield is increased? Change p_1 to 7. Can you select a price that uses almost all of the reserve?

17.2–2 Will the operation get started if you begin with almost no capital (M = 0)? How would it be different if you had twice as much capital as the original program? Try M = 2.

17.2–3 What is the effect of finding more reserves after the operation has been going for 100 months? Type a new statement to add 300 more units of Q: 730 IF T*T0 = 100 THEN Q = Q + 300.

17.2–4 What would happen if the initial reserve were small and less available? Make Q = 300.

◆

17.3   MODEL OF LOANS, INTEREST, AND BANKING (BANK)

The BANK model (Figure 17.3a) represents financial processes involved with banking. Depositors on the left supply money J_d

TABLE 17.2
Program for Economic Use of Reserves
(TANKSALE.bas)
(Figure 17.2)

```
10  REM PC: TANKSALE (Economic use of reserves)
15 CLS
30 SCREEN 1, 0
35 COLOR 7, 0
40 LINE (0, 0)-(319, 180), 3, B
100 REM Scaling factors
110 DT = .1
130 Q0 = .13
140 M0 = .2
150 T0 = 10
200 REM Starting storages
210 Q = 1000
220 M = 1
300 REM Sources
310 E = 1
320 P1 = 5
330 P2 = 1
400 REM Coefficients
410 K1 = .001
420 K2 = .000001
430 K3 = .0005
440 K4 = 1
450 K5 = .0003
500 REM Equations
510 F = K4 * M / P2
520 DQ = K1 * E - K2 * Q - K3 * Q * F
530 DM = P1 * K5 * Q * F - K4 * M
600 REM Change equations
660 Q = Q + DQ * DT
670 M = M + DM * DT
680 T = T + DT
700 REM Plotting
710 PSET (T * T0, 180 - Q * Q0), 1
720 PSET (T * T0, 180 - M * M0), 3
800 REM Iteration
810 IF T * T0 < 319 GOTO 500
```

to the bank's storage of money M. Dollars given to the bank go into the money storage M, and the amounts are also tallied in the deposit account (tank D). D is not a storage of money; it is a record of the money owed to depositors. The quantity D decreases when the deposits are repaid. The bank pays interest to the depositors according to a depositor's interest rate. Money is also drawn out by depositors.

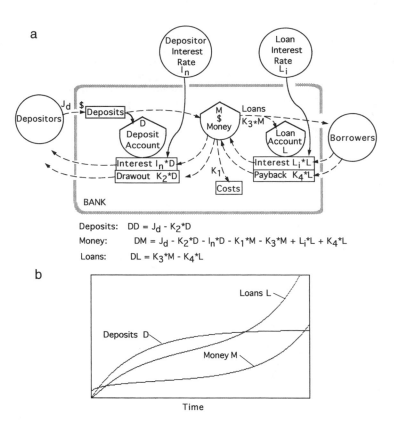

Deposits: $DD = J_d - K_2*D$

Money: $DM = J_d - K_2*D - I_n*D - K_1*M - K_3*M + L_i*L + K_4*L$

Loans: $DL = K_3*M - K_4*L$

FIGURE 17.3.
Model of banking (program
BANK.bas). (a) Energy systems
diagram and equations;
(b) simulation of profitable
regime.

Shown to the right in the diagram, the bank loans its money
(in excess of a required cash reserve), and the amounts are tallied
in the loan account L. From its borrowers, the bank receives
interest at a higher rate than it pays its depositors. Money also
goes for costs of bank operations. The money in the bank M is
a balance between the money it gets from its loans and the money
it pays its depositors. It makes a profit by paying less interest to
its depositors than it charges its borrowers.

The simulation (Figure 17.3b) represents the start-up of a
bank. Deposits and the deposit account build to a steady level
limited by the number of depositors. The money increases with
deposits. There is a profit if the rate at which money is deposited
plus that earned from loans is greater than the rate at which
money is loaned and paid to depositors. If there is a net profit
(Figure 17.3b), there is an accelerating growth of money. However, if there is not a net profit, the money spirals downward
reaching zero, stopping the program and printing the word
"BANKRUPT."

EXPLANATION OF EQUATIONS

The change in the bank's money M is a balance between receipts and payments. Receipts include the money deposited J_d, the interest on its loans L_i*L, and borrowers paying back the loans K_4*L. Money paid to depositors in interest is the product of the going interest rate and deposit account I_n*D. In proportion to the account, money flows back to the depositors K_2*D as they withdraw their deposits by writing checks. In this model loans are made by the bank in proportion to its money K_3*M. Borrowers pay the bank interest as a product I_i*L of the loan interest rate L_i and the loan account L. In the model payback of loans K_4*L is in proportion to the quantity loaned L. The change in the deposit account at each iteration is a balance of the deposits J_d and the drawout of deposits K_2*D. The change in the loan account is the difference between the loans K_3*M and the payback K_4*L.

EXAMPLES OF THE BANK MODEL

Banks accept people's deposits to make mortgage loans to finance buying of homes. If you sometimes borrow and also lend money, you fit into this model. Although you may act as if the money you borrowed is a bonanza to spend, you also need to keep track of your debt, since you have to pay it back with the interest added.

"What If" Experimental Problems for the Program in Table 17.3

17.3–1 What happens to the bank's deposits if the interest rate paid to depositors IN goes up? Increase IN to 0.07 and see. To see the original graph and the change on the screen at the same time, first run the program; then change IN to 0.07 and run the new program on top of the first by typing RUN 70. This skips the statements that clear and reset the screen.

17.3–2 Suppose the rate of new deposits goes down by half. What will happen to the bank's money and to its debt? Make JD = 1.5 E4 and rerun. Can the bank accelerate its profit with few or no depositors (assuming initial capital to loan)?

TABLE 17.3
Program for Simple Banking (BANK.bas) (Figure 17.3)

```
10 REM PC
20 REM BANK.bas (Deposits, debt account, and loans)
30 SCREEN 1,0
40 COLOR 0,0
50 LINE (0,0)-(320,180),,B
60 REM Sources and Starting Values
70 JD = 300000!
80 M = 1000000!
90 D = 10000
95 L = 1
100 IN = .04:REM Depositors interest rate
105 LI = .06:REM  Bank loan interest rate
200 REM Scaling
210 T0 = 2
220 M0 = .00001
230 D0 = .00001
235 L0 = .00001
240 DT = .5
250 REM Constant Coefficients
255 K1 = .01:REM Cost, Taxes, Losses
260 K2 = .03:REM Drawout Rate
265 K3 = .14:REM Loan Rate:
270 K4 = .05:REM Payback Rate
300 REM Equations
320 DM = JD - K2 * D - IN *D - K3*M + LI*L + K4*L-K1*M
325 DD = JD - K2 * D
330 DL = K3*M - K4*L
340 M = M + DM * DT
345 IF M <0 THEN M =0
350 IF M = 0 THEN PRINT "Bankrupt":END
360 D = D + DD * DT
370 L  = L + DL*DT
380 T = T + DT
385  PSET (T * T0,180 - M * M0)
390  PSET (T * T0,180 - D * D0)
395 PSET (T*T0,180 - L*L0)
400 IF T * T0 < 320 GOTO 300
```

17.3–3 What happens if the bank is unable to make as many loans? Reduce the loan income rate K3 by half. Suppose there is a limited number of borrowers available so that there is a ceiling. Insert the lines:

```
375   IF L> 8000000! THEN K3 = 0
376   IF L< 8000000! THEN K3 = .14
```

17.3–4 What happens if the going interest rate for loans is reduced? What can the bank do to stay profitable? (Assume
depositors and borrowing are unchanged).

◆

17.4 MODEL OF PRODUCTION, CONSUMPTION, AND CIRCULATING MONEY (ECONP&C)

In the model ECONP&C (Figure 17.4a), circulating money
is related to the production and consumption of real wealth. Real
wealth (such as food, fiber, wood, and fish) is first produced in
the rural area using the renewable energy of the environment
interacting with goods and services from the city. Products are

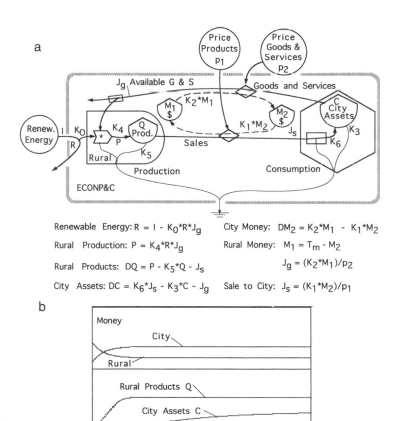

Renewable Energy: $R = I - K_0*R*J_g$ City Money: $DM_2 = K_2*M_1 - K_1*M_2$

Rural Production: $P = K_4*R*J_g$ Rural Money: $M_1 = T_m - M_2$

Rural Products: $DQ = P - K_5*Q - J_s$ $J_g = (K_2*M_1)/p_2$

City Assets: $DC = K_6*J_s - K_3*C - J_g$ Sale to City: $J_s = (K_1*M_2)/p_1$

FIGURE 17.4.
Model of the countercurrent
link between money circulation
and energy use (program
ECONP&C.bas). (a) Energy
systems diagram and equations;
(b) simulation.

sold to the city consumers who generate city assets. Goods and services of the city feed back to the rural area (to the left).

Money that is used to buy and sell products and goods and services flows as a circular countercurrent in the opposite direction. As shown by the diamond-shaped symbols, the flow of products and goods and services are linked to the flow of money by the prices. In this model the prices are determined by the larger system's markets outside this system's boundary. The working capital in the bank accounts of the producers and consumers are the small tanks indicated with $ signs. Money circulates between rural money (tank M_1) and city money (tank M_2). The spending of money is proportional to the money in the bank accounts—the more money, the more spending. Because the energy source is a renewable inflow, limited at its source, the model is an appropriate simplification for an agrarian economy before the industrial revolution.

The simulation (Figure 17.4b) started with small storages of products Q and consumer assets C and then shows a growth of rural products Q and city assets C as money goes around and around (Figure 17.4b). Since the money supply T_m is constant, the same money buys more as production increases until the growth is limited by the renewable energy source.

EXPLANATION OF EQUATIONS

Change of money in the city with each iteration is a balance between expenditures from the rural area K_2*M_1 and expenditures by the city K_1*M_2. Since the total money supply T_m is held constant, the rural money is the difference between the total and that in the city $(T_m - M_2)$. The expenditure of the city money K_1*M_2 buys products from the rural stock $Q(J_s = K_1*M_2/p_1)$ according to their price p_1. The expenditure of rural money K_2*M_1 buys goods and services $J_g = K_2*M_1/p_2$ from the city C with the price p_2.

Available renewable energy R is the unused remainder of the input I. The energy in use K_0*R*J_g depends on the goods and services from the city J_g. The resulting production P is K_4*R*J_g. The change in stock of rural products DQ is a balance between the production P, losses due to depreciation K_5*Q and sales K_1*M_2/p_1. The change in stock of city assets is a balance between the consumers' work K_6*J_s using the inflow of rural products J_s, depreciation K_3*C, and the outflow of goods and services to the rural area $J_g = K_2*M_1/p_2$.

EXAMPLES

An example is agricultural production of food and fiber, which consumers in the city use to provide the labor and capital assets needed for agricultural production. In Paris in 1860, hay for horses was sold to the city and manure returned to the surrounding farms.

"What If" Experimental Problems for the Program in Table 17.4

17.4-1 In 1815 dust from a volcanic explosion shielded the sun, and in that summer there were crop failures in New England with frost all summer. What is the effect of reducing the source input by half?

17.4-2 What is the effect of increasing the money supply TM without changing the resource availability or the prices?

17.4-3 Suppose the price of human goods P2 were increased, other things being the same? What is the effect on stored products and city assets? Explain.

◆

17.5 MODEL OF ECONOMIC USE OF RENEWABLE RESOURCES (ECONUSE)

With the model ECONUSE (Figure 17.5a) environmental products based on renewable energy are used for economic production and sale. The product Q (such as forest wood, fish, or nuts) is produced by an autoctalytic environmental production process that uses the renewable energies of sun, rain, and wind. The products accumulate in storage Q where some depreciates and the rest is used in economic production (interaction symbol) and sold to the main economy on the right. The yield of product ready for sale E depends on both the quantity of original product Q and the amount of processing equipment and labor A. The money obtained from the sale is in tank M where it is used to

TABLE 17.4
Program for Economic Countercurrent of Money and Energy (ECONP&C.bas) (Figure 17.4)

```
10 REM PC ECONP&C (Countercurrent of money and energy)
20 CLS
40 SCREEN 1,0:COLOR 0,1
50 LINE (0,0)-(320,180),3,B
55 LINE (0,90)-(320,90),3
60 REM Starting Conditions and Scaling Values
70 I = 1: REM Renewable Resources
80 C = 1: REM Consumer assets
90 Q= 1: REM Product storage
100 TM = 100: REM  Total money supply
120 M1 = 80: REM Producers' money
130 Q0 = .7
140 M0 = .5
145 C0 = 2
150 T0 = .5
155 DT = 1
160 REM Constant Coefficients
165 K0 = 45
170 K1 = .01
175 K2 = .02
180 K3 = .005
185 K4 = 200
190 K5 = .05
195 K6 = .5
205 P1 = 1! :REM Price of products
210 P2 = 2.5:REM Price of goods and services
220 REM Equations
230 R = I  / (1 + K0 * JG)
235 P = K4*R*JG
240 Jg = (K2 * M1)/ P2
242 IF C < JG THEN JG = 0
245 M2 = TM - M1
250 Js = (K1 * M2)/P1
255 IF Q < JS THEN JS = 0
260 DQ = P - K5 * Q - JS
270 DM = K1 * M2- K2 * M1
277 DC = K6*JS - K3*C - JG
280 Q = Q + DQ*DT
290 IF Q < 0 THEN Q = 0
294 C = C + DC*DT
296 IF C < 0 THEN C = 0
300 M1= M1 + DM * DT
305 IF M1 <0 THEN M1 = 0
310  PSET (T * T0,180 - Q * Q0),1
320  PSET (T * T0, 180 - C * C0),3
322 PSET (T*T0, 90 - M1*M0),1
324 PSET (T*T0, 90 - M2*M0),3
330 T = T + DT
340 IF T * T0 < 320 GOTO 220
```

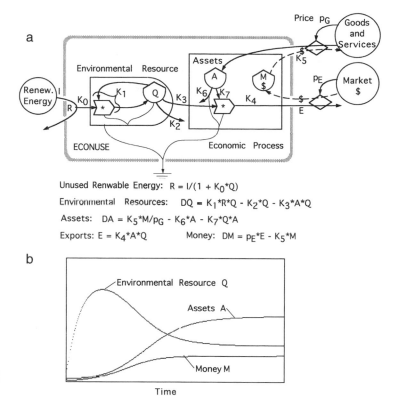

Unused Renewable Energy: $R = I/(1 + K_0*Q)$

Environmental Resources: $DQ = K_1*R*Q - K_2*Q - K_3*A*Q$

Assets: $DA = K_5*M/p_G - K_6*A - K_7*Q*A$

Exports: $E = K_4*A*Q$ Money: $DM = p_E*E - K_5*M$

FIGURE 17.5.
Model of economic use of
environmental resources
(program ECONUSE.bas).
(a) Energy systems diagram and
equations; (b) simulation.

buy necessary goods and services to sustain assets A for the economic production process.

The simulation (Figure 17.5b) starts with small quantities of environmental product Q, equipment assets A, and money M. First the quantity of environmental product (e.g., wood) Q grows. Sales increase, and more money M becomes available to buy more goods and services like saws, tractors, trucks, and roads to build up economic assets A. This economic growth uses up some of the stocks of wood. Later a balance of production and use is reached at a rate of use based on the natural limitation of tree growth on the renewable energies.

EXPLANATION OF EQUATIONS

In this model, R is the unused renewable energy after the use K_0*R*Q for autocatalytic environmental production K_1*R*Q. The change of storage in each iteration is the difference between the production, the depreciation K_2*Q, and that used for economic processing K_3*Q*A. The economic product exported from the system for sale $E = K_4*Q*A$ generates income p_E*E. The change in money available DM is the difference between the money brought in from the sale of the product p_E*E minus the money paid out for goods and services K_5*M. The change in economic processing assets DA is the difference between the purchased goods and services K_5*M/p_G, the depreciation K_6*A, and assets used up in the economic production K_7*Q*A.

EXAMPLES OF ECONOMIC USE

This model is appropriate for any salable product that grows without human assistance or reinforcement—wildflowers, natural honey; forest products like rubber, resin, turpentine, wood, fruits, nuts; marine products like fish, lobsters, clams, coral, shells.

"What If" Experimental Problems for the Program in Table 17.5

17.5–1 What difference does a one-half increase in the price of goods and services make? Find the statement for PG and increase it by a half.

17.5–2 If the environmental resource is scarce, sales are less. Can the user stay in business? Change I0 to half and run the program again.

17.5–3 What difference does it make if you start with 10 times more money?

17.5–4 How vulnerable is your business to the price you can get for your product? To see, halve PE and then double it.

17.5–5 As an economic product gets scarce, its price tends to increase. Can you add a statement to the program with PE inverse to Q? For example, what happens with $PE = K8/Q$ and $K8 = 150$?

◆

17.6   MODEL OF THE ECONOMIC ROLE OF BIODIVERSITY RESERVE AREAS (RESERVE)

Because most agriculture and intensive plantation forestry wears out soils, sustainable production depends on a cycle that

TABLE 17.5
Program for Economic Use of Environmental Resource (ECONUSE.bas) (Figure 17.5)

```
20 REM PC: ECONUSE (Economic use of environmental source)
30 CLS
40 SCREEN 1, 0: COLOR 7, 0
50 LINE (0, 0)-(319, 180), 3, B
60 I0 = 100
70 Q = 10
80 A = 10
90 M = 10
100 PE = 1
110 PG = 2
120 K0 = .19
130 K1 = .002
140 K2 = .005
150 K3 = .00005
160 K4 = .0001
170 K5 = .01
180 K6 = .003
190 K7 = .000005
200 DT = 10
210 T0 = .1
220 Q0 = .5
230 A0 = .3
240 M0 = .3
250 R = I0 / (1 + K0 * Q)
260 E = K4 * Q * A
270 DQ = K1 * R * Q - K2 * Q - K3 * Q * A
280 DA = K5 * M / PG - K6 * A - K7 * Q * A
290 DM = PE * E - K5 * M
300 A = A + DA * DT
310 Q = Q + DQ * DT
320 M = M + DM * DT
330 PSET (T * T0, 180 - Q * Q0), 1
340 PSET (T * T0, 180 - A * A0), 2
350 PSET (T * T0, 180 - M * M0), 3
360 T = T + DT
370 IF T * T0 < 319 GOTO 250
```

includes a period of time with successional restoration by a natural ecosystem. This restoration of fallow land is automatic and rapid if the area is surrounded by a complex, forest with high diversity of species to transport and seed the restoration process. The restoration land cycle is modeled as RESERVE in Figure 17.6. If there are few biodiversity reserves, seeding is delayed and economic output is less in the long run. If the areas reserved

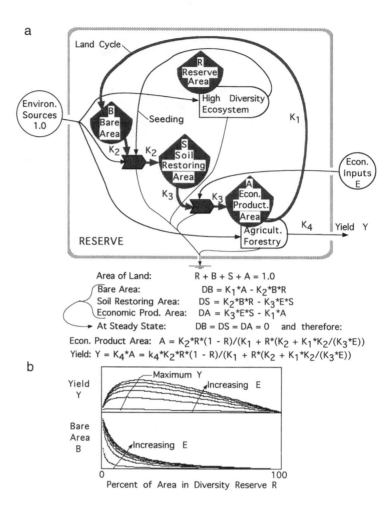

a

Area of Land: $R + B + S + A = 1.0$
Bare Area: $DB = K_1{*}A - K_2{*}B{*}R$
Soil Restoring Area: $DS = K_2{*}B{*}R - K_3{*}E{*}S$
Economic Prod. Area: $DA = K_3{*}E{*}S - K_1{*}A$
At Steady State: $DB = DS = DA = 0$ and therefore:
Econ. Product Area: $A = K_2{*}R{*}(1 - R)/(K_1 + R{*}(K_2 + K_1{*}K_2/(K_3{*}E)))$
Yield: $Y = K_4{*}A = k_4{*}K_2{*}R{*}(1 - R)/(K_1 + R{*}(K_2 + K_1{*}K_2/(K_3{*}E)))$

FIGURE 17.6.
Model of the economic role of
reserve areas of high diversity
(program RESERVE.bas).
(a) Energy systems diagram and
equations; (b) simulation of
yield and bare area as a
function of percent of land
in reserve.

b

for complex forest are large, the area for economic output is small and the economic output less.

The program has been arranged to simulate the effect on the yield Y of changing the reserve area R. The resulting graph (Figure 17.6b) plots yield Y and bare area B as a function of R. The curve has a maximum. On the left where there is little reserve, the bare land is huge. On the right where the reserve is large, the land for economic yield is less.

The program runs a family of curves. After each run, the economic input E is increased and the program plots another curve. With more economic input the yields are larger, but more economic production also increases the land deterioration rate, causing more bare land. The peak of the upper curves shifts to the right, since the reserve required to generate maximum yield increases.

Unlike most of the programs in this book, the coefficient calibrations are included in the program (instead of being in a separate table or spread sheet). In lines 100–145 you enter fractional areas of the four categories in the land cycle (total should be 1.0). In lines 160–190 the program calculates the coefficients that would produce a cycle with that land distribution. As Lotka showed years ago, a storage in a circular steady state is inverse to the outflow coefficients. In other words, storages accumulate ahead of a bottleneck.

EXPLANATION OF EQUATIONS

Environmental energy sources in this model are unlimited (constant 1.0), with land area limiting each environmental production process. The change equation for each land use DA, DB, DS is a balance between a flow from the previous state and the transfer to the next. At steady state the change rates are zero. Solving these equations for one variable and algebraically substituting, a steady-state equation for economic area A is obtained as a function of the reserve area R. Thus time was eliminated. In the program, instead of iterating with time, R is changed on each iteration. After the plot is complete the program continues to lines 550–570 where R is reset to a low value, economic input E is increased, and the simulation is sent back to iterate the value of R again.

"What If" Experimental Problems for the Program in Table 17.6

17.6–1   Insert STOP in the program after line 190 just after the observed steady-state areas are converted into coefficients. Then activate the command line and type ?K1, K2 (? means to print what follows). Then type ?K3, K4. Write down these coefficients. Looking at the model diagram, which coefficient indicates the rate at which the land in economic use wears out? Increase this coefficient to 0.3 and run the simulation again. What is the effect on economic production, the amount of bare land, and the reserve needed for maximum yield?

TABLE 17.6
Program for Simulating the Effect of Biodiversity Reserves on Economic Yield from Agriculture and Forestry (RESERVE.bas) (Figure 17.6)

```
5 REM   PC RESERVE.BAS (Optimum diversity reserve for maximum economic
         yield)
10 CLS
30 SCREEN 1, 0: COLOR 7, 0
40 LINE (0, 0)-(319, 180), 3, B
50 LINE (0, 80)-(319, 80), 3
60 DR = .001
70 T0 = 1
80 T = 60: REM time for cycle
90 REM Put in values of R, B, S, A, and J so that coefficients are calibrated
100 J = 1 / T
110 A = .4: REM .4
120 S = .1: REM .3
130 R = .2: REM .1
140 B = 1 - S - R - A
145 IF B < .05 THEN B = .05
150 Y = 1E+07
155 E = .01
160 K1 = J / A
170 K2 = J / (R * B)
180 K3 = J / S
190 K4 = Y / A
200 REM Scaling
205 R0 = 1 / 320
210 A0 = .01
220 B0 = .01
230 R0 = .2
340 Y0 = 130000!
350 REM below this Y is plotted as a function of R
400 R = .01
420 A = (K2 * R * (1 - R)) / (K1 + R * (K2 + K1 * K2 / (K3 * E)))
430 B = 1 - A - R - (K1 / (K3 * E)) * A
450 Y = K4 * A
500 PSET (R / R0, 80 - Y / Y0), 3
510 PSET (R / R0, 180 - B / B0), 3
520 R = R + DR
530 IF R / R0 < 319 GOTO 420
550 E = E + .05
560 R = .01
570 IF E < .3 GOTO 420
```

17.6–2 Suppose an area believed to be in a steady-state rotation were observed with small fraction of land area bare $B = 0.1$ and biodiversity reserve $R = 0.1$. Recalibrate and run the simulation with these values (also $A = 0.4$ and $S = 0.4$). For this economic yield to be sustained, what can you say about the rate of reseeding of this area $K2$. What parts of an ecosystem might be responsible?

Chapter Eighteen

◆

MODELS OF MACROECONOMIC OVERVIEW

The minimodels discussed in this chapter are simple overviews of a whole economy, including the environmental sources of real wealth and the circulation of money. Models of a whole economy are called *macroeconomic models*. MONEYGRO adds money supply, accelerating economic growth. BUYPOWER relates production and prices to stages in use of nonrenewable resources. After diagramming the textbook equations used in macroeconomics, MACROEC compares growth with limited and unlimited resources. ROTATION simulates the allocation of land area between environment and urban economic development.

18.1 MODEL OF MONEY-DRIVEN GROWTH (MONEYGRO)

A simple minimodel that represents the standard view of growth from economics is MONEYGRO (Figure 18.1). Money is circulating (dashed lines) as a countercurrent to the flow of products and to the feedback of goods and services into production. As the diagram shows, the resource availability E is held constant, which means that resources are never limiting, regardless of the demand on them. Because new money is added to the money supply M in proportion to the development of real wealth (assets A) without causing inflation, the result is exponential growth without limit (see EXPO, Figure 9.4). The rate of growth is sensitive to the depreciation rate of the assets. This is a version of the Domar–Harrod model of economic growth introduced 60 years ago. To see what happens when the sources are limited, see models BUYPOWER (Figure 18.2) and MACROEC (Figure 18.3) that follow.

EXPLANATION OF EQUATIONS

In the model, new money is added to the money supply M in proportion to economic assets K_1*A, making the growth inflation free. Economic production K_0*E*M is a product of the money circulating M and the unlimited energy supply (E = constant).

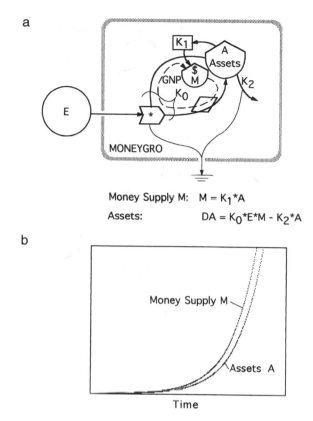

Money Supply M: $M = K_1 * A$

Assets: $DA = K_0 * E * M - K_2 * A$

FIGURE 18.1.
Model of money driven growth
(program MONEYGRO.bas).
(a) Energy systems diagram and
equations; (b) simulation.

Depreciation of real assets is linear $K_2 * A$. Taken together the
two equations are autocatalytic.

EXAMPLES OF EXPONENTIAL GROWTH DRIVEN BY MONEY CIRCULATION

The model in Figure 18.1 applies to the growth of an industry
that is utilizing a large new resource, or to the growth of a
country, where the concentration of resource is constant. Because
many people believe it is always possible to substitute one resource for another, they believe that the resource E is unlimited
and that exponential growth as shown here is a general model
for the whole economy and its component businesses. Resource
scientists know that resources are not unlimited and that such
exponential growth is only temporary.

*"What If" Experimental Problems for the Program
in Table 18.1*

18.1–1 If availability of resources is less, how will the growth
        of money and assets differ? Change E to 0.9.

18.1–2 What happens if the depreciation rate of the assets is
        decreased? Change K2 to 0.02.

18.1–3 What if the economy changes the rate of adding money?
        Increase K1 to 0.02; then decrease it to 0.012.

18.1–4 If a company starts with more assets, how does its growth
        compare to a start with fewer assets?

◆

18.2 MODEL OF PRICE IN RELATION TO UNSUSTAINABLE GROWTH (BUYPOWER)

Where growth is based on a large nonrenewable source (fuels)
as well as renewable sources (sun, wind, rain), there is a pulse of

TABLE 18.1
**Program for Money-Driven Exponential
Growth (MONEYGRO.bas) (Figure 18.1)**

```
20  REM PC: MONEYGRO (Money-driven growth)
30  CLS
40  SCREEN 1, 0: COLOR 7, 0
50  LINE (0, 0)-(240, 180), 3, B
60  REM STARTING QUANTITIES
70  E = 1
80  A = .2
90  REM  COEFFICIENTS
100 K0 = 3.7
110 K1 = .0133
120 K2 = .033
130 REM SCALING FACTORS
140 M0 = 100
150 A0 = 1
160 DT = 1
170 T0 = .5
180 REM  PLOTTING
190 PSET (T * T0, 180 - A * A0), 1
200 PSET (T * T0, 180 - M * M0), 2
210 REM  EQUATIONS
220 M = K1 * A
230 DA = K0 * E * M - K2 * A
240 A = A + DA * DT
250 T = T + DT
260 IF T * T0 < 240 GOTO 180
```

growth before the economy returns to a lower state. BUYPOWER (Figure 18.2a) is like the 2SOURCE model of Figure 13.4 but includes the circulation of money. The model shows the behavior of prices of products generated during the growth cycle. The model has autocatalytic production generating economic assets Q with three pathways, one using fuels and minerals, and two based only on renewable resources.

In the simulation shown in Figure 18.2b money supply was held constant, and an index of prices p_r (money per unit produce) was graphed. Prices reached a minimum when the production rate was highest before assets reached their maximum. Then, as nonrenewable sources were used, prices increased again, showing

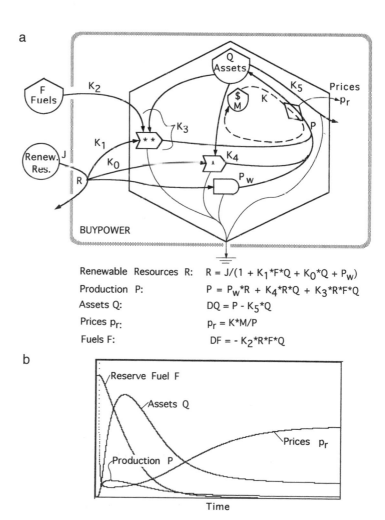

Renewable Resources R: $R = J/(1 + K_1*F*Q + K_0*Q + P_w)$

Production P: $P = P_w*R + K_4*R*Q + K_3*R*F*Q$

Assets Q: $DQ = P - K_5*Q$

Prices p_r: $p_r = K*M/P$

Fuels F: $DF = - K_2*R*F*Q$

FIGURE 18.2.
Model of prices resulting from growth on nonrenewable reserves (program BUYPOWER.bas). (a) Energy systems diagram and equations; (b) simulation.

inflation (fewer assets per dollar). Productivity P was also graphed, showing a maximum when assets and nonrenewable resources F were both moderately high. In the real economy, money is usually added to the money supply each year.

EXPLANATION OF EQUATIONS

Renewable source inflow is J, and the availability of nonrenewable sources is F. The externally limited inflow of environmental renewable sources is represented by the equation for available unused inputs remaining R. There are three production processes, including linear conversion of environmental energies P_w*R, an autocatalytic process K_4*R*Q without the nonrenewable resource, and $K_3*R*F*Q$, an autocatalytic process with both sources. The change of assets at each iteration is a sum of these production inputs minus depreciation and consumption K_5*Q. Money M is circulating in proportion to the money supply $K*M$. Prices p_r is the ratio of the money circulating (gross economic product) divided by production $K*M/P$. In other words, product price p_r goes down and up again as the buying power increases and decreases.

EXAMPLES OF MACROECONOMIC MODELS LIKE BUYPOWER

This model is a very simplified view of the world economy, but it can represent any country that is dependent on fuels and renewable resources and not very dependent on imports and exports. Examples are large nations like the United States and Russia.

"What If" Experimental Problems for the Program in Table 18.2

18.2–1   What happens to prices if the system starts with more fuel? Make F = 150 and run both graphs.

18.2–2   If the system converts fuels more efficiently, how will that affect assets and prices? Increase K3 to 0.006.

18.2–3   How does the system react if it finds a way to use its renewable energies more efficiently? Increase K4 to 0.08.

TABLE 18.2
Program for Prices Determined by Energy for Growth (BUYPOWER.bas) (Figure 18.2)

```
20 REM PC: BUYPOWER (Prices determined by energy for growth)
30 CLS
40 SCREEN 1, 0: COLOR 0, 1
50 LINE (0, 0)-(319, 180), 3, B
60 F = 100
70 J = 1
80 PW = .01
90 Q = .01
100 M = 2
110 K0 = .02
120 K1 = .00041
130 K2 = .00066
140 K3 = .005
150 K4 = .07
160 K5 = .05
170 DT = 1
180 T0 = .5
190 Q0 = 1
200 F0 = 1.6
210 P0 = 3
220 R = J / (1 + K1 * F * Q + K0 * Q)
230 P = K3 * R * F * Q + K4 * R * Q + PW
240 PR = 100 / P
250 DF = -K2 * F * R * Q
260 DQ = P - K5 * Q
270 Q = Q + DQ * DT
280 F = F + DF * DT
290 PSET (T * T0, 180 - P * P0), 1
300 PSET (T * T0, 180 - Q * Q0), 2
310 PSET (T * T0, 180 - PR), 3
320 PSET (T * T0, 180 - F * F0), 1
330 T = T + DT
340 IF T * T0 < 319 GOTO 220
```

◆

18.3 MODEL OF TEXTBOOK MACROECONOMICS (MACROEC)

Introductory texts in economics often describe macroeconomics with a set of equations for income Y, consumption C, savings S, and investment I, as listed and diagrammed in Figure 18.3a. Note the circulation of money in dashed lines in two loops. One is the money circulating C through consumers. In the other loop, money is saved S into the money supply M and flows out as investment I. The income, Y, is where the two loops are joined.

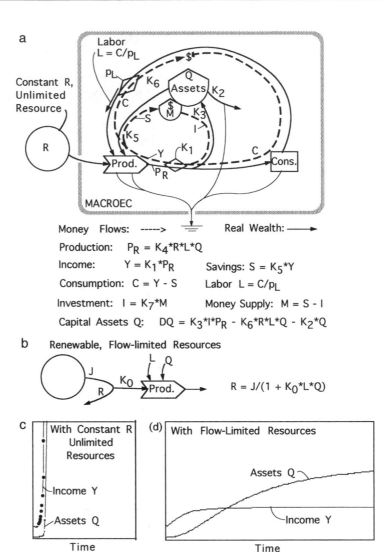

Money Flows: ----> $\equiv$ Real Wealth: ⟶

Production: $P_R = K_4 * R * L * Q$

Income: $Y = K_1 * P_R$ Savings: $S = K_5 * Y$

Consumption: $C = Y - S$ Labor $L = C/p_L$

Investment: $I = K_7 * M$ Money Supply: $M = S - I$

Capital Assets Q: $DQ = K_3 * I * P_R - K_6 * R * L * Q - K_2 * Q$

FIGURE 18.3.
Model relating textbook
macroeconomic equations to
energy resources (program
MACROEC.bas). (a) Energy
systems diagram and equations
with unlimited resources (R =
constant); (b) design of source
that is limited by inflow;
(c) simulations with unlimited
resource; (d) simulation with
flow-limited resources.

Flows of real resources, products, and services are drawn as
solid lines flowing in opposite direction from the money. In this
model the income is generated by the real production (although
the income money is flowing in a countercurrent in the opposite
direction). Production is a product of the available resources, the
capital assets, and labor. Savings is a fraction of the income;
saved money is invested. Labor is in proportion to the money
paid to consumers.

In Figure 18.3c and d two simulations are shown, the first

with unlimited energy sources (R = constant) and the second with flow limited renewable energy sources. Compare the design for the limited production process in Figure 18.3b with the unlimited one in Figure 18.3a. With limited resource inflow, the growth levels reach a steady state, but with unlimited source explosive exponential growth of the real assets and of the circulating money occurs. Exponential growth with unlimited energy was used with the MONEYGRO simulation also (Figure 18.1b). The constant resource case can only apply for a short time, because no energy source is large enough to be constantly available indefinitely.

Although the equations of macroeconomics were used in both runs, the type of resource controlled the nature of growth. This exercise helps show how economic systems are similar to general models for autocatalytic systems, but with additional flexible money mechanisms that facilitate human participation in the patterns that may maximize performance.

EXPLANATION OF EQUATIONS

The storage of real wealth is Q (capital assets). The energy resource R is a factor in the production process $P_R = K_4 * R * L * Q$ along with labor L and capital assets Q. Income is generated by the production $Y = K_1 * P_R$, where K_1 is the ratio of the flows of money to production of real wealth P_R. A fraction of this goes into savings $S = K_5 * Y$. The rest goes to consumption $C = Y - S$. Labor depends on the consumption money C according to the wage rate p_L ($L = C/p_L$). At each iteration money supply M is increased by the savings flow S and decreased by money flowing out to investment proportionately, $I = K_7 * M$. Capital assets Q is a balance between the part of production that goes into new assets $K_3 * I * P_R$ minus assets used in production process $K_6 * R * L * Q$ and the linear losses to depreciation $K_2 * Q$. With unlimited resources (a), R is constant. With flow-limited resources (b), R is the unused remainder of the flow junction in Figure 18.3b, where $R = J - K_0 * R * L * Q$, and therefore $R = J/(1 + K_0 * L * Q)$.

EXAMPLES OF MACROECONOMIC SYSTEMS

The macroeconomic models are often applied to nations, states, and regions. The provisions for source control added to these minimodels make the macroeconomics equations dynamic and realistic.

"What If" Experimental Problem for the Program in Table 18.3

18.3–1 What is the effect of an increase in depreciation rate of the growing assets? Double K2.
18.3–2 What is the effect of increasing the percent of the income saved and reinvested? Increase K5.
18.3–3 What is the effect of increasing the availability of energy in the exponential case of Figure 18.3a? Increase R.
18.3–4 What is the effect of increasing the inflow of energy in the flow-limited system of Figure 18.3b? Increase J.

18.4   MODEL OF AN ECONOMY IN RELATION TO LAND USE (ROTATION)

Agricultural-based human economies have to rotate their land back and forth between a resting state and farming. During years of farming, the soil usually loses structure, organic matter, and nutrients, so that farm productivity declines. In the resting (fallow) state, plant succession is allowed to run for several years, rebuilding the vegetation and soil.

The model ROTATION, shown in Figure 18.4a, relates land area to development in an agrarian time without fossil fuels. The human economic assets Q depend on the land in economic use A_2 and the stored products of environmental work S, such as wood and soil. The environmental support system has biomass structure B, which depends on the environmental sources I and the area of fallow land A_1. Land rotation is modeled by letting land cycle back and forth between fallow and economic use.

The simulation (Figure 18.4b) starts with a natural landscape, such as a forest cover, with large biomass structure B (e.g., trees) but low economic assets Q. As the economic system develops, land shifts from A_1 to A_2, and forest biomass is used up as it is transferred into trees ready to be processed S and then into economic assets Q.

TABLE 18.3
Program for Macroeconomic Equations (MACROEC.bas)
(Figure 18.3)

```
10 REM PC
20 REM MACROEC (Macroeconomics)
30 REM Initial run is with limited resources: Then reset X in Line 70
40 SCREEN 1, 0
50 COLOR 0, 0
60 LINE (0, 0)-(319, 180), , B
70 X = 0: REM X=0 for limited renewable source; X=1 for unlimited constant
        source
80 REM SCALING FACTORS
90 DT = 1
100 T0 = 1
110 y0 = .3
120 Q0 = .3
130 REM COEFFICIENTS
140 K0 = 9.000001E-02
150 K1 = 10
160 K2 = .01
170 K3 = .02
180 K4 = .1
185 K5 = .1
190 K6 = .01
195 K7 = .1
200 REM SOURCES
220 J = 15
240 M = 100
250 Y = 100
260 S = K5 * Y
265 I = K7 * M
268 C = Y - S
270 PL = 90
275 L = C / PL
280 Q = 10
290 R = 8
300 REM PLOTTING
305 IF Y * y0 > 180 GOTO 530
320 PSET (T * T0, 180 - Y * y0)
325 IF Q * Q0 > 180 GOTO 530
330 PSET (T * T0, 180 - Q * Q0)
335 REM EQUATIONS
340 IF X = 1 GOTO 360
350 R = J / (1 + K0 * L * Q): REM RESIDUAL RESOURCES AVAILABLE
360 PR = K4 * R * L * Q
365 Y = K1 * PR
370 S = K5 * Y
375 C = Y - S
380 L = C / PL
385 I = K7 * M
390 DM = S - I
395 DQ = K3 * I * PR - K6 * R * L * Q - K2 * Q
440 Q = Q + DQ * DT
445 M = M + DM * DT
450 IF Q < .00001 THEN Q = .00001
510 T = T + DT
520 IF T * T0 < 319 GOTO 300
530 END
```

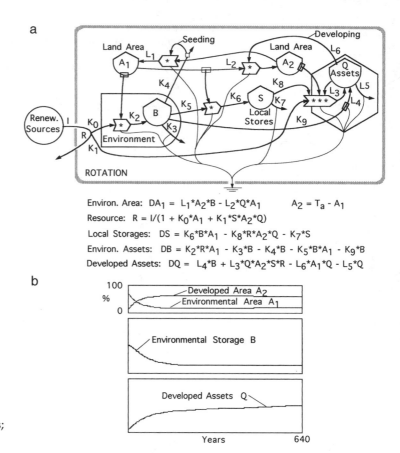

Environ. Area: $DA_1 = L_1*A_2*B - L_2*Q*A_1$ $A_2 = T_a - A_1$

Resource: $R = I/(1 + K_0*A_1 + K_1*S*A_2*Q)$

Local Storages: $DS = K_6*B*A_1 - K_8*R*A_2*Q - K_7*S$

Environ. Assets: $DB = K_2*R*A_1 - K_3*B - K_4*B - K_5*B*A_1 - K_9*B$

Developed Assets: $DQ = L_4*B + L_3*Q*A_2*S*R - L_6*A_1*Q - L_5*Q$

FIGURE 18.4.
Model of land allocation in development (program ROTATION.bas). (a) Energy systems diagram and equations; (b) simulation.

EXPLANATION OF EQUATIONS

Production of environmental biomass B is proportional to the available renewable energy resources R (such as sun, wind, rain) and the undeveloped area A_1 (K_2*R*A_1). Economic production of human assets is a product of the local storages S, the developed area A_2, the renewable energy R, and the assets Q ($L_3*R*S*A_2*Q$). When human assets increase, they cause more environmental land area A_1 to go into developed use (pathway L_2*A_1*Q). Land area is returned to ecological succession with the help of the seeding K_4*B from environmental biomass structure B interacting (L_1*A_2*B) with developed land area A_2. The small rectangles on seedflow pathway K_4 and on land development flow L_2 indicate influences derived from these flows, forces that go to other interactions.

TABLE 18.4
Program for Land Allocation in Development
(ROTATION.bas) (Figure 18.4)

```
20 REM PC: ROTATION (Land rotation)
30 CLS
40 SCREEN 1, 0: COLOR 0, 0
45 LINE (0, 0)-(319, 50), 3, B
50 LINE (0, 60)-(319, 110), 3, B
55 LINE (0, 120)-(319, 190), 3, B
60 DT = 1
70 A0 = .5
80 T0 = .5
90 S0 = .5
100 Q0 = 3
110 B0 = .3
120 I = 100
130 TA = 80
140 A1 = 10
150 B = 100
160 S = 2
170 Q = 10
180 K0 = .18
190 K1 = .00008
200 K2 = .0032
210 K3 = .004
220 K4 = .002
230 K5 = .0002
240 K6 = .0002
250 K7 = .004
260 K8 = 3.2E-07
270 K9 = .002
280 L1 = .0002
290 L2 = .0002
300 L3 = .0000002
310 L4 = .001
320 L5 = .001
330 L6 = .00004
340  PSET (T * T0, 110 - B * B0), 1
350  PSET (T * T0, 50 - A2 * A0), 2
360  PSET (T * T0, 50 - A1 * A0), 2
370  PSET (T * T0, 190 - Q * Q0), 3
380 A2 = TA - A1
390 R = I / (1 + K0 * R * A1 + K1 * S * Q * A2)
400 DB = K2 * R * A1 - K3 * B - K4 * B - K5 * B * A1 - K9 * B
410 DS = K6 * B * A1 - K7 * S - K8 * S * R * A2 * Q
420 DA = L1 * A2 * B - L2 * Q * A1
430 DQ = L3 * Q * A2 * S * R + L4 * B - L5 * Q - L6 * A1 * Q
440 A1 = A1 + DA * DT
445 IF A1 < 0 THEN A1 = 0
450 B = B + DB * DT
460 S = S + DS * DT
470 Q = Q + DQ * DT
480 T = T + DT
490  IF T * T0 < 319 GOTO 340
```

EXAMPLES OF LAND ROTATION SYSTEMS

This model illustrates what happens when a new area is colonized without much outside imports. It is appropriate for early slash-and-burn agriculture in the rain forest. There, a very small part of the forest was in agriculture, while most was in very slow succession.

The model also represents the whole human civilization in relation to environmental resources. In that example, B is the product of geologic production such as fuels and minerals. We are using these products much faster than the small amount of resting land can produce them.

"What If" Experimental Problems for the Program in Table 18.4

18.4–1 What would happen to the land rotation if so much of the original vegetation were destroyed that there were no seeds for reseeding?

18.4–2 What is the effect of starting with larger storage of environmental products? Set S = 100.

18.4–3 What is the effect of having less total land area? Set TA = 50.

Chapter Nineteen

◆

MODELS OF INTERNATIONAL RELATIONS AND TRADE

Although this chapter has minimodels of relationships between nations, the models may apply to other systems where there are exchanges between units. The model ENTRADE shows how exchanges convert competition to cooperation. The models FREEMARK and EMEXCHNG show the effects of a global economy on a nation. With the model WAR, diversion of energy for defense and war is compared with other international relationships. The model INFOBEN shows coexistence by free sharing of world information. Some of the relationships of competition and cooperation in Chapter 14 are also relevant.

An open question is this: What relationships maximize the whole system's performance? In addition to plots of each of the competing or cooperating units, the total output power of the combined system is plotted by adding the outputs of the production processes (interaction symbols).

19.1 MODEL OF TRADE AND EMPOWER (ENTRADE)

In the model ENTRADE, two populations are competing for the common resource R. The production process K_2*R*A for unit A is slightly more productive K_4*R*B than for B; K_2 is greater than K_4. Other coefficients of A and B are identical. In the simulation (Figure 19.1b), A tends to displace B, as in the EXCLUS model of competitive exclusion in Figure 14.1. After an initial period, trade is turned on between the two, and the populations become a symbiotic system of coexistence. With trade, the total production of both together is greater than with competition. Their total assets are also greater when they are trading.

EXAMPLES OF TRADE

Examples of trade between populations are nations like Japan with the United States, and groups of nations like the European

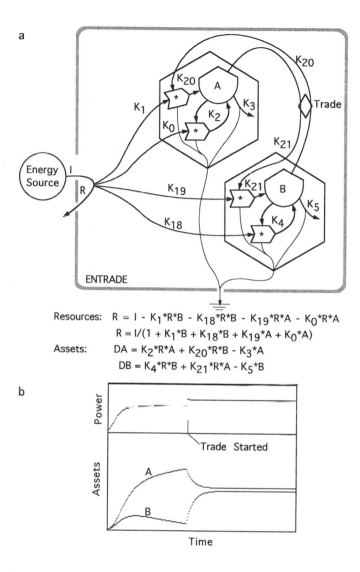

Resources: $R = I - K_1 * R * B - K_{18} * R * B - K_{19} * R * A - K_0 * R * A$
$R = I/(1 + K_1 * B + K_{18} * B + K_{19} * A + K_0 * A)$
Assets: $DA = K_2 * R * A + K_{20} * R * B - K_3 * A$
$DB = K_4 * R * B + K_{21} * R * A - K_5 * B$

FIGURE 19.1.
Model of trade and coexistence
(program ENTRADE.bas).
(a) Energy systems diagram and
equations; (b) simulation.

Union with the Asian rim countries. Animal populations trade
too, like large ocean fish with their smaller cleaner fish—the large
fish have their parasites cleaned off and the smaller fish get food.

EXPLANATIONS OF EQUATIONS

In Figure 19.1 the coefficient for each mathematic pathway
expression is labeled on each line. To find the whole expression
for a line, find the coefficient in its equation below the diagram.

At any time the available resource R is that part of the renewable inflow I remaining after that used by four production processes. Two of the production processes are the autocatalytic production flows without trade ($K_2 *R*A$ and K_4*R*B), and two ($K_{20}*R*B$ and $K_{19}*R*A$) operate using the availability of assets from the other unit when trade is turned on (when Z in the program of Table 19.1 is changed from 0 to 1).

"What If" Experimental Problems for the Program of Table 19.1

19.1–1 If you increase the productivity K2 of population A, what happens to the assets of A and B, and to the total power? Explain.

19.1–2 If you increase K3 (the death rate of animal population A or the pollution depreciation of nation A), what happens to the assets and to power? Explain.

◆

19.2   EFFECTS OF A GLOBAL ECONOMY ON A NATION (FREEMARK AND EMEXCHNG)

The global economy impacts a nation through many kinds of pathways including investments, loans, sales, and purchases. The minimodel FREEMARK shows some of these exchanges in Figure 19.2a. Whether the money economy is stimulated or drained depends on the prices of exports and imports and on interest rates. The model was normalized by calibrating the inflow, the environmental storage, and the assets storage with the value 100, a much smaller number than the actual values. The flows were then determined from an estimate of turnover time. Figure 19.2b is a typical result of simulating this program without exchange, and Figure 19.2c is a typical result when prices are favorable for stimulating growth of assets A.

Perhaps more important is the exchange of real wealth as measured by the balance of emergy. The universal value concept, emergy (spelled with an "m"), was explained and simulated in Chapter 11 as a way to measure real wealth on a common basis. The standard of living goes down if economic exchange carries out more real wealth than it brings in. Emergy calculations are necessary to determine real wealth. A longer version, program

TABLE 19.1
Program for Simulating the Symbiotic Effects of Trade
(ENTRADE.bas) (Figure 19.1)

```
20 REM PC
40 REM ENTRADE (Symbiotic effects of trade)
43 CLS
45 SCREEN 1, 0
50 COLOR 0, 0
60 REM Trade is on when Z=1
70 LINE (0, 0)-(240, 180), 3, B
80 LINE (0, 60)-(240, 60), 3
90 A0 = .01
100 B0 = .01
110 S0 = .025
120 P0 = .0025
130 DT = 1
160 Y = 1
170 I = 1
180 A = .01
190 B = .01
200 K0 = 4
210 K1 = 4
220 K2 = .52
230 K3 = .1
240 K4 = .48
250 K5 = .1
390 K18 = 4
400 K19 = 4
410 K20 = .5
420 K21 = .5
470 PSET (T, 180 - A / A0), 1
480 PSET (T, 60 - P / P0), 1
500 IF A > 179 THEN A = 179
510 PSET (T, 180 - B / B0), 3
520 R = I / (1 + K0 * A + K1 * B + Z * K18 * B + Z * K19 * A)
530  IF R < 0 THEN R = 0
540 DA = (K2 + .1 * S) * R * A - K3 * A + Z * K20 * R * B
550 DB = (K4 + .1 * S) * R * B - K5 * B + Z * K21 * R * A
590 A = A + DA * DT
600 B = B + DB * DT
700 P = (K2 + .1 * S) * R * A + (K4 + .1 * S) * R * B + Z * K20 * R * B + Z * K21
       * R * A
710 T = T + DT
715 IF T > 100 THEN Z = 1
720  IF T < 240 GOTO 470
```

EMEXCHNG.bas, has emergy equations added. These diagrams and equations are shown in Figure 19.2d with a typical simulation of the emergy storages in Figure 19.2e.

For models like this one with normalized calibration, the emergy calculation can be made by including the normalization factors (the ratios of the real energy value to 100). These factors (N_{fJ}, N_{fQ}, and N_{fA}) are included in emergy equations in Figure 19.2d.

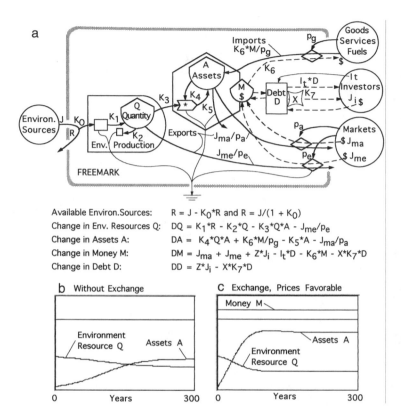

FIGURE 19.2.
Model of effect of global exchange on the economy of a nation. (a) Energy systems diagram and equations for program FREEMARK.bas; (b) simulation without exchange; (c) exchange with favorable prices; (d) emergy diagram and equations for program EMEXCHNG.bas; (e) simulation with favorable prices but net loss of emergy (real wealth).

Available Environ.Sources: $R = J - K_0{*}R$ and $R = J/(1 + K_0)$
Change in Env. Resources Q: $DQ = K_1{*}R - K_2{*}Q - K_3{*}Q{*}A - J_{me}/p_e$
Change in Assets A: $DA = K_4{*}Q{*}A + K_6{*}M/p_g - K_5{*}A - J_{ma}/p_a$
Change in Money M: $DM = J_{ma} + J_{me} + Z{*}J_i - I_t{*}D - K_6{*}M - X{*}K_7{*}D$
Change in Debt D: $DD = Z{*}J_i - X{*}K_7{*}D$

EXPLANATIONS OF EQUATIONS

The stock of environmental resource Q (Figure 19.2a) is a balance between the production $(+)$ and three uses $(-)$. The consumption within the environmental unit is $K_2{*}Q$, the use by the human development is $K_3{*}Q{*}A$, and the export is J_{me}/p_e. The developed assets storage A is a balance between economic production $K_4{*}Q{*}A$ plus imported goods–services–fuels $K_6{*}$ M/p_g minus local depreciation–consumption $K_5{*}A$ and export of assets J_{ma}/p_a. The propensity of outside markets to purchase (J_{me} and J_{ma}) are a major influence on the nation's economy.

The available international money M is a balance between the money from purchases by the outside economy (J_m and J_a) and the payout for imports $K_6{*}M$. Prices on international markets are p_e for environmental products, p_a for assets, and p_g for goods–services–fuels. If there is investment from outside J_i, money flows into storage of international dollars M, and accompanying debt D also results. Interest outflow is $I_t{*}D$ where I_t is the interest

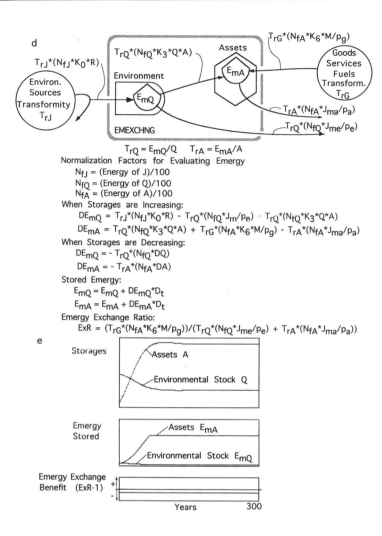

$$T_{rQ} = E_{mQ}/Q \qquad T_{rA} = E_{mA}/A$$

Normalization Factors for Evaluating Emergy

$$N_{fJ} = (\text{Energy of } J)/100$$
$$N_{fQ} = (\text{Energy of } Q)/100$$
$$N_{fA} = (\text{Energy of } A)/100$$

When Storages are Increasing:
$$DE_{mQ} = T_{rJ}*(N_{fJ}*K_0*R) - T_{rQ}*(N_{fQ}*J_m/p_e) - T_{rQ}*(N_{fQ}*K_3*Q*A)$$
$$DE_{mA} = T_{rQ}*(N_{fQ}*K_3*Q*A) + T_{rG}*(N_{fA}*K_6*M/p_g) - T_{rA}*(N_{fA}*J_{ma}/p_a)$$

When Storages are Decreasing:
$$DE_{mQ} = - T_{rQ}*(N_{fQ}*DQ)$$
$$DE_{mA} = - T_{rA}*(N_{fA}*DA)$$

Stored Emergy:
$$E_{mQ} = E_{mQ} + DE_{mQ}*D_t$$
$$E_{mA} = E_{mA} + DE_{mA}*D_t$$

Emergy Exchange Ratio:
$$ExR = (T_{rG}*(N_{fA}*K_6*M/p_g))/(T_{rQ}*(N_{fQ}*J_{me}/p_e) + T_{rA}*(N_{fA}*J_{ma}/p_a))$$

FIGURE 19.2.
(*Continued*)

rate. Repayment outflow $X*K_7*M$ is turned on when X (in the switch symbol) is changed from 0 to 1.

By aggregating the first diagram (Figure 19.2a), a simple overview of the emergy contributions is obtained (Figure 19.2d). There are only two inflows of emergy, and these are calculated as the product of outside transformities T_{rJ} and T_{rG} and energy inflows (corrected with the normalization factor). Emergy flows out in sales of environmental products $T_{rQ}*(N_{fQ}*J_{me}/p_e)$ such as wood, minerals, and agricultural products, and in export sale of assets $T_{rA}*(N_{fA}*J_{ma}/p_a)$. Emergy at any time is the balance of inflows and exports represented in simulations by the emergy exchange ratio ExR = emergy imported/emergy exported [plotted in Figure 19.2e as exchange ratio minus 1 so as to graph gain (+) upward and loss (−) downward].

"What If" Experimental Problems for the Programs in Table 19.2

19.2–1 First run the program FREEMARK.bas (Table 19.2a) without any outside exchanges by setting JME, JMA, and K6 = 0. The results should be the graph in Figure 19.2b. Then turn on investment by setting INV = 1 and K6 = 1. Note that the program (lines 510–560) allows investments for 50 years before starting the payback. Interest is paid as long as there is debt. What happens to the debt D plotted in the upper panel? Is the early growth of assets A changed? Turn off the investment sequence (set INV = 0 in line 370) before doing the next problem.

19.2–2 Next run the program with environmental stock sales and purchases of outside goods–services–fuels by setting JME = 0.5 and K6 = 1 with favorable prices: PG = 0.2 and PE = 10 (JMA = 0). The result should be the graph in Figure 19.2c. With only the internal system running, there is no international money (M = 0). Next, what is the effect of lower price (PE = 1) for sale of environmental products Q? Next, what is the effect of a large increase in price of purchased goods–services–fuels (set PG = 2)? Why? How is this scenario like some undeveloped countries? What is the effect of borrowing with these price conditions (run with INV = 1 and JI = 0.5) compared with the result in Problem 20.3–1?

19.2–3 Use the program EMEXCHNG.bas (Table 19.2b) to simulate emergy and the ratio of real wealth received from the global economy to that exported. Run the program with favorable prices (PG = 0.2; PE = 10; JI = .5; K6 = 1). The benefit or loss from the foreign exchanges is plotted in the lower panel in Figure 19.2e (emergy exchange ratio defined in Figure 19.2d). Is there a net benefit or loss in real wealth (exchange ratio ExR greater than 1)? Can you explain how developing countries can accelerate their growth with trade and loans but experience net loss of real wealth?

Next increase the price of purchases (PG = 2) or decrease prices of sales (PE or PA) and run again. Explain the effect of world prices on the emergy measures of real wealth and the ratio of inflow to outflow. How much increase in the sale price and/ or decrease in the price of fuels–goods–services is required to

TABLE 19.2a

Program FREEMARK.bas Simulating Global Impacts on a Nation (Figure 19.2a)

```
10 REM PC: FREEMARK.BAS (Global Impacts on a Nation's Economy)
20 CLS
25 SCREEN 1, 0: COLOR 7, 0
30 LINE (0, 0)-(300, 190), , B
70 LINE (0, 50)-(300, 50)
71 LINE (0, 50)-(300, 50)
100 REM Constant coefficients
115 K0 = 9
120 K1 = .5
130 K2 = .035
140 K3 = .00015
150 K4 = .0005
160 K5 = .05
170 K6 = 1: REM 1
180 K7 = .1: REM .01
200 REM Sources
205 J = 100: REM Renewable source flow
210 JME = .1: REM 0.5 Purchases of env. prod. by outside market
215 JMA = 0: REM .1 Assets purchases by outside markets
220 JI = .5: REM  .5 Investments
230 IT = .05: REM Interest rate
240 PE = 10: REM 10  Price of env. exports
250 PG = .2: REM .2  Price of imports
255 PA = 10: REM  Price of assets exports
300 REM Starting Values
310 Q = 142
320 A = 10:
330 M = 1
340 D = 0:
350 X = 0: REM Switch to 1 to payback
360 Z = 0: REM Switch to 1 to start investment Ji
370 INV = 0: REM Set Inv to 1 to include investment, debt, and payback
400 REM Scaling
410 DT = .5
420 T0 = 1
430 Q0 = 2
440 A0 = 1.5
450 M0 = .01
460 D0 = .5
500 REM Equations
510 IF INV = 0 GOTO 600: REM If invest = 1 loan & payback sequence
520 Z = 1: REM Z = 1 starts investment and debt
530 X = 0: REM X = 1 starts payback
540 IF T > 50 THEN Z = 0
550 IF T > 50 THEN X = 1
560 IF T > 50 THEN INV = 0
600 R = J / (1 + K0)
610 DQ = K1 * R - K2 * Q - K3 * Q * A - JME / PE
620 DA = K4 * Q * A + K6 * M / PG - K5 * A - JMA / PA
630 DM = JME + JMA + Z * JI - K6 * M - IT * D - X * K7 * D
```

(continues)

TABLE 19.2a (Continued)

```
640 DD = Z * JI - X * K7 * D
800 REM Change Equations
810 A = A + DA * DT
820  IF A < .1 THEN A = .1
830 Q = Q + DQ * DT
840  IF Q < .1 THEN Q = .1
855 M = M + DM * DT
860  IF M < .001 THEN M = .001
870  D = D + DD * DT
880  IF D < 0 THEN D = 0
945 T = T + DT
950 REM Plotting
952  PSET (T / T0, 50 - M / M0)
953  PSET (T / T0, 50 - D / D0)
955  PSET (T / T0, 190 - A / A0)
960  PSET (T / T0, 190 - Q / Q0)
980  REM Iteration
9900  IF T / T0 < 300 GOTO 500
```

obtain a net benefit in real wealth as measured by the emergy exchange ratio ExR?

19.2–4 In both programs, consider the effect of selling assets (finished products) after setting the inflow of outside purchasing money JMA = 0.2. Then increase prices for sale of assets PA larger than the price of asset purchases PG. What is the effect of the price changes on growth of assets and on emergy exchange?

◆

19.3 MODEL OF DEFENSE AND WAR (WAR)

Whereas Chapter 14 presented models with simple patterns of competition between two populations, the model WAR has defense and war, features of competing countries. In Figure 19.3 two systems A and B are drawing from the same limited environmental resources. Each competitor builds a defense storage. When at war, the two competitors have three pathways of negative interaction. Each system has a pathway controlling destruction of the assets of the other. Then there is the pathway where each unit interacts to decrease the defense of the other.

In the simulation in Figure 19.3b, unit A prevails at the end of the war period, but the total useful productivity of the whole system P for the time is not changed.

TABLE 19.2b
**Program EMEXCHNG.bas Simulating Emergy and Economic
Effects of a Global Economy on a Nation (Figure 19.2)**

```
10 REM PC: EMEXCHNG.BAS (Global Impacts on a Nation's Economy)
20 CLS
25 SCREEN 12, 0: REM  COLOR 7, 0
30 LINE (0, 0)-(300, 150), 3, B
40 LINE (0, 175)-(300, 275), 3, B
50 LINE (0, 300)-(300, 350), 3, B
60 LINE (0, 325)-(305, 325), 3
100 REM Constant coefficients
115 K0 = 9
120 K1 = .5
130 K2 = .035
140 K3 = .00015
150 K4 = .0005
160 K5 = .05
170 K6 = 1: REM 1
180 K7 = .1: REM .01
200 REM Sources
205 J = 100: REM Renewable source flow
210 JME = .5: REM 0.5 Env. purchasing from outside
215 JMA = 0: REM 0.1 Assets purchasing from outside
220 JI = .5: REM  .5 Investments
230 IT = .05: REM Interest rate
240 PE = 10: REM 10  Price of exports
250 PG = .2: REM .2 Price of imports
255 PA = 10: REM 10 Price of assets exported
260 REM Emergy and Transformities
265 NfQ = 10000: REM 10000:REM Normalization factor Q = 1E6/100 = 1E4
270 NfA = 500: REM 500:REM Normalization factor A = 50000/100 = 500
275 NFJ = 20000!: REM 20000:REM Normalization factor J = 2E6/100 = 2E4
280 TRJ = 20000: REM Transformity Environmental source flow
285 TRG = 100000!: REM Transformity Flow of Goods, Services, Fuels
288 TRQ = 1000000!: REM Transformity of Environmental Stock
290 TrA = 2E+07: REM Transformity of Assets Stock
300 REM Starting Values
310 Q = 142
320 A = 10:
330 M = 1
340 D = 0
345 REM Emergy including Energy Scaling Correction
347 EMQ = TRQ * (NfQ * Q)
349 EmA = TrA * (NfA * A)
350 X = 0: REM Switch to 1 to payback
360 Z = 0: REM Switch to 1 to start investment Ji
370 INV = 0: REM set Inv to 1 to include investment, debt, and payback
400 REM Scaling
410 DT = .5
420 T0 = 1
430 Q0 = 2
440 A0 = 1.5
```

(continues)

TABLE 19.2b (*Continued*)

```
450 M0 = .05
460 D0 = 1
470 EXR0 = .1
480 EMQ0 = .2 * EMQ
490 EMA0 = .2 * EmA
500 REM Emergy Equations
510 IF INV = 0 GOTO 600: REM If Inv = 1, loan sequence starts
520 Z = 1: REM Starts investment Ji
530 X = 0: REM Starts payback
540 IF T > 50 THEN Z = 0
550 IF T > 50 THEN X = 1
560 IF T > 50 THEN INV = 0: REM ends loan sequence
600 R = J / (1 + K0)
610 DQ = K1 * R - K2 * Q - K3 * Q * A - JME / PE
620 DA = K4 * Q * A + K6 * M / PG - K5 * A - JMA / PA
630 DM = JME + JMA + Z * JI - K6 * M - IT * D - X * K7 * D
640 DD = Z * JI - X * K7 * D
700 REM Emergy evaluation
705 TRQ = EMQ / (Q * NfQ)
710 TrA = EmA / (A * NfA)
715 IF DQ > 0 THEN DEMQ = TRJ * (NFJ * K0 * R) - TRQ * (NfQ * JME / PE) - TRQ
      * (NfQ * K3 * Q * A)
720 IF DA > 0 THEN DEMA = TRQ * (NfQ * K3 * Q * A) + TRG * (NfA * K6 * M / PG)
      - TrA * (NfA * JMA / PA)
725 IF DQ = 0 THEN DEMQ = 0
730 IF DA = 0 THEN DEMA = 0
735  IF DQ / Q < .005 * DT THEN DEMQ = 0
740  IF DA / A < .005 * DT THEN DEMA = 0
745 IF DQ < 0 THEN DEMQ = TRQ * (DQ * NfQ)
750 IF DA < 0 THEN DEMA = TrA * (DA * NfA)
800 REM Change Equations
810 A = A + DA * DT
820  IF A < .1 THEN A = .1
830 Q = Q + DQ * DT
840  IF Q < .1 THEN Q = .1
850 M = M + DM * DT
860  IF M < .001 THEN M = .001
870  D = D + DD * DT
880  IF D < 0 THEN D = 0
900 REM Emergy Change Equations
910 EMQ = EMQ + DEMQ * DT
915 IF EMQ < .00001 THEN EMQ = .00001
920 EmA = EmA + DEMA * DT
930 IF EmA < .00001 THEN EmA = .00001
940  EXR = (TRG * (NfA * K6 * M / PG)) / (TRQ * (NfQ * JME / PE) + TrA * (NfA
      * JMA / PA))
945 T = T + DT
950  REM Plotting
955  PSET (T / T0, 150 - A / A0), 3
960  PSET (T / T0, 150 - Q / Q0), 3
965  PSET (T / T0, 275 - EMQ / EMQ0), 3
970  PSET (T / T0, 275 - EmA / EMA0), 3
975  PSET (T / T0, 325 - (EXR - 1!) / EXR0), 3
980  REM Iteration
990 IF T / T0 < 300 GOTO 500
```

EXAMPLES OF SYSTEMS AT WAR

Examples are two pre-industrial nations at war, since they are not importing anything from outside. In modern times a country's assets could include fuels from within the country. This model

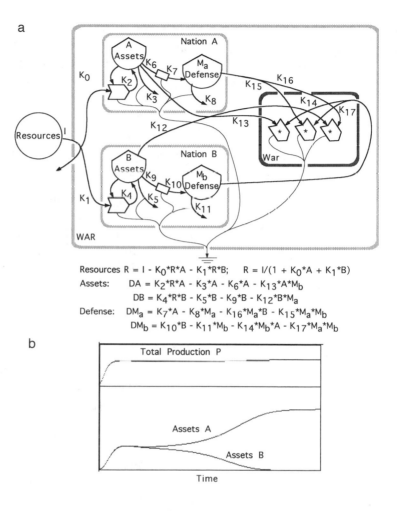

Resources $R = I - K_0*R*A - K_1*R*B$; $R = I/(1 + K_0*A + K_1*B)$

Assets: $DA = K_2*R*A - K_3*A - K_6*A - K_{13}*A*M_b$

$DB = K_4*R*B - K_5*B - K_9*B - K_{12}*B*M_a$

Defense: $DM_a = K_7*A - K_8*M_a - K_{16}*M_a*B - K_{15}*M_a*M_b$

$DM_b = K_{10}*B - K_{11}*M_b - K_{14}*M_b*A - K_{17}*M_a*M_b$

FIGURE 19.3.
Model of defense and war in international relations (program WAR.bas). (a) Energy systems diagram and equations; (b) simulation.

could also represent two groups of animals fighting over territory, like elephants, or fish in your fish tank. Animals have some of their resources put into defense mechanisms held in readiness.

EXPLANATION OF EQUATIONS

At any time, the resource R is the unused remainder of the renewable inflow I after part is used by the competing production processes K_0*R*A and K_1*R*B. The assets of both systems are a balance of positive autocatalytic production flows K_2*R*A and K_4*R*B minus depreciation flows K_3*A and K_5*B, minus assets used to support defense K_6*A and K_9*B and minus destruction from war $K_{13}*A*M_b$ and $K_{12}*B*M_a$. The storages for defense are a balance between the allocations from assets K_7*A and $K_{10}*B$ minus the depreciations K_8*M_a and $K_{11}*M_b$, and minus the destructive drains by war $K_{16}*M_a*B$, $K_{15}*M_a*M_b$, $K_{14}*M_b*A$ and $K_{17}*M_a*M_b$. The coefficients K are the same for both units except that the allocation for defense in unit A (K_6 and K_7) is slightly greater than for unit B (K_9 and K_{10}).

"What If" Experimental Problems for the Program in Table 19.3

19.3–1 Adjust the coefficients so that neither nation is destroyed. Explain what you did and why it did or did not work.

19.3–2 When you increase the environmental resources, how does this affect each nation and the timing of the destruction of B? Explain.

19.3–3 Change one coefficient so that B wins the war. Explain in terms of two warring countries.

◆

19.4   MODEL OF GLOBALLY SHARED INFORMATION (INFOBEN)

As the earth has developed new information and new means of information processing, the possibility of a new pattern of peace among nations is suggested by a model of global information sharing. As diagrammed in Figure 19.4, two nations are linked by shared information, which is mutually beneficial to production of both countries. Simulation of the program in Table 19.4 shows how countries share information, maximize power, and develop coexistence (Figure 19.4b). Shared information S contributes to production of both populations, making them mutually beneficial.

It may be an assumption, but information moves easily throughout the world with television and the Internet, not much constrained by the many efforts to restrict information access with economic profit making. Information has high transformities. High-transformity products are rapidly transmitted and flexible in use. It will be a new era if the self-organization of information finally displaces the old global system of power politics and territoriality.

TABLE 19.3
**Program for Simulating Competition with Negative
Interactions (WAR.bas) (Figure 19.3)**

```
10 REM PC: WAR (Competition with negative interactions)
30 CLS
40 SCREEN 1,0: COLOR 0,1
70 LINE (0,0)-(319,180),3,B
80 LINE (0,60)-(319,60),3
90 A0 = .01
100 B0 = .01
110 K13 = .01
120 P0 = .0025
130 DT = 1
140 I = 1
150 A= .01
160 B =.01
162 MA = 0
163 MB = 0
170 K0 = 4
180 K1 = 4
190 K2 = .5
200 K3 = .1
210 K4 = .5
220 K5 = .1
230 K6 = .011
240 K7 = .055
250 K8 = .05
260 K9 - .01
270 K10 = .05
280 K11 = .05
290 K12 = .1
300 K13 = .1
310 K14 = .01
320 K15 = .01
330 K15 = .01
340 K16 = .01
350 K17 = .01
360 REM EQUATIONS
370 PSET (T,180-A/A0),1
380 PSET (T,60-P/P0),1
390 IF A >179 THEN A=179
400 PSET (T,180-B/B0),3
410 R = I/(1+K0*A+K1*B)
420  IF R < 0 THEN R = 0
430 DA = K2*R*A - K3*A -K6*A - K13*A*MB
440 DB = K4 * R * B - K5 * B - K9*B - K12*B*MA
450 DMA= K7*A - K8*MA - K16*MA*B - K15*MA*MB
460 DMB = K10*B - K11*MB -K17 *MB*MA -K14*A*MB
470 A = A + DA*DT
480 B = B + DB*DT
490 MA = MA + DMA*DT
500 MB = MB + DMB*DT
510 P = K2*R*A +K4*R*B
520 T = T + DT
530  IF T < 320 GOTO 360
```

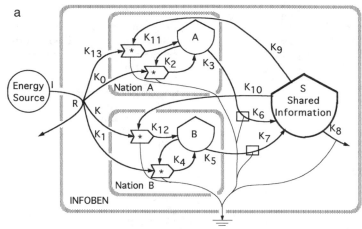

Resources Available: $R = I - K_0*R*A - K_1*R*B - K_{11}*R*S - K*R*S$
                       $R = I/(1 + K_0*A + K_1*B + K_{11}*S + K*S)$

Assets, A & B: $DA = K_4*R*S + K_2*R*A - K_3*A$
                       $DB = K_4*R*B + K_{12}*R*S - K_5*B$

Shared Information S: $DS = K_6*A + K_7*B - K_9*R*S - K_{10}*R*S - K_8*S$

Total Power P: $P = K_2*R*A + K_{11}*R*S + K_4*R*B + K_{12}*R*S$

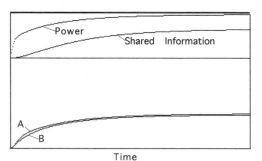

FIGURE 19.4.
Model of benefits of globally shared information (program INFOBEN.bas). (a) Energy systems diagram and equations; (b) simulation.

EXAMPLES OF SYSTEMS SHARING INFORMATION

Medical information shared around the world has decreased deaths and increased productivity. As information about birth control spreads, productivity could increase even more. Political information about what is happening in one country, via TV, can affect activities in another country, encouraging protests or elections toward political systems that are internationally cooperative. Animal populations share information that allows populations to operate in herds, schools, and rookeries. Interspecies

TABLE 19.4
Program for Shared Information (INFOBEN.bas)
(Figure 19.4)

```
20 REM PC: INFOBEN (Shared information)
30 CLS
40 SCREEN 1, 0: COLOR 0, 1
50 REM Type CONT for phase plane.
70 LINE (0, 0)-(319, 180), 3, B
80 LINE (0, 60)-(319, 60), 3
90 A0 = .02
100 B0 = .02
110 P0 = .0033
120 S0 = .03
130 T0 = 2
140 DT = 2
150 I = 1
170 A = .01
180 B = .01
190 K = 1
200 K0 = 8
210 K1 = 8
220 K2 = .52
230 K3 = .1
240 K4 = .48
250 K5 = .1
260 K6 = .01
270 K7 = .01
280 K8 = .01
290 K9 = .05
300 K10 = .05
310 K11 = 1
315 K12 = 1
318 K13 = 1
320 IF X = 1 THEN PSET (2 * A / A0, 180 - 2 * B / B0): GOTO 380
330 PSET (T / T0, 180 - A / A0), 2
340 PSET (T / T0, 60 - P / P0), 3
350 PSET (T / T0, 60 - S / S0), 1
360 IF A > 179 THEN A = 179
370 PSET (T / T0, 180 - B / B0), 1
380 R = I / (1 + K0 * A + K13 * S + K1 * B + K * S)
400  IF R < 0 THEN R = 0
410 DA = K2 * R * A - K3 * A + K11 * R * S
420 DB = K4 * R * B - K5 * B + K12 * R * S
430 DS = K6 * A + K7 * B - K8 * S - K9 * S * R - K10 * S * R
440 A = A + DA * DT
450 B = B + DB * DT
460 S = S + DS * DT
470 P = K2 * R * A + K4 * R * B + K * R * S + K11 * R * S
480 T = T + DT
490  IF T / T0 < 320 GOTO 320
```

information is shared for mutual benefit, such as when squirrels make warning chatter at the sight of a snake.

EXPLANATIONS OF EQUATIONS

Shared information S is increased by linear pathways from A and B minus losses K_8*S and that used in feedback loops K_9*R*S and $K_{10}*R*S$. The maintenance of shared information requires continuous duplication and distribution like that which occurs in child rearing, education, and television.

With each iteration, the change of assets in each nation (A and B) is the sum of the autocatalytic production K_2*R*A and K_4*R*B plus the production increased by the information $K_{11}*R*S$ and $K_{12}*R*S$ minus outflows K_3*A and K_5*B. In this model these linear outflow pathways generate shared information and also include the depreciation losses from A and B.

"What If" Experimental Problems for the Program in Table 19.4

19.4–1 If the energy source I is decreased, what happens to the total productivity? Explain with an example.

19.4–2 What happens if the shared information, S, is zero? Explain. Add a line 465   S = 0.

19.4–3 Suppose one nation does not contribute to shared information. Set K6 = 0. What happens?

Chapter Twenty

◆

MODELS OF THE GLOBAL GEOBIOSPHERE

M odels assembled in this chapter present overview perspectives for the whole earth. The model WORLDCO2 simulates carbon dioxide resulting from global metabolism fuel uses and oceanic buffering. The model PEOPLE, for world population, includes energy resources for public health and epidemic disease. The model EARTHGEO overviews the geological processes in the development of the earth. The STATECON minimodel presents a state economy driven by a world minimodel of availability of inputs. The macroeconomic models in Chapter 18 and the growth minimodels in Chapter 13 can also be used for global perspectives.

20.1 MODEL OF WORLD CARBON DIOXIDE (WORLDCO2)

As the consumption of nonrenewable fuels has increased in this century, the global carbon dioxide concentration of the atmosphere has increased, affecting climate. As shown by the pathways in and out of the CO_2 storage in Figure 20.1, the carbon dioxide is increased by consumers on land and in the sea and by burning of fossil fuels. Carbon dioxide is decreased by land and marine photosynthesis and the buffering reaction with the carbonates in the sea, deserts, and elsewhere. For the long time periods of this model (years) the separate, rapid, chemical reactions of the buffering chemical reactions can be aggregated to those appropriate.

As illustrated by the simulation in Figure 20.1b, the model binds the excess carbon dioxide into the carbonate buffer system until the global system comes into better balance in later years as fuel use decreases. *Ppmv* means parts per million by volume. In this model balance between global production and use of carbon dioxide can be achieved by reforestation. Reforestation is represented by the coefficient of terrestrial production K_1.

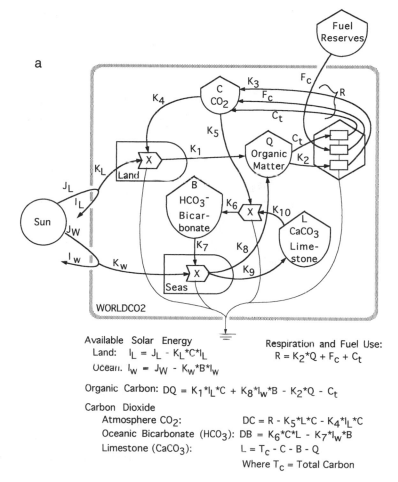

a

Available Solar Energy
Land: $I_L = J_L - K_L*C*I_L$
Ocean: $I_w = J_W - K_w*B*I_w$

Respiration and Fuel Use:
$R = K_2*Q + F_c + C_t$

Organic Carbon: $DQ = K_1*I_L*C + K_8*I_w*B - K_2*Q - C_t$

Carbon Dioxide
  Atmosphere CO_2: $DC = R - K_5*L*C - K_4*I_L*C$
  Oceanic Bicarbonate (HCO_3): $DB = K_6*C*L - K_7*I_w*B$
  Limestone ($CaCO_3$): $L = T_c - C - B - Q$

Where T_c = Total Carbon

FIGURE 20.1.
Model of world carbon dioxide
(program WORLDCO2.bas).
(a) Energy systems diagram and
equations; (b) simulation.

EXPLANATION OF EQUATIONS

The available sunlight is the unused remainder I_l and I_w after solar energy use by the production processes on land K_L*I_L*C and in the sea K_w*I_w*B. The photosynthetic production (land and seas production symbols) goes into the storage of organic carbon Q, where its change in a unit time DQ also depends on two outflows: respiration K_2*Q and the cutting and burning of forest C_t.

Change in atmospheric carbon dioxide DC is a balance between positive inputs from respiration K_3*Q, fuel consumption F_c, and cutting and fires C_t and two negative outflows, to land

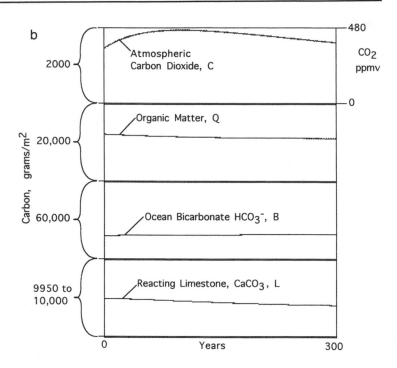

FIGURE 20.1.
(Continued).

photosynthesis $K_4 * I_L * C$ and oceanic photosynthesis $K_5 * L * C$. As listed in Table 20.1, the rate of carbon dioxide contributed by fuel combustion F_c was calculated as the starting value F_{cs} decreased by one percent each year $(0.00035 * YR)$. Decline in global fuel consumption is anticipated as fuels become less abundant and more expensive.

Bicarbonate in oceans is a balance between that produced when carbon dioxide is buffered with limestone $K_6 * C * L$ and that consumed directly and indirectly by marine photosynthesis $K_7 * I_w * B$.

The reacting limestone L is that in chemical exchange with these processes in the biogeosphere. The change in limestone ($CaCO_3$) DL was increased by photosynthetic-driven elevation of pH increasing skeletal deposition, for example, coral reefs and coccoliths $K_9 * I_w * B$. Limestone is decreased by the acidification of carbon dioxide diffusing into the sea $K_{10} * L * C$.

The model was given stability by providing a conservation of carbon. Total organic and inorganic carbon T_c was given an initial value that was added to by the fossil fuel combustion F_c on each iteration $T_c = T_c + F_c * dt$. The organic matter was

TABLE 20.1
Program for World Carbon-Dioxide Changes (WORLDCO2.bas) (Figure 20.1)

```
10 CLS
30 REM PC: WORLDCO2.bas (Global CO2, fuel consumption, and ocean buffering)
40 SCREEN 12, 0
45 REM COLOR 0,0
50 LINE (0, 0)-(300, 400), 3, B
60 LINE (0, 100)-(300, 100)
70 LINE (0, 200)-(300, 200)
80 LINE (0, 300)-(300, 300)
130 REM Scaling factors
140 DT = 20
150 YR = 100
160 T0 = 365 * YR / 300
180 CO = 10
190 B0 = 1000
195 CG = 1000
200 L0 = 1
210 LG = 9950
220 Q0 = 200
230 REM Starting conditions
240 Q = 12300
250 L = 10000
260 B = 31475
270 C = 1455
280 TC = Q + B + C + L
290 REM Forcing Sources
300 J = 100
310 FCS = .035: REM Fuel consumption
320 CT = 0: REM Tropical forest cutting
330 REM Coefficients
340 K1 = .000048: REM Forest production rate
350 K2 = .000035
360 K3 = K2
370 K4 = K1
380 K5 = 1.237E-08
390 K6 = 1.237E-08
400 K7 = .0000019
410 K8 = .0000019
420 K9 = 1.05E-08
430 K10 = .001 / 10000 / 1455
440 KW = .00072
450 KL = .0059
460 REM Equations
465 YR = T / 365
495 J = 100 + 50 * SIN(T * .017)
500 IL = J * .29 / (1 + KL * C)
505 FC = FCS - .00035 * YR
507 IF FC < 0 THEN FC = 0
510 IW = .71 * J / (1 + KW * B)
520 P = K1 * IL * C + K8 * IW * B
530 R = K2 * Q + FC + CT
540 DC = R + FC - K5 * L * C - K4 * IL * C
550 DB = K6 * L * C - K7 * IW * B
560 DL = K9 * IW * B - K10 * L * C
565 DQ = K1 * IL * C + K8 * IW * B - K2 * Q - CT
570 C = C + DC * DT
580 IF C < .0001 THEN C = .0001
590 B = B + DB * DT
600 IF B < .0001 THEN B = .0001
610 L = L + DL * DT
620 TC = TC + FC * DT
630 Q = TC - C - B - L: REM  grams Carbon/m2
650 PSET (T / T0, 400 - (L - LG) / L0), 2
660 PSET (T / T0, 300 - B / B0), 1
670 PSET (T / T0, 100 - ((C - CG) / CO)), 3
680 PSET (T / T0, 200 - Q / Q0), 3
690 T = T + DT
700 YR = T / 365
710 CO2 = C * .2336: REM parts per million
730 IF T / T0 < 300 GOTO 460
```

able R is the remainder of the inflow J not yet in use. (See the RENEW model, Figure 13.1, for more explanation of the limitations of constant, renewable sources.) One production flow, $K_3*R*F*N*A$, is based on available nonrenewable fuels F interacting with population, assets, and renewable resources. Fuels are used up by this production flow $-K_0*R*F*N*A$. Another production term, K_4*R*A, becomes important only when fuel reserves F are small. The quantity of economic assets A is a balance between the two productive flows (positive terms in the equation) and the outflows (negative terms). The outflows include depreciation, K_5*A, the economic assets used to develop populations, $K_6*(A/N)*N$, the economic assets used for regular health and medicine, $L_0*(1 - K_9*A)$, and those used for epidemic disease, $L_0*N*N*(1 - K_9*A)$. The calibration was made with epidemic mortality at 10% of ordinary mortality.

The population N is the balance between births and deaths. The birthrate, $L_1*(A/N)*N$, is directly proportional to assets per individual (A/N) and the population N. Two pathways of mortality are included, regular deaths and deaths from epidemic disease. Regular mortality, $K_7*N*(1 - K_9*A)$, is in proportion to the population N, but diminished in proportion to the economic assets used in health care $1 - K_9*A$. Epidemic mortality is increased in proportion to the square of the population K_8*N*N, but diminished in proportion to those assets used in health care and medicine $1 - K_9*A$. A square of population, $N*N$, is appropriate because epidemics spread when people are crowded, in proportion to population interactions, which is mathematically the square of the number.

The program (Table 20.2) was calibrated with values for 1980. When the program is run, it generates the graph shown in Figure 20.2b. As the nonrenewable fuels are used, economic assets pass through a maximum and start to decrease. Not many years later the population crests and decreases rapidly, a result of declining birthrates and higher mortalities.

The curves in the population simulation (Figure 20.2b) show that the assets per person during growth are higher than those during the decline period. However, later with lower populations, the assets per person improve again. In one sense this model is an optimistic one, implying that a reasonable standard of living is possible in lower energy times, providing population levels decline in proportion with resources available for use.

TABLE 20.2
Program for World Population (PEOPLE.bas) (Figure 20.2)

```
10 REM PC: PEOPLE (World population)
20 CLS
30 SCREEN 1, 0: COLOR 15, 0
50 LINE (0, 0)-(240, 180), 3, B
60 LINE (0, 60)-(240, 60), 3
70 J = 8.560001
80 DT = 1
90 N = .5
100 A = 1
110 R = 12.5
120 F = 1000
130 K0 = .00003
140 K1 = .0000176
150 K2 = .01
160 K3 = .0000176
170 K4 = .01
180 K5 = .05
190 K6 = .0135
200 K7 = .126
210 K8 = .004116
220 K9 = .009
230 L1 = .001054
240 L0 = .0433
250 T0 = 2.8
260 N0 = .08
270 F0 = 16
280 A0 = 1.8
290 X = 1
310 R = J / (1 + K1 * F * N * A + K2 * A)
320 DF = -K0 * R * F * N * A
330 DA = K3 * R * F * N * A + K4 * R * A - K5 * A - K6 ^ A - L0 * N * N * (1 - K9
   * A)
340 B = L1 * (A / N) * N
350 D = K7 * N * (1 - K9 * A) + K8 * N * N * (1 - K9 * A)
360 IF D < 0 THEN D = 0
370 F = F + DF * DT * X
380 N = N + (B - D) * dt
390 IF N < .01 THEN N = .01
400 A = A + DA * DT
410 IF A < 1 THEN A = 1
420 BR = B * 100 / N
430 DR = D * 100 / N
460 PSET (T / T0, 180 - N / N0), 1
480 PSET (T / T0, 180 - A / A0), 3
500 PSET (T / T0, 60 - F / F0), 1
510 IF F < 1 THEN F = 1
520 T = T + DT
560 IF T / T0 < 240 GOTO 310
```

"What If" Experimental Problems for the Program in Table 20.2

20.2–1 If the fossil fuel tank F is set to zero, the model depicts a human population growing on renewable energies alone. How will the graph differ from the original? Why?

20.2–2 Increase K9 to demonstrate the results of increased medical care. How does this change the long-range population?

20.2–3 How would the graph change if, after about 250 years, the birthrate were cut 90%—what would be the changes in the immediate and long-range futures? (Add line 515 IF T = 250 THEN L1= 0.0001.) Also explain the difference in the graph of the fuels.

20.2–4 If an epidemic of a disease like AIDS gets worse, how would this change the population prediction? Increase K8 10 times and run the simulation. Then, to make things worse, also cut the medical care K9 to zero.

◆

20.3 OVERVIEW MODEL OF GEOLOGICAL PROCESSES IN EARTH EVOLUTION (EARTHGEO)

Figure 20.3a represents a model of the main inputs and processes that developed the pattern of oceans and continents of the earth requiring a billion years or more. There are two energy sources, solar energy and the concentrated heat deep in the earth. The earth processes, as they have become organized in cycles over many years, have rapid atmospheric processes helping drive the ocean. These contribute to the hydrologic cycle, which contributes to land formation and earth cycles, coupled to deep heat-driven emissions of materials to the surface. Water emerges from the deep earth in volcanic flows and leaves the earth when decomposed in the high atmosphere, and the light hydrogen molecules escape into space. The pathways of circulating water are shaded darkly on the left, while the flow of land in the sedimentary and igneous–metamorphic cycles is shown on the right with speckled shading.

Figure 20.3b uses the values of flow and storage to calibrate the model. Whereas the processes of atmosphere, ocean, and hydrologic cycle have turnover times of a few thousand years, the turnover time involved in building and sustaining the continents is around a billion years. Running a model with such a wide range of turnover times would be difficult if A and L were both represented as storages. By representing the fast turnover items as a "flow use junction," a running average of available sunlight R and atmosphere-ocean energy A are calculated in each computer iteration (see Section 6.4). The average contributions are provided to the slower large-scale land production process

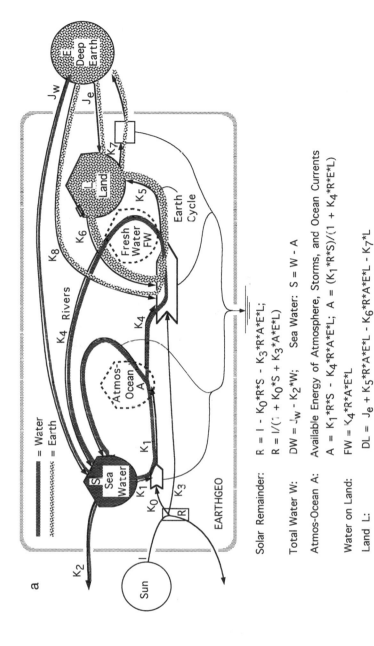

Solar Remainder: $R = I - K_0 * R * S - K_3 * R * A * E * L;$

$R = I / (1 + K_0 * S + K_3 * A * E * L)$

Total Water W: $DW = J_w - K_2 * W;$ Sea Water: $S = W - A$

Atmos-Ocean A: Available Energy of Atmosphere, Storms, and Ocean Currents

$A = K_1 * R * S - K_4 * R * A * E * L;$ $A = (K_1 * R * S) / (1 + K_4 * R * E * L)$

Water on Land: $FW = K_4 * R * A * E * L$

Land L: $DL = J_e + K_5 * R * A * E * L - K_6 * R * A * E * L - K_7 * L$

FIGURE 20.3.

Model of earth development (program EARTHGEO.bas).

(a) Energy systems diagram and equations; circulating materials are shaded; (b) calibration values;

(c) simulation.

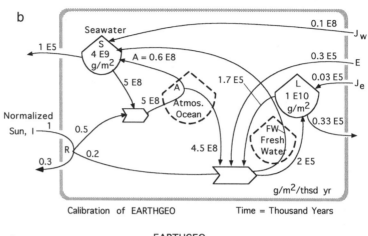

Calibration of EARTHGEO   Time = Thousand Years

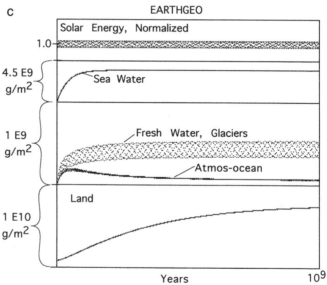

FIGURE 20.3
(*Continued*).

on the diagram to the right. The values used in the calibration are written in storages and on pathways in Figure 20.3b. Values of earth and water are in grams per square meter above a plane 4000 meters below sea level. The unit of time is 1000 years.

The simulation shown in Figure 20.3c starts with a smaller amount of seawater and continental land than now. A sine wave represents the solar energy with a 20% oscillation of energy on a cycle of about 130,000 years. The zone of the oscillation shows up in the upper graph as a stippled bar. As the seawater grows, the atmospheric–oceanic processes increase, contributing to land development and development of a freshwater hydrologic cycle.

The plot of freshwater glaciers in the middle panel oscillates in response to variations in solar energy, suggesting ice ages. The increase of the freshwater hydrologic cycle occurs by diverting some of the atmospheric–oceanic cycle. The energy of the fast flows on the left are accumulated in the slow development of the continents.

EXPLANATION OF EQUATIONS

In Figure 20.3a, solar energy I is shown heating and vaporizing water from the sea S to generate the water and energy in atmospheric circulation, storms, and ocean currents A. In the program (Table 20.3) I_i is the average solar insolation around which it is varied plus and minus 20%. On the right are shown upflows from the deep earth bringing water J_w, energy E, and earth matter in volcanic flows and intrusions J_e and $K_8*R*A*E*L$. In the middle there is a large productive interaction that represents the coupling between the atmosphere–ocean energy A, available energy from the deep earth E, existing land L, and the available solar energy R. A hydrologic cycle of freshwater F_w is generated with lakes, rivers, and glaciers on land driving the erosion of the land $K_6*R*A*E*L$, which is cycled to form new land $K_5*R*A*E*L$ using the sedimentary cycle plus new materials from the deep earth $K_8*R*A*E*L$. Some land sinks back into the deep earth K_7*L.

"What If" Experimental Problems for the Program in Table 20.3

20.3–1 Double the solar energy flow I. What happens? Why?

20.3–2 Set the water flow from deep earth JW to zero. What happens? Why?

20.3–3 Reduce the energy E from the deep earth to 0.1 and then 0.01. What happens? Why?

◆

20.4 MODEL OF A STATE DRIVEN BY A WORLD MINIMODEL (STATECON)

The state-miniworld model (STATECON) in Figure 20.4 simulates the response of a state to changes in the resources of the

TABLE 20.3
Program for Simulating Development of Ocean and Land with (EARTHGEO.bas) (Figure 20.3)

```
2 REM PC: EARTHGEO.bas (Long-term earth processes)
3 CLS
4 SCREEN 1, 0: COLOR 1, 0
6 LINE (0, 0)-(320, 300), , B
10 LINE (0, 50)-(320, 50)
12 LINE (0, 100)-(320, 100)
15 LINE (0, 200)-(320, 200)
20 REM Scaling Factors; Time in 1000 years
25 DT = 1000
30 T0 = 3300
40 W0 = 1E+08
55 L0 = 1.2E+08
60 I0 = .05
65 FW0 = 1E+07
70 REM Coefficients
75 K0 = 4.2E-10
80 K1 = .423
85 K2 = .000025
90 K3 = 1.11E-18
95 K4 = 2.5E-09
100 K5 = 1.1E-12
110 K6 = 9.43E-13
115 K7 = .0000033
120 K8 = 5E-14
140 REM Outside Sources
145 II = 1: REM Average solar energy
150 E = 1: REM Deep Earth energy availability
155 JE = 3000!: REM Earth seep  from the deep earth
160 JW = 100000!: REM Water outflow from the deep earth
170 REM Starting Conditions
175 W = 4E+07: REM Total Water
177 A = 1E+07: REM Atmospheric & oceanic storms, currents
180 S = W - A: REM Seawater
185 L = 1E+09: REM land
190 FW = K4 * R * A * E * L
300 REM Plotting
305 PSET (T / T0, 50 - I / I0): REM Solar insolation
307 PSET (T / T0, 100 - S / W0):
310 PSET (T / T0, 200 - A / FW0): REM Atmospheric water and energy
320 PSET (T / T0, 200 - FW / FW0): REM Total water
325 PSET (T / T0, 300 - L / L0): REM Continents, land
400 REM Equations
405 I = II - .2 * SIN(.001 * T): REM
410 R = I / (1 + K0 * S + K3 * A * E * L)
422 DW = JW - K2 * W
425 DL = JE + (K5 - K6) * R * A * E * L - K7 * L
427 A = (K1 * R * S) / (1 + K4 * R * E * L)
430 IF A < 1 THEN A = 1
435 W = W + DW * DT
450 S = W - A
455 IF S > W THEN S = W
490 L = L + DL * DT
495 FW = K4 * R * A * E * L
500 T = T + DT
510 IF T / T0 < 320 GOTO 300
```

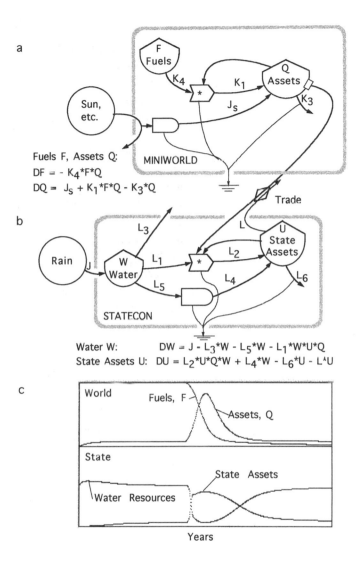

a

Fuels F, Assets Q:
DF = - K_4*F*Q
DQ = J_s + K_1*F*Q - K_3*Q

MINIWORLD

b

STATECON

Water W:　　　　DW = J - L_3*W - L_5*W - L_1*W*U*Q
State Assets U:　DU = L_2*U*Q*W + L_4*W - L_6*U - L^AU

c

FIGURE 20.4.
Minimodel of global
development affecting local
development (Program
STATECON.bas). (a, b) Energy
systems diagram and equations;
(c) simulation.

outside world. The world economy represented by MINIWORLD
in Figure 20.4a grows slowly before discovery and use of fossil
fuels; then with abundant fuels turned on it has a pulse of growth
of assets and products, which are available to stimulate local
economies. Finally global assets decrease again as fuel resources decline.

The state represented by the lower model in Figure 20.4b uses
some of the world's economic assets Q to produce its own assets
U. The production of the state's assets U is in proportion to the

world assets. The more world assets there are, the more the state can grow. Water W is used as an index of the state's natural resources, which interact with the world assets for the state growth.

In the simulation, growth of the state's assets follows that of the world. The upper graph of Figure 20.4c shows curves for world economic assets Q and the introduction of fossil fuels F. In the lower graph are curves for the state's environmental resources W and its economic assets U. The simulation goes halfway across the screen, representing a colonial period without fossil fuels. Then after 300 years fossil fuels F are turned on by technology, which starts the Industrial Revolution. As these nonrenewable resources are used, economic assets go through a maximum in the world, followed almost as rapidly in the state. The state's assets decline because local natural resources W and world assets decline. As growth decreases, the storages of natural resources start to recover.

EXAMPLES THAT FIT THE STATE-WORLD MODELS

STATECON can be used for any nation or state that depends on imports and exports from the world economy. For example, Florida imports fuels and goods, like machines and cars, and exports oranges, vegetables, and tourist services. Japan is an example of a country that must import and export to keep its economy going. It imports fuels and raw products and exports finished electronic goods. Its economy is quite dependent on the rest of the world.

DESCRIPTION OF EQUATIONS

The five equations in Figures 20.4a and 20.4b are stated as the following changes for each unit of time that passes:

$$DF = -K_4 * F * Q.$$

Change of nonrenewable resource reserves F (fuels, etc.) is equal to the use in the economic production resulting from product interaction of fuel reserves F and global assets Q.

TABLE 20.4
Program (STATECON.bas) for a Regional Economy Driven by A Minimodel of the World Economy (Figure 20.4)

```
10 REM PC
20 REM STATECON.bas (State Economy driven by Miniworld model)
30 CLS
40 SCREEN 1,0:Color 15,1
50 LINE (0,0)-(320,180),,B
60 LINE (O,80)-(320,80)
80 J = 2
90 Js = 3
100 Q = 1
110 W = 10
120 F = 0
130 U = .1
140 K1 = .0001
150 K2 = .0000005
160 k3 = .04
170 K4 = .00004
220 L = .2
230 L1 = .00016
240 L2 = .0003
250 L3 = .17
260 L4 = .0013
270 L5 = .006
280 L6 = .04
300 Q0 = .08
310 F0 = .08
320 DT = 1
330 T0 = .45
340 U0 = 3
350 W0 = 5
360 REM WORLD IS TOP OF SCREEN
370 IF T = 300 THEN F = 1000
380 PSET (T * T0,80 - Q * Q0)
400 PSET (T * T0,80 - F * F0)
410 PSET (T * T0,180 - W * W0)
420 PSET (T*T0,180 - U*U0)
430 DQ = Js + K1 * F * Q -k3*Q
450 DF = - K4 * F * Q
460 DW = J - L1 * Q * U * W - L3 * W - L5 * W
470 DU = - L * U + L2 * U * Q * W + L4 * W - L6 * U
480 Q=Q+DQ*DT
500 F = F + DF*DT
510 W = W + DW*DT
520 IF W < .1 THEN W = .1
530 U = U + DU*DT
540 T = T + DT
550 IF T * T0 < 320 GOTO 360
```

$$DQ = J_s + K_1{}^*F^*Q - K_3{}^*Q.$$

Change of global assets equals steady inflow of products from environmental work J_s, plus the economic production resulting from fuel use interacting with global assets Q ($K_1{}^*F^*Q$), minus assets-dependent losses $K_3{}^*Q$.

$$DW = J - L_3*W - L_5*W - L_1*W*U*Q.$$

Change of water storages DW equals inflow of rain J, minus outflow L_3*W, minus linear use in developing state assets L_5*W, minus water use by state economic production $L_1*W*U*Q$.

$$DU = L_2*U*Q*W + L_4*W - L_6*U - L*U.$$

Change of state assets DU is increased by the autocatalytic production based on imported global assets Q interacting with local resources W $(L_2*U*Q*W)$, plus local production L_4*W, minus depreciation L_6*U, minus state products traded $L*U$.

"What If" Experimental Problems for the Program in Table 20.4

20.4-1 What would happen to this state if it developed with 10% less use of resources obtained by trade with the world? To make a 10% lower rate of use of world resources, decrease L2 by 10%.

20.4-2 If the world's sunlight is greater, how does this affect the state's growth? After running the program with its current settings, increase JS.

20.4-3 If a new rich fuel source were discovered in about the year 2000, how would the world and state economies react? Add a statement as line 375 to make F = 1000 when T = 400.

PART FOUR

APPLICATION

Part IV applies simulation to real systems. After considering fundamental models and their simulation with several easy methods in Parts I through III, you are ready for a modeling project. Chapter 21 presents the steps for simulating an ecosystem example and discusses the controversial question of what complexity is appropriate. Chapter 22 briefly introduces other kinds of simulation modeling including spatial models, statistical models, mathematical methods for complex networks, and other simulation software.

Chapter Twenty-One

◆

MODELING PROJECTS AND COMPLEXITY

$\mathbf{A}$ modeling project is the next step after learning principles and means for simulation. Chapter 21 introduces the process of modeling and simulation of a real system. Examples in previous chapters were *minimodels* with relatively few units and relationships (2 to 15 coefficients). Each minimodel shows a few important principles, but the patterns over time in the real world are more complex, with all the principles and scales operating together. Investigators want all the main features of a system to show in the model. The matching of observed data usually requires models with more units and interconnections. But the controversial question is this: How much complexity is desirable?

21.1 OVERVIEW

COURSE PROJECTS

In our systems courses, after working with principles and minimodels, students do special projects on systems of special interest to them. To limit difficulty, complexity, and the time required for a first application and to develop ability to aggregate, the assignment requires that the whole system of the declared interest and its main sources be included. However, for the first application the simulation model is limited to four storages and four interactions. LAGOON (see Section 21.2) is an example of an aggregated model of an ecosystem, only slightly more complex than the minimodels in Part III. Students are discouraged from isolating a small part of the system of interest, since to do so puts the model at the wrong scale for understanding the defined system and its problems. An oral presentation is made to the group as soon as there are graphs to show from the initial simulation runs. After benefiting by the discussion, the model is revised and runs completed.

A written report is required to include and explain the following items:

1. The system under study and its questions of special interest
2. A complex diagram of the real system
3. Diagram of the aggregated simulation model and coefficients
4. The derived equations (on a page facing the model diagram)
5. Diagram of the model with calibration numbers
6. Table or spreadsheet used for calculating coefficients
7. Listing of the simulation program
8. Simulation graphs for the calibration run
9. Simulation graphs for "What If" experiments that explore the questions of special interest
10. Summary of insights gained from the exercise

After the course project is completed, students are ready to develop models for their other courses, theses, dissertations, and jobs.

MODELING A REAL SYSTEM

To model a real system, first define a boundary and the time scale of interest (hours, days, years, etc.). Then use words to make three lists: (1) the main external factors (sources), (2) the main parts inside, and (3) the main processes (interactions) believed to be important at that place and time scale. Then draw the systems diagram and write the equations. Find data on flows and storages for the diagram and calibration table. Numbers used for calibration need to be in the right order of magnitude and adjusted to be internally consistent with other flows in the diagram (example: Figure 21.1b). At this stage precise data are not needed. In one simple method, estimate average storage quantities, and then use estimated turnover times to estimate the flows on the pathways in and out, determine the coefficients from the flows and storages, write the program, and start runs.

DEBUGGING

Mistakes (bugs) are almost always present in the programming and have to be removed. Bugs include errors in thinking and careless typing. The first runs may jump off scale, show nothing, give negative values, or induce error messages from the computer. Debugging is a trial-and-error process. Reduce the iteration interval DT to eliminate any artificial chaos. Cut the input sources to a small value to get the output range to stay within the screen. Decrease off-scale amplitudes with scaling factors. Keep the energy systems diagrams in view. After each run use the command line to print values to see if they are in the right order of magnitude. If not, use the diagram to identify the equations and coefficients affecting that variable. Start on the left, checking each input and storage quantity. Where interactions are drawn from storages, they can be pulled below zero, which causes the whole program to generate nonsense. Insert statements to prevent these values from going below zero (e.g., IF Q < 0.00001 THEN Q = .00001).

If steady-state numbers are used for calibration, if the program is not oscillating, and if the first runs are made with the calibration numbers as the starting values, the graphical output is horizontal straight lines, which may mean that bugs are eliminated. Reset the starting conditions to 10% of steady state and rerun to observe growth and leveling. If the model is oscillating, check to see if amplitudes and frequencies are realistic.

Controversy exists as to how much complexity and detail are desirable in simulation models and in their validation. Many modelers go immediately to the level of complexity planned for simulation. However, a better way is to program a model with important features and get it running to gain insight on its performance in comparison to the observed system. Then add additional parts and mechanisms. By making the preliminary model, debugging is easier, the investigator keeps structure and function better related, and he or she enjoys the process more.

APPLICATIONS OF MINIMODELS

Minimodels aid development of more complex models according to scale:

1. At its *own scale* a minimodel and its simulation show a few principal relationships and their consequences over time. The simulation of a minimodel that has the main essence in overview should be a smoothed version of the simulation of the

more complex model. For example, a minimodel of a microbial population might show a smoothed curve produced by the aggregated of mechanisms of growth and limitations.

2. Relative to the *smaller scale,* a minimodel aggregates more complex details. It provides a top-down overview, which may be enough if the details of the smaller scale are not the primary concern. At the scale of the minimodel the smaller scale often contributes rounded averages. For example, most of the details of the rapid processes of biochemistry within the bacteria are not needed on the time scale of studies of microbial population growth.

3. The minimodels are building blocks for developing more complex models. Knowledge of the behavior of the simpler configurations helps understand the more complex model in which these are embedded. For example, the models of the bacterial population can be inserted into a model of a garbage dump.

21.2 LAGOON: AN EXAMPLE OF AN INITIAL PROJECT MODEL

The model LAGOON (Figure 21.1) illustrates the process of developing a preliminary simulation of a real system in a project. Figure 21.1a has the main features of an ecosystem in a tropical marine lagoon in Puerto Rico. The broad shallow lagoon 1 m deep was surrounded by mangroves and connected with the sea by a narrow tidal channel. Shrimp and larger fish were harvested by nets in the channel. The detritus from the mangroves and algae was consumed by larger organisms such as young shrimp and fishes. The lagoon water exchanged with the sea through the channel had a greater outflow than inflow because of the freshwater received from land runoff.

The modeling procedure started with a written list of sources, components, and processes. Then these were diagrammed and equations written (Figure 21.1a). Next, values of flows and storages were collected and written on the diagram (Figure 21.1b). Constant coefficients were calculated with a calibration table (using a spreadsheet). The simulation program was written, run, and debugged. To save time, an old BASIC program

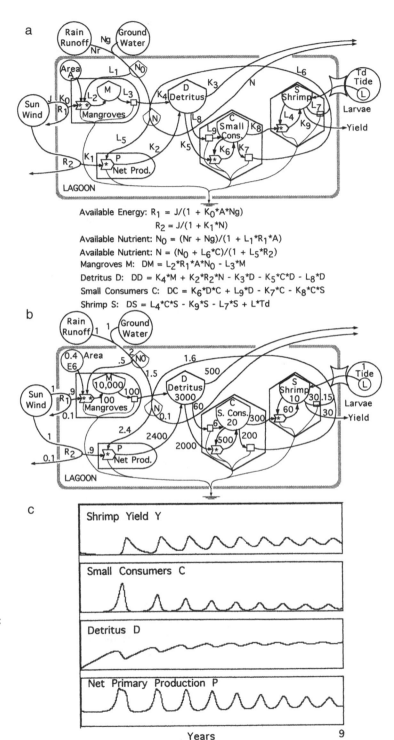

Available Energy: $R_1 = J/(1 + K_0*A*Ng)$
$$R_2 = J/(1 + K_1*N)$$
Available Nutrient: $N_0 = (Nr + Ng)/(1 + L_1*R_1*A)$
Available Nutrient: $N = (N_0 + L_6*C)/(1 + L_5*R_2)$
Mangroves M: $DM = L_2*R_1*A*N_0 - L_3*M$
Detritus D: $DD = K_4*M + K_2*R_2*N - K_3*D - K_5*C*D - L_8*D$
Small Consumers C: $DC = K_6*D*C + L_9*D - K_7*C - K_8*C*S$
Shrimp S: $DS = L_4*C*S - K_9*S - L_7*S + L*Td$

FIGURE 21.1.
Model of a tropical marine
lagoon connected by a channel
with the sea (program
LAGOON.bas). (a) Energy
systems diagram and equations;
(b) calibration values for
storages (per square meter) and
flows (grams per square meter
per year) written on the
diagram; sun, wind and tide
were normalized as unity;
(c) typical simulation of the
LAGOON model.

of similar complexity was loaded and statements changed. Then simulations were rerun with various changes in inputs or coefficients, relating the responses to what was known about the lagoon.

This aggregated ecosystem model (Figure 21.1a) illustrates food chain patterns in ecosystems and the recycle of mineral nutrients. The external sources, sun, nutrient inflow, and tidal patterns are source-limited inputs that determine the carrying capacity of the system to support higher organisms.

It was known that populations pulsed in this lagoon. In typical simulations (Figure 21.1c) of LAGOON.bas, consumers pulse almost with an annual time scale even when no seasonal inputs are applied. The buildup of detritus storage permits the consumer food chain to develop an oscillation. The simple prey–predator interaction of the shrimp and small consumers causes the pulsing, which also causes the primary production, detritus, and nutrient levels to oscillate. Pulsing is facilitated by having linear and auto-catalytic pathways to the small consumers. This lagoon model illustrates the ways larger species can control the periodicities of ecosystems.

EXPLANATION OF MODEL AND EQUATIONS

To represent the main features of the lagoon, the model LAGOON (Figure 21.1a) was defined with the boundary at the edge of the wetland border of the lagoon. The model includes a producer unit for mangroves with stored biomass M occupying area A. Marked with a producer symbol, the net photosynthetic production P in the open water and on the shallow bottom was modeled as a product of available light R_2 and the nutrient concentration N. Because the scale of time chosen was years, the very fast turnover components algae and nutrients at position N_0 and N were modeled without storages. The "flow use" design (Section 6.4) provides a running average of the nutrient concentrations, simplifying the simulation. In Figure 21.1a and the program (Table 21.1) both K and L are used for constant coefficients; L alone is the storage of larvae and juveniles.

Nutrients available to the mangroves N_0 are the sum of inflows N_r and N_g minus that used by mangroves $L_1*R_1*A*N_0$. Nutrients N available to the freewater aquatic production P are those passing through the mangrove border N_0 plus that recycled L_6*C

TABLE 21.1
Program for Marine Lagoon (LAGOON.bas)
(Figure 21.1)

```
2 REM PC
3 REM LAGOON: MARINE LAGOON
5 SCREEN 12, 0: REM No Color
6 LINE (0, 0)-(320, 60), , B
7 LINE (0, 70)-(320, 130), , B
8 LINE (0, 140)-(320, 200), , B
9 LINE (0, 210)-(320, 270), , B
20  REM  Scaling Factors
21 DT = .002
23 T0 = 35
27 C0 = .4
30 P0 = .02
35 D0 = .02
37 M0 = .005
40 Y0 = 4
43 N0 = 10
44 N = .1
47 REM Sources and Starting Conditions
52 L = .1
55 NG = 1
60 NR = 1
65 J = 1
66 A = .1
67 M = 10000
68 D = 100
69 C = 2
70 S = 2
80 TD = 1
100  REM   Coefficients
110 K0 = 45
120 K1 = 90
130 K2 = 200000!
140 K3 = .167
150 K4 = .01
155 K5 = .067
157 K6 = .0167
160 K7 = 10
165 K8 = 1.5
168 K9 = .33
170 L1 = 50
175 L2 = 5000!
178 L3 = .01
180 L4 = .3
190 L5 = 240
195 L6 = .05
197 L7 = 3
198 L8 = .02
199 L9 = .002
200  REM  Plotting
220  REM PSET (T*T0, 270 - M*M0)
```

(continues)

TABLE 21.1 (*Continued*)

```
230  PSET (T * T0, 270 - P * P0)
250  PSET (T * T0, 200 - D * D0)
285  PSET (T * T0, 60 - Y * Y0)
290  PSET (T * T0, 130 - C * C0)
300  REM  Equations
305  R1 = J / (1 + K0 * A * N0)
310  R2 = J / (1 + K1 * N)
317  N0 = (NR + NG) / (1 + L1 * A * R1)
320  N = (N0 + L6 * C) / (1 + L5 * R2)
325  IF N < .00001 THEN N = .00001
330  P = K2 * R2 * N
340  Y = K9 * S * TD
350  REM  New storage values
352  DD = K2 * N * R2 - K3 * D - K5 * C * D + K4 * M - L8 * D
355  DC = K6 * D * C - K7 * C - K8 * C * S + L9 * D
360  DS = L4 * C * S - K9 * S - L7 * S + L * TD
365  DM = L2 * A * N0 * R1 - L3 * M
370  D = D + DD * DT
375  C = C + DC * DT
377  IF C < .0001 THEN C = .0001
380  S = S + DS * DT
390  M = M + DM * DT
392  T = T + DT
400  IF T * T0 < 320 GOTO 200
```

minus that used in production L_5*R_2*N. A pool of organic accumulation characteristic of most ecosystems is included as the detritus storage D. The animals are included as a pair of autocatalytic consumer units connected with prey–predator type interaction.

In Figure 21.1a the storage of detritus organic matter D receives the net photosynthetic production flow of the algae K_2*R_2*N and some of the litter from the mangroves K_4*M. Some detritus washes out K_3*D, and some is used by the small consumers on two pathways, one linear L_8*D and one an autocatalytic use K_5*D*C. The small consumers (zooplankton and bottom microzoa) are increased by the food conversion pathways L_9*D and K_6*D*C and decreased by linear losses K_7*D and the autocatalytic consumption $K*C*S$ by larger consumers (labeled as shrimp). The larger consumers S are increased by conversion of their food L_4*C*S and influx of juveniles and larvae L with the tide. Storage S is decreased by linear losses L_7*S and fishing yield K_9*S.

"What If" Experimental Problems for the Program in Table 21.1

21.1-1 What happens to the storages and the oscillations if nutrients running off the land through the mangrove border are increased? Change NR from 1 to 3.

21.1-2 What is the effect of cutting the mangroves on detritus and consumers? Reduce the mangrove area A from 0.1 to 0 and change the initial value of M to 0.

21.1-3 What happens if larvae and juveniles of larger consumers from offshore reproduction are not available to exchange back into the lagoon with the tide? Set $L = 0$ and $S = 0$.

◆

21.3   COMPLEXITY OF MODELS

On every scale the real world is vastly more complex than humans can visualize, and no model can represent all of its detail. But humans can visualize main features in simplified models. Traditional science studies pieces. For understanding, systems and simulation combine the pieces into models. Associated with O'Neill is the principle that the complexity of ups and downs of patterns over time is proportional to the complexity of the parts and pathways. In other words, minimodels generate smoothed graphs of phenomena, whereas more complex models of the same system have more of the detailed ups and downs. Reality modeling is intermediate in complexity and accuracy between the minimodels and the real world.

The complexity of models that is desirable is controversial, and depends on scales. Complexity results as modelers attempt to include more of what they observe in the real world. Many scientific disciplines are dedicated to one scale and sometimes these investigators want to include everything they know. Many people believe that combining more and more parts will generate a correct simulation of the system, even though the main features of design that control are from mechanisms of a larger scale outside of the model. Yet a model at one scale may not have the controlling influences of the larger scale necessary for simulation of that scale. The human mind acting like a zoom microscope observes each scale in a system of interest with the same detail and often wants to include them equally. Yet fine detail of smaller scales may not affect the scale of main interest.

MORE COMPLEX LAGOON MODEL

To illustrate the process of increasing complexity, model LAGOONII (Figure 21.2a) was made by adding details to LAGOON. The program contains 120 coefficients and required several months to develop and debug (Odum and Munroe, 1993). It has many of the uncertainties and arbitrary choices of moderately complex models. Like the simpler LAGOON, it has producers, a food web of consumers, nutrient recycle, and tidal seeding. Like the real lagoon, the simulation has many complex pulsing sequences which are sensitive to population levels, nutrients, and other factors (example: Figure 21.2b). In one simulation overfishing caused blooms of fast turnover carnivores (jellyfish).

The complex version still could not duplicate the fine detail sequences of each species population because they were not in the model separately. Nor were the mechanisms of population detail included such as age classes, special species physiology, and special relationships to other species. However, additional understanding was obtained of important aspects of the ecosystem. For example, increased organic matter from the land facilitated pulsing. Neither model can be expected to generate all of the observed fluctuations of the smaller scales that had been aggregated.

SIMULATING SMALL SCALE WITH MAIN FEATURES OF A LARGER SCALE

Where small-scale detail is of interest, it can be included within a model that for the most part has larger scale structure and processes. For example, details on a species, to the satisfaction of the species scientist, can be imbedded in an otherwise simpler ecosystem model like LAGOON in order to have its inputs appropriately driven by interactions with the ecosystem around it. For example, one of the consumer units in Figure 21.2c can be expanded to include both mass and population numbers, eggs, larvae, age classes, and physiological effects of temperature.

a

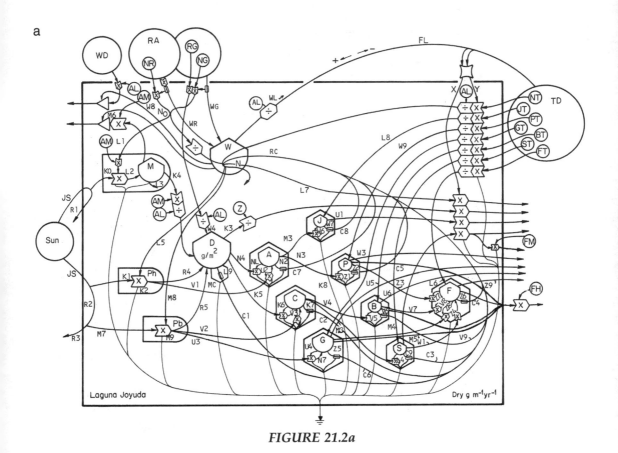

FIGURE 21.2a

LIMITS ON COMPLEXITY

It is nonsense to try to model all the detail of the real world. To do so would be to lose understanding in the confusion. It takes more information to model something than to be that something. To find out what the real world does, measure the real world. To understand and predict the main features of the real world, develop models on a scale appropriate to human understanding, with frequent checks with observed data so that the model shows the main features, but not all the details of scales outside of main interest.

When large models are generated by combining many relationships piece by piece, even the modeler-programmer who did it may not comprehend the product, often forgetting what is in the code. Opportunity for programming errors increases with the complexity. It is very difficult to know if all the errors have been

FIGURE 21.2.

(a) More complex model LAGOONII of a tropical marine lagoon used to explore oscillations in a food web in relation to productivity and nutrients (Odum and Munroe, 1993). Equations:

Limiting Flows:

$R_1 = JS - K_0{}^*R_1{}^*N_0/AM$ Therefore $R_1 = JS/(1 + K_0{}^*N_0/AM)$

$R_2 = JS - K_1{}^*N^*R_2$ Therefore $R_2 = JS/(1 + K_1{}^*N)$

$R_3 = R_2 - M_7{}^*N^*R_3$ Therefore $R_3 = R_2/(1 + M_7{}^*N)$

$R_4 = PH - V_1{}^*A^*R_4$ Therefore $R_4 = PH(1 + V_1{}^*A)$

$R_5 = PB - V_2{}^*C^*R_5 - U_3{}^*G^*R_5$ Therefore $R_5 = PB/(1 + V_2{}^*C + U_3{}^*G)$

Water:

$WR = RA^*AL$; WL $(W/AL) - Z$; $FL = W_2{}^*(TD - WL)$; $EV = W_8{}^*WD^*AL$

If $TD > WL$ then $Y = 1$: $X = 0$; If $TD < WL$ then $Y = 0$: $X = 1$

$DW = WR + WG + FL - W_8{}^*WD^*AL - M_6{}^*M^*AM$

Phosphorus:

$N_0 = WR^*NR + WG^*NG - L_1{}^*R_1{}^*AM^*NO.$

Therefore $N_0 = (WR^*NR + WG^*NG)/(1 + L_1{}^*R_1{}^*AM)$

$N = N_0/AL + Y^*M_1{}^*NT^*(FL/AL) + RC - X^*L_7{}^*N^*FL/AL - L_5{}^*R_2{}^*N - M_8{}^*R_3{}^*N$

Therefore $N = (N_0/AL + Y^*M_1{}^*NT^*FL/AL + RC)/(1 + L_5{}^*R_2 + M_8{}^*R_3 + L_7{}^*FL/AL)$

Production:

$PH - K_2{}^*R_2{}^*N$; $PB = M_9{}^*R_3{}^*N$; $GP = PH + PB$; $RS = 200^*RC$

Detritus: $DD = R_4 + R_5 + X^*K_3{}^*(D/Z)^*FL/AL - K_5{}^*C^*D + K_1{}^*M^*AM/AL - N_4{}^*D^*A - L_9{}^*D + W_4{}^*WG^*RG/AL$

Benthos: $DC = K_6{}^*D^*C - K_7{}^*C - K_8C^*S - V_3{}^*R_5{}^*C - V_4{}^*C^*B$

Shrimp: $DS = L_4{}^*C^*S - X^*M_5{}^*S - C_9{}^*S - W_1{}^*F^*S + Y^*M_4{}^*ST^*FL/AL$

     $YS = X^*M_2{}^*S^*FM$

Mangroves: $DM = L_2{}^*R_1{}^*N_0/AM - L_3{}^*M - K_4{}^*M$

Zooplankton: $DA = N_1{}^*D^*A - N_2{}^*A - N_3{}^*A^*P - M_3{}^*A^*J + U_2{}^*R_4{}^*A$

Grazing fish: $DG = U_4{}^*R_5{}^*G - N_7{}^*G - N_8{}^*G - Z_5{}^*G^*F + Y^*N_9{}^*GT^*FL/AL - U_7{}^*FH^*G$

Plankton fish: $DP = Z_1{}^*A^*P - Z_2{}^*P - W_3{}^*P - Z_3{}^*P^*F + W_9{}^*PT^*FL/AL - N_5{}^*FH^*P$

Bottom fish: $DB = V_5{}^*C^*B - V_6{}^*B - V_7{}^*B^*F - U_6{}^*B + Y^*U_5{}^*BT^*FL/AL - N_6{}^*FH^*B$

Jellyplankton: $DJ = W_6{}^*A^*J - W_7{}^*J + Y^*L_8{}^*JT^*FL/AL + X^*U_1{}^*(J/Z)^*(FL/AL)$

Higher consumers:

$DF = Z_4{}^*S^*F + Z_7{}^*P^*F + Z_8{}^*G^*F + V_8{}^*B^*F - Z_6{}^*F - Z_9{}^*F - K_9{}^*FH^*F + Y^*L_6{}^*FT^*FL/AL$

Fishing: $YF = K_9{}^*FH^*F + N_5{}^*FH^*P + N_6{}^*FH^*B + U_7{}^*FH^*G$

debugged. With large numbers of coefficients, the calibration is difficult. It is difficult to understand large models, although it helps to provide an energy systems diagram on a poster board.

Many examples are available of the failure of modeling and simulation which started with a false ideal of having the computer combine everything that could be learned about the parts of a system. The IBP (International Biological Program) of the late 1960s

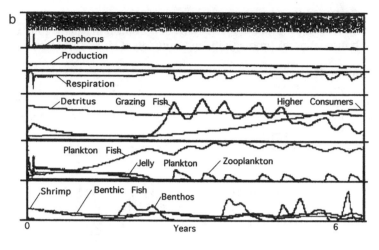

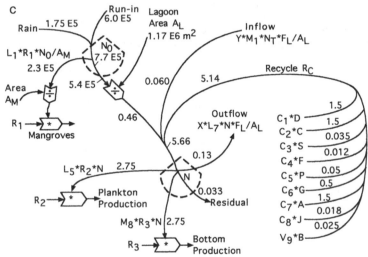

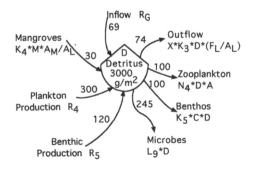

FIGURE 21.2

(*Continued*). (b) One of the simulations of the LAGOONII model with large detritus input; (c) sections of the systems model isolated to show calibrations. Letters used for constant coefficients are C, K, L, M, N, and V.

sought to "understand ecosystems for the benefit of human society." The traditional science ideal was adopted of studying all the parts, going from the scale of ecosystems down into the scales of physiology of organisms and below, leaving synthesis to a few people with mathematical background. Much was learned and published about the parts of ecosystems, but the models were so complex that they were never finished, not realistic, and mostly never published. What understanding was obtained about the ecosystem scale came from a few simplified models.

COMPLEXITY AND UNIQUENESS

In real systems uniqueness and creativity emerge with complexity. There are more possible arrangements and connections between parts than there is energy to arrange them. Thus, different combinations result for the same number of parts and pathways. Complex ecosystems like coral reefs and rain forests tend to be unique—never exactly the same. The same is true for the complex models. When more detail is added, the choices tend to be arbitrary, representing the special knowledge and interests of the modelers.

The more complex a model is, the more possible variations there are. Even at moderate levels of complexity, two modelers are likely to design models that have different units and connections. This doesn't mean they are wrong. The models reflect the different choices on the parts of the real world included, aggregated, or omitted. Creativity and originality emerge with complexity both in nature's creation and in the understanding of unique models.

REPRESENTING COMPLEX MODELS
WITH DIAGRAMS

As illustrated in this book, simulation models should always be diagrammed, preferably with the participation and approval of the authors. Each diagram should be accompanied by the change rate and logic equations extracted from computer codes and represented by the symbol network. The diagramming catches inconsistencies and dangling pathways. Making models visible and more easily understood encourages use by more people, more discussion of the structure and functions in previous models, and more building of one effort on another. People can trust a model better if they understand what is in it. Then they can suggest the changes they require for additional use in other situations.

In wide usage since 1966, and explained in Part I of this book, the symbols and diagramming rules of the energy systems language (Chapter 2 and Appendix D) contain mathematical equivalents, energy constraints, and means of representing information. When symbols are placed on the page from left to right in order of their energy transformity, a model drawn by one person is congruent with those of another. Whereas the language is also used qualitatively in a soft way as a first step in converting verbal, mental models to network form, this language can be used precisely to represent equations and program code.

Figure 21.2a is an example of representing the essentials of a moderately complex simulation model with energy systems diagramming and equations. To represent the calibration data, confusion can be reduced by separating parts on which the calibration values and the equations are shown (sample in Figure 21.2c).

Very complex models can require several pages. With the symbols in memory as "macros," complex diagrams can be drawn in a few hours with computer drawing programs such as Canvas and Corel Draw.

USE OF COMPLEX MODELS IN NEW SITUATIONS

Some complex models with large investments of time, developed for one area, are being systematically applied in other areas and ecosystems by integrative organizations such as the network of Long Term Ecological Research sites (LTERs) of the National Science Foundation. Without systems diagrams of the details, participants in these group projects have no easy way to know their content. Even the authors of models using them later do not remember all the details. Even when printouts of computer code are available, it is not easy to see what is being represented, what should be modified, or how an investigator dealing with a particular part can relate.

So-called "canned" models are adapted to new conditions by

resetting coefficients with data, but the overall structure and features may not be appropriate. No model or its output should be used until its equations and systems diagram are available and understood.

21.4   JUDGING THE SUCCESS OF SIMULATION

At the least, simulation shows the quantitative consequences of ideas about relationships, often to the surprise of the person making causal inferences with verbal thinking. At best, simulation fits observed data and predicts future patterns.

VERIFYING MODEL PERFORMANCE

Verification is the process of showing that the model has appropriate mechanisms and output to represent the concepts believed to be important in the real system. For example, a model of productivity should respond to variations in the availability of necessary inputs to production. Study the pathways and configurations of the systems diagram to see if the simulation results are consistent with the mechanisms that were introduced. Are these consistent with beliefs about the real system? If not, make changes. The minimodels in Part III provide insight on responses that follow from model structures. They should be rerun until the patterns are plausible. Recalibration may be part of this process, an easy effort if a spreadsheet program has been used for calculating coefficients.

The equations for a component unit can be examined to write the mathematical function for its storage or production. Then observed data can be plotted on the same graph, which has the mathematically generated line. Where possible, coordinates should be chosen so that the theoretical line from the equation is straight. For example, Figure 16.4c has a straight line relating species number to the square root of the resource area. Then a statistical correlation or regression equation can be calculated to evaluate the probability of the fit between line and data. For another example, a semi-log graph can be used to test for exponential growth (Figure 9.4) or decay (Figure 6.7) because these functions are straight on coordinates log Q as a function of time T.

A whole model might generate realistic looking time series curves with incorrect mechanisms. In addition to comparison of data with runs of the whole model, comparisons can be made between the performance of component units and observed data. For example, the data on production rate of plants (model output and observed data) can be graphed as a function of light intensity. A good model not only generates the time series, but has the components of the model performing realistically. A part of the model can be isolated to test its response. In BASIC this is easily done by using GOTO to bypass program lines or to disable lines by typing REM just after the line number.

COMPARING OBSERVATIONS
WITH SIMULATIONS

Comparing the simulation graphs with data from the real system is sometimes called *validation*. The simulation results can be compared with the data that were used to calibrate the model. Then the model can be compared with data from other times and situations. However, it would not be appropriate to compare the model with data for a different system known to have major factors not included in the model.

For those using models to represent real systems, success is often tested by comparing observed data with the simulation of the whole model. Actual records of sources like sunlight and water inflow are used as inputs to the simulation. Then observed data for the storages are plotted on a graph of properties with time generated by the model. In a typical example, Campbell and Newell (1998) simulated a system of mussel culture with the model shown in Figure 21.3a. Note the way the mussel production and respiratory functions include the exponential effect of temperature (e^T). Validation was done by comparing observed data with simulated graphs. Figure 21.3b is one of these comparisons.

If changes have resulted from experiments on the real system (or natural experiments), the model should show the same changes when the model is experimentally modified. This does not mean recalibrating. If the model is a good one, it will change its output to that observed without recalibration.

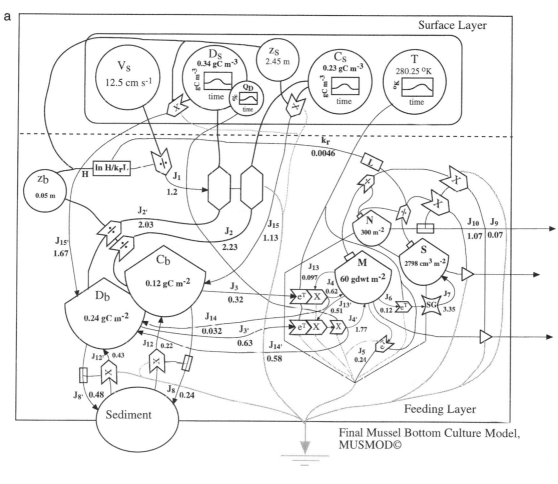

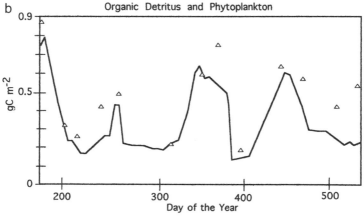

FIGURE 21.3.

Example of validation, culture of bottom mussels (Campbell and Newell, 1998). (a) Energy systems model; (b) simulation graph of organic detritus–phytoplankton biomass with observed data points plotted as small triangles; redrawn from Campbell and Newell (1998).

Statistical methods can be used to measure the fit between simulation and data. Observed curves can be statistically compared with a simulated curve using the chi-squared test, which indicates probability of fit. The values of a variable generated by the simulation may be plotted on a graph as a function of the observed values of that variable. For a perfect fit, points would be in a straight line. The degree of fit can be evaluated with a correlation or regression program. If the comparison shows a poor match, the model and its calibration can be changed until there is a better match.

Perfect fit is not to be expected in any simulation, since no model has all the factors and scales operable in the real world. Nor are observed data abundant enough and accurate enough to prove a fit. The model is successful if it reproduces the main features of the system of interest. It is useful if it helps understand the problems at issue in the real system.

Chapter Twenty-Two

◆

SIMULATION APPROACHES

Chapter 22 surveys other approaches to simulation including spatial models, mathematical approaches, statistical models, techniques for larger models and software packages. We can only explain the key ideas and where to go to learn more. Published translations of models from the symbols of mathematics and other fields into energy diagrams are available (Odum, 1983, 1994).

22.1   SPATIAL MODELS

Typical spatial simulation divides the geometric area of a system into component blocks, each of which has its own simulation model running, but connected by pathways between the blocks. For example, in Figure 22.1a the landscape is divided into nine blocks. Each block has its own minimodel, receiving inputs from outside sources. The model in each block has inputs and output pathways joining adjacent blocks (small squares in Figure 22.1a). After equations are written for each block, all of the block models are simulated as one program. The output can be nine sets of graphs, one for each block. More often, the values of each block are given a shading or color so that an attractive map is displayed with shadings changing with time. Sometimes three-dimensional graphics are used to show the values of each block as a vertical column. For applications to real landscapes, the blocks are usually not square but follow natural features of the landscape.

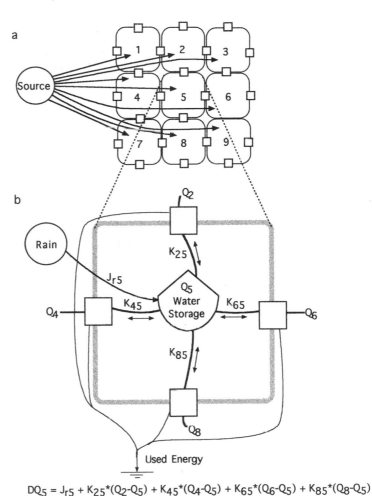

FIGURE 22.1.
Example of spatial simulation, water distribution on a flat landscape with the area divided into nine blocks. (a) Pattern of blocks showing rain inflow from an outside source and small squares indicating flows between adjacent blocks; (b) details of block 5 showing the unit model with energy systems symbols, including coefficients and change equation.

$$DQ_5 = J_{r5} + K_{25}*(Q_2-Q_5) + K_{45}*(Q_4-Q_5) + K_{65}*(Q_6-Q_5) + K_{85}*(Q_8-Q_5)$$

SIMPLE EXAMPLE: LINEAR WATER FLOWS

A very simple example is the model in Figure 22.1b for the flow of water in a flat, sandy landscape. In this case each block has the same minimodel. The repeating unit model in this case is a simple water storage, which exchanges water with the next block according to the difference in water quantity (Figure 22.1b). Rain enters each block and flows toward those blocks with less water, dispersing some energy from the frictional processes. The model is appropriate where flow is through a porous medium (such as sand) and thus appropriate for D'Arcy's law where flow is proportional to difference in pressure (a linear pathway), and pressure is proportional to the water stored Q.

The equation for each block is the same, although the coefficients (K) may be different (different friction). For the model in Figure 21.1 the program for simulation has nine equations, one for each of the nine storages.

A unit model with more pathways is required for simulating water flow on most landscapes, one including rain, percolation into deeper levels of the earth, evaporation, transpiration by the vegetation, runoff and other interblock flows, and quadratic flow where there is turbulent flow in channels. In the field of hydrology, commercial software packages help simulate water distribution and flow.

SPATIAL SELF-ORGANIZATION BY PULSING UNITS

With a more complex unit model (PULSE from Figure 15.4), Richardson (1987) studied the spatial patterns of energy production and consumption generated by spatial simulation. The unit model (Figure 22.2b) was simulated in 100 blocks (Figure 22.2a) with different kinds of interblock pathways, starting with the simple linear flow used in Figure 22.1. A digital-analog hybrid computer was used for the computations. Shading was used to show the magnitude of simulated quantities, with darker shading for larger quantities. Shown as a time sequence from left to right in Figures 22.2c and 22.2d are two sequences of shaded maps, one for the accumulated storage of producer products and the other for the storage of consumer assets during their surge of consumption and growth. This power-maximizing design automatically develops an organized center in spatial simulation. The consumer frenzy develops a hierarchical center because more flows converge to the blocks in the center than to those on the perimeter. Figure 22.2e has the graphs for one of the middle blocks. The peaks of the curves correspond to the dark shadings in Figures 22.2c and 22.2d. In the field of embryological development, Meinhardt (1982) simulated the early development of structures and patterns of organisms with spatial models of chemical concentrations.

SPATIAL SIMULATION SOFTWARE

In application to real systems, there may be many blocks and more complex unit models, so that the amount of computer

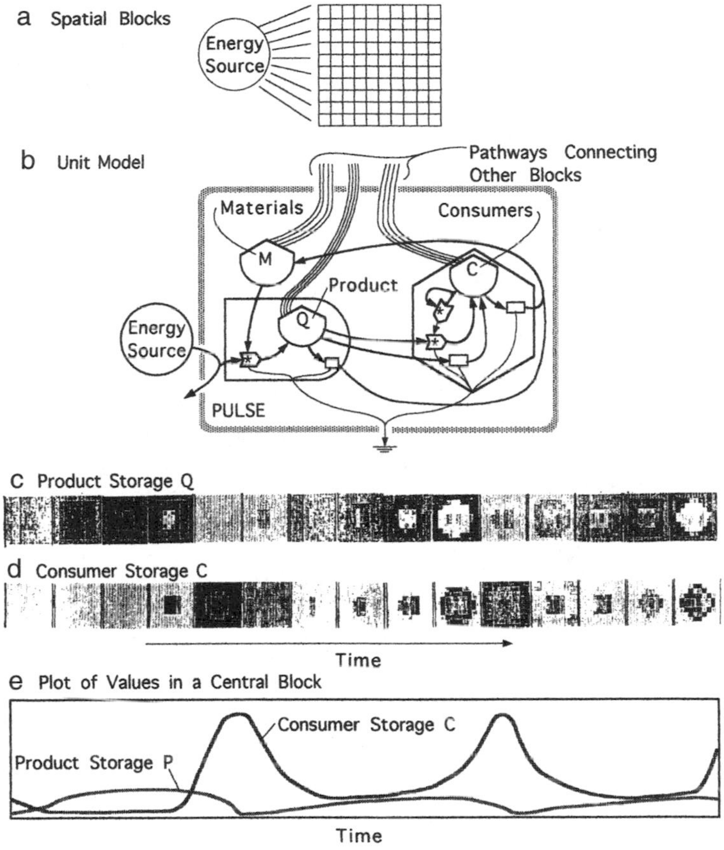

a Spatial Blocks

b Unit Model

c Product Storage Q

d Consumer Storage C

e Plot of Values in a Central Block

FIGURE 22.2.
Example of spatial simulation with the unit model PULSE (Figure 15.4) by Richardson (1987). (a) 100 blocks; (b) unit model; (c) 15 successive patterns of accumulated product arranged in sequence from left to right (quantity indicated by the intensity of shading); (d) time sequence of consumer storage; (e) time sequence of accumulated products and consumer storage in one of the blocks.

computation is large. Commercially available software packages for spatial simulation are used such as MAP and MFWORKS. These are part of the tools available to study geographic information (GIS = geographic information systems). ARCINFO is one of the computer graphics programs widely used to represent landscape data. The landscape is divided into polygons (irregular blocks) to which data for each area are assigned. Color-coded maps of various properties and combinations of data are derived. GRID is a version of that program that can be used for simulating changes in the blocks over time.

RAIN FOREST SUCCESSION

Using GRID (ARCINFO), Weber (1994) developed a unit model of rain forest succession in the Tabonuco forest of Puerto

Rico (Figure 22.3a) and simulated its spatial patterns for a square forest area with 1000 blocks. Values of storages were represented with intensity of shading (Figure 22.3b). The *Cecropia*-Tabonuco unit model (Figure 22.3a) has *Cecropia* representing successional vegetation drawing on sunlight and nutrients competing in parallel with mature tree species represented by Tabonuco (*Dacryodes excelsa*). Organic matter passes from the tree biomass to a litter storage, from which animal consumption and microbial decomposition release nutrients that recycle. Each block receives sunlight, nutrients (from rain, rocks, and other blocks), and seeds from other blocks. Biomass is removed from the blocks by herbivore animals that occupy larger territory. The switch symbols on seed pathways and light indicate threshold actions.

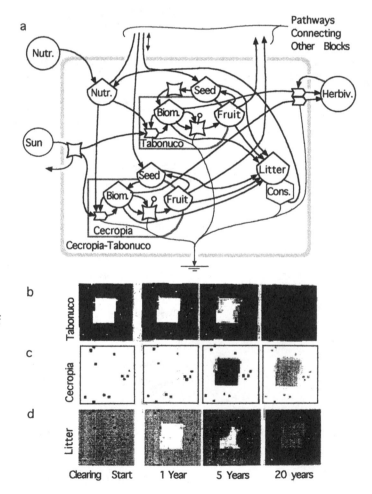

FIGURE 22.3.
Example of spatial simulation of rain forest succession in Puerto Rico by Weber (1994) relating successional tree species (*Cecropia*), mature Tabonuco forest tree species and litter. (a) Unit model; (b–d) time sequence of regrowth following forest clearing in the center of the area; (b) mature forest trees (Tabonuco); (c) successional *Cecropia*; (d) litter.

The successional species, *Cecropia,* is a weed tree with high net production, but short life (fast turnover time), developing rapidly where seeds have been dispersed. In the program seeds were distributed from a few trees scattered in the forest to surrounding blocks with probability decreasing with distance. Also depending on fruit dispersal, mature species develop as soon as shade levels reach a threshold. The mature species develop more slowly, with higher quality represented in the model as a longer turnover time. The fruits of these trees are fewer and larger, dispersing slowly.

In the simulation scenario shown in Figure 22.3b–d, central area of forest area was cleared and maps plotted of regrowth patterns. First, the open area filled in with *Cecropia* (Figure 22.3c). Then the mature vegetation developed closing the canopy in 20 years (Figure 22.3b). In the cleared area there was less litter production, and ground litter decreased (Figure 22.3d). As the vegetation redeveloped, organic litter returned to normal. These patterns are similar to those observed where patches of forest were experimentally cleared (Smith, 1970).

To represent the varying factors of light, nutrients, and topography on the steep mountain rain forest of the Luquillo forest, Hall et al. (1992) simulated a unit model of production, consumption, organic matter, and recycle in response to the destructive action of Hurricane Hugo. Irregular blocks were defined by ridge-ravine topography, which also affected solar energy because of the variation in slopes of land facing the sun.

As the examples here show (Figures 22.1–22.3), spatial simulations depend on the unit model, the interblock connections, and the impact of large-scale influences affecting all the blocks. Sometimes spatial simulations are presented without reporting the unit model or demonstrating its validity. Without an understanding of what the simulation is doing, the result cannot be used for policy or trusted for prediction.

FRACTALS

Fractal forms have the same shapes on each scale, large patterns on larger scale, and the same pattern but in a miniaturized version on the smaller scale (Mandelbrot, 1983; Schroeder, 1991). Simulation can generate the small shapes within the larger ones, cascading through many levels as space is divided into fractions, each with the shape of the original and so on. For example, a branching network of a tree results if half of a limb is diverted into a branch and each branch again split into half and that one split again and so on. Fantastic shapes result when various geometric operations are propagated in this way. Such patterns are common in nature (examples: snowflakes, flowers, deltas, highways, tidal channels, galaxies), apparently because of the hierarchy of energy transformations. Similar energy processes occur on each scale, but with different magnitude. The small are coupled to the large. For example, in turbulence large eddies interact with smaller eddies and these with even smaller eddies, etc. Finding similar designs on different scales is a property expected where each scale has the same model. Fractal algorithms can be included in simulations with digital compression programs and graphic plotting to represent observed patterns in space and time (Barnsley and Hurd, 1993).

22.2   VARIATION AND STATISTICAL MODELS

Statistics is the study of variation. Concepts and equations are developed for representing the distribution of values in space and time. Observations in nature show variation involved in nearly everything. There are very different views about the causes of variation and how this should be modeled and simulated.

INHERENT RANDOMNESS

One approach regards variation as inherently indeterminate, referred to as *randomness*. Mathematical equations are developed that start with estimations of chance which are represented as probability. Probability of a result is the fraction of its expected occurrence among possible occurrences. In flipping a coin, the chance of getting heads is one out of two, a probability of 0.5.

Bennett (1998) summarizes the history of mathematical thought and many controversies about randomness. The distribution of errors (X) in measurements is often represented with the Gaussian distribution, the well-known bell-shaped curve, where the number of values (Y) with deviations (X) from a mean are distributed according to a negative exponential squared: Y =

$K*e^\wedge - k*x^2$ (Figure 22.4a). Distributions that represent this concept of inherent variation are generated by mechanical devices for repeating chance actions or by similar manipulation of numbers by computer algorithms. For example, tables of random numbers exist (generated by these chance selections of numbers). With this approach to variation, randomness can be added to simulation models with random number tables or some computer algorithm that generates variation with a symmetrical distribution of variation about the curves generated by dynamic equations.

VARIATION AS SMALLER SCALE OSCILLATIONS

An opposing concept regards variation as the dynamic oscillations on a smaller scale that are seen as variations in curves of

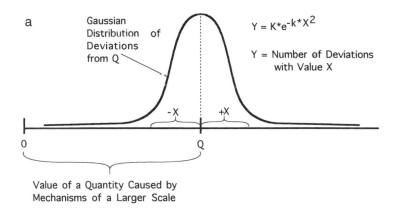

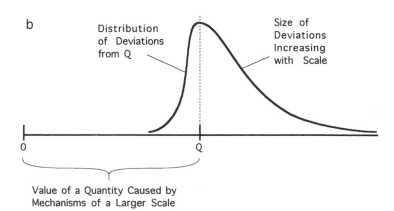

FIGURE 22.4.
Frequency distribution (Y) of deviations (X) from the values expected for phenomena of a larger scale. (a) Symmetrical distribution of variation described by the Gaussian equation; (b) skewed distribution of variation expected from fluctuations dependent on the scale of oscillations.

models on a larger scale. To simulate this variation, models of small scale can be added to larger scale models, making a more realistic model (Chapter 21). This point of view does not accept inherent randomness as a valid concept. Errors in measurements are regarded as dynamic fluctuations in the phenomena being measured and/or in the devices and persons making the measurements. This dynamic variation concept recognizes the hierarchy of the real world and its values (ultimately caused by the hierarchical distribution of energy transformations). Most variation in the real world is not distributed symmetrically (Figure 22.4a), but increases with scale (Figure 22.4b). Such curves are said to be *skewed*. The same percentage variation is found at each level of size, because each level has the similar effects of the variations from the next smaller scale. Measures of variations (such as variance) increase with the means. Weibull, gamma, and lognormal equations are used when a skewed distribution of variation exists.

Whether you believe that there is inherent randomness or that variation is the result of a hierarchy of dynamic oscillations or both, dynamic curves may be given somewhat more realistic variation with statistical devices for generating variation. Values generated by the equations (dynamic model) are varied according to a statistical distribution (Figure 22.4). A model with a source of variation included is a *stochastic* model (Nelson, 1995). For example, stochasic models fit marketing, inventory, and optimization (Ross, 1992). Simulations are used to estimate risk (Vose, 1996; Evans, Olson, 1998).

VALIDATION WITH STATISTICAL MEASURES

If variation fits a statistical distribution somewhat, the tests for probability derived from that mathematics are convenient for determining how well a simulation curve fits observed data. The probability of curves being different is estimated. In Chapter 21 we suggested use of chi-squared tests to compare simulated and observed time series. Another way of validating a model plots a graph of simulated data as a function of observed data. Then the fit can be evaluated with correlation measures. Or vertical bars can be placed on points representing some measure of the variation. Bars on observed data plots can be compared with the bars on curves generated by a stochastic model. Where plus and minus the standard error of the mean is used for the bars and the two sets of bars don't overlap, the curves have a high probability of being different (a property based on Student's t distribution for small samples). Refer to texts on biometry and statistics. However, all of these statistical measures and procedures are based on assumed distributions of variation (normal distributions, Student's t distribution, chi-squared distribution, Poisson distribution) and are not accurate where the variation does not follow the mathematical equations used for that distribution. Sometimes the data are transformed to achieve a distribution of variation that matches the equations on which the statistical measure is based.

STATISTICAL METHODS OF FINDING AN APPROPRIATE DYNAMIC MODEL

With statistical modeling, the equations for dynamic simulation are derived by statistical procedures for fitting lines or curves to data. For example, a regression equation can be obtained to represent observed data for growth. Grant (1986) provides explanations and wildlife examples.

To derive equations from observed data sounds realistic. However, descriptive equations derived without understanding of the mechanisms provide no basis for using them for a different range of values, different time, or different situation. For example, an observed growth curve might be part of a system that will develop a steady state, or it might be part of a pulse that will turn down. Extrapolating the observed curve has no basis.

Determining coefficients by empirical fitting to past data cannot predict the future if the coefficients are actually variables affected by changes. For example, in a complex model of economic growth, Meadows et al. (1972) determined many coefficients relating one variable to another from existing data. These coefficients were determined in a time of cheap available energy but were applied to a future time when less fossil fuel energy is expected. Because energy augments nearly all production processes in our economy, output coefficients decline when energy is more expensive. When the same procedures were used to simulate the economy 25 years later (Meadows et al., 1992), limits to growth were found 80 years sooner than in the previous simulations made before the improved coefficients had changed.

◆

22.3   APPROACHING MODELS THROUGH MATHEMATICS

Procedures in this book translate verbal thinking into network diagrams after which equations are written for simulation programs. Emphasis is on visualizing and calibrating the whole model in a network view before writing equations for the computer to do the various calculations. Little mathematical training is required. Students may simulate systems without realizing that they are doing calculus operations.

However, many people approach simulation through mathematics, writing equations for each part of the system using mathematical expressions for various mechanisms and processes from analytic geometry and other fields (Beltrami, 1997). For example, mathematical models are used for growth of microbes (Hurst, 1996) and for animal behavior (Lendrem, 1986; Mangel and Clark, 1988). Many published papers and books on simulation give the models in mathematical form and use its associated vocabulary. For example, the word *graph* refers to pathway diagrams. A branched network without a loop is a *tree*. Simulation is solving equations by numerical integration. Students may want to take additional math courses to use this literature.

For simpler systems, families of curves are often derived from mathematical computation rather than from simulation. Equations derived from mathematical integration do not have cumulative errors of numerical simulation. (Examples Figures

In the widely used state-change way of thinking, the set of conditions at one time are the states (values of storages). There is a set of coefficients changing states at time t to that at the next time step. Each state change coefficient is multiplied by one or more state variables to obtain part of the change in itself or other state variables during the time interval. The set of coefficients is a transition matrix. The coefficients in many cases turn out to be the same pathway coefficients we derive from the systems diagrams. Even where equations for the parts are written first, overview can be aided and consistency checked by diagramming the set of equations (Chapter 6).

STELLA and the library of blocks supplied with EXTEND require the writing of equations first. This contrasts with our use of EXTEND in Chapters 4 and 5 where simulation is arranged with the symbols, which automatically apply the appropriate equations.

MATHEMATICS SOFTWARE

Routine teaching of mathematics now uses commercial mathematical software such as MATHEMATICA, MAPLE, MATLAB, etc., in which the student can enter equations, values, math symbols, and plot the results of the calculations including attractive finished graphs of one variable as a function of others. These software packages for calculations can be used for simulation. For example, Roughgarden (1998) uses MATLAB as the basis for a course to introduce concepts of mathematical ecology and related fields, emphasizing, as in this book, that understanding requires models and simulation. In his approach, after explaining the rationale for an equation, it is typed into the program in text form, after which the computer plots the graph with time or some other variable. The emphasis is on population equations elaborating on mathematical relationships of some of the same minimodels given in Part III of this book. The Roughgarden book, like this one, started as a workbook of models and exercises, but requires more mathematics background.

SIMULATING FLUID DYNAMICS

Air, water in lakes and oceans, and fluid earth move according to the balance of forces on each parcel. The usual forces affecting lateral motion include pressure gradient force, Coriolis force of earth rotation, centrifugal force where fluids are spinning, eddy transfer of momentum, and friction. For up and down motions, there is also gravity. For spatial simulation these appropriate forces are included in each unit model. After the area is divided into blocks, a point in the center of each block is designated as the center of mass of the fluid parcel contained in that block. The equation for each block is the sum of the forces, called the equation of motion. It is an acceleration equation. Fluids are accelerated along the pathway from one block's center to those adjacent according to the difference in forces, the density, and the distance between the centers. Where mass of fluid accumulates in a block, it has to expand vertically, increasing the pressure below. A "continuity" equation calculates these changes. The unit models become more complex when equations for heat, moisture changes, and density are added. In applications to the real atmosphere in three dimensions over the earth there are so many blocks and equations that super computers are required. Jacobsen (1998) provides an introduction.

22.4 METHODS FOR LARGER SIMULATION PROGRAMS

Special techniques are useful for larger simulation programs.

COMPILING

With the fast computers now available to everyone, minimodel simulations run almost too fast for the eye to see as points plot in sequence across the screen. Larger models may take longer, either because they are more complex or because they have so many duplicate units. With appropriate converter software, a simulation model may be compiled to run faster. The compiler converts the many separate steps into one process in binary machine language (the 1's and 0's of computer memory). Many simulation software packages automatically compile for each run.

SENSITIVITY ANALYSIS

In the chapters in Part III, "What If" experiments were suggested to explore the responses of the systems. In a sensitivity analysis, a systematic procedure is used to test the response of the system to changes in all of the variables, one by one. For example, the test can record the responses to a doubling of each source and state variable. Learning where a system is sensitive can help us anticipate impacts. Large responses in simulation can be compared with a range of variations in observed data. Simulation software packages such as STELLA and EXTEND have subprograms for doing a general sensitivity test in one operation.

USING MATRIX EQUATIONS FOR SIMULATION

Where there are many units, all with similar equations, and where the equations are simple, the simulation programs can compute the iterations efficiently using matrix operations. To get the idea, consider the spatial simulation with many blocks in Figure 21.1. For one block on each iteration a new storage value Q_{t+1} is calculated as the old Q_t plus the inflow J plus the change due to flows between the blocks DQ. The change due to interblock flows is the product of the K coefficients and the appropriate Q storages (= states) in the adjacent blocks. For block 5 in Figure 21.1a:

$$Q_{5t+1} = Q_{5t} + J_{r5} + DQ_5, \quad \text{where}$$
$$DQ_5 = K_{25}{}^*Q_2 - K_{25}{}^*Q_5 + K_{45}{}^*Q_4 - K_{45}{}^*Q_5$$
$$+ K_{65}{}^*Q_6 - K_{65}{}^*Q_5 - K_{85}{}^*Q_8 - K_{85}{}^*Q_5$$

The equations for all of the blocks together can be represented with matrix mathematics so that all the blocks are calculated simultaneously on each iteration. The matrix expression for the system simulation is analogous in form to the equation for a single block and is shown In Figure 22.5 with block 5 highlighted and its coefficients shown.

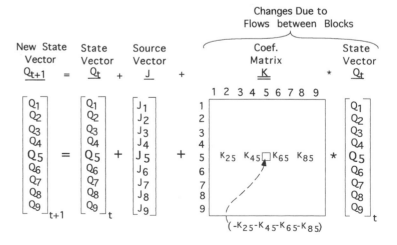

FIGURE 22.5.
Matrix expressions for an iteration with the water model in Figure 22.1.

Multiplying the state vector Qt by the coefficient matrix K generates a vector of changes DQ due to the flows between blocks. (In this operation the column vector is changed to a row vector and multiplied by each row of the matrix, starting at the top. Then the products in the row are added across to get the new Q values.) On each iteration, the program calculates new state values Q_{t+1} by adding the vector of source inputs J and the changes due to exchanges with surrounding blocks $K*Q$ to the state vector Q_t for time t. The state values Q_t for each iteration can be graphed or assigned by spatial graphics to show the distribution on a map.

Here the use of matrix equations in simulation was explained using a spatial example. However, matrix operations are used for other kinds of simulation wherever there are duplicate, repetitive iterations. For example, the Leslie matrix is a table of population numbers with age classes on one axis and the processes affecting each age class on the other axis. It is readily simulated by multiplying on each iteration the vector of age class numbers by the matrix of age-specific processes to obtain a new vector of age class numbers for each new time step. Grant (1986) explains with wildlife examples.

Programming languages (including BASIC) contain fundamental mathematics operations such as matrix addition, subtraction, multiplication, and inversion (analogous to division). Matrix operations are useful in spatial hydrology and fluid dynamics simulations where the same set of equations is used for large numbers of blocks.

On the right side of Figure 22.5, a state vector (list of Q storage values) was multiplied by a transition matrix (table of K coefficients) to obtain a new state vector. In many cases matrix algebra can calculate a substitute matrix that has the same result when multiplied by the vector, but has numbers only on the diagonals. These are called eigenvalues. Called an eigenvector, a set of Q values can be defined that will generate a multiple of itself when multiplied by the substitute matrix with its eigenvalues. This set of values in the eigenvector is the steady state. For example, the eigenvalues and eigenvector for the population numbers in a Leslie matrix are the stable age distribution (the distribution of numbers by age classes that generates itself in steady state).

For systems where the state change coefficients are linear (multiplied by only one state variable), matrix operations are used to find steady states and criteria for stability. However, except when available energy is very small, real systems develop autocatalytic designs (maximum power principle) and are not linear. When equations are solved for the steady state (changes equal zero), many nonlinear terms become linear, but then the simplified equations only apply to small changes near equilibrium. Since all storages have depreciation (second law), there are negative terms on the diagonal of system matrices, a mathematical criterion of stability. Logofet (1993) provides matrix views of many ecological examples.

The rules for use of pathways and numbers on some other kinds of network diagrams are quite different from the energy systems language. For example, in one usage the arrowhead is directed from consumer to food, opposite to the energy flow. In energy systems language the arrowhead is always in the direction of the flows of energy, matter, and information—from food to consumer. The action of the consumer to get the food is shown as an autocatalytic feedback interaction (Figures 2.7 and 2.13a).

INPUT–OUTPUT SIMULATION

Simulations of systems can be done by matrix operations that extrapolate input–output relationships. The flows of materials or money between the units of a network can be represented with an *input–output table* (sometimes called a *transaction matrix*) (Figure 22.6a). Flows of materials are from the items listed at

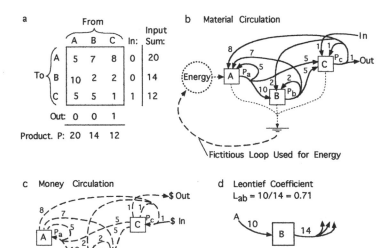

FIGURE 22.6.
Example of an input–output table and its representation as an energy systems diagram. (a) Table of flows between three units; (b) network diagram with energy pathways added; (c) network diagram with reversed directions of flow, which are appropriate if flows are money; (d) calculation of a Leontief coefficient describing the flow relationship of two units.

the top to the items listed along the side. These flows can also be represented with an energy systems diagram that can show the energy and hierarchical relationships (Figure 22.6b). The diagonals of the matrix are flows circulating within the unit. The total of numbers added across is the total of inflow, whereas the total of columns adding down is the productive output. At steady state the sums of rows and columns are the same.

Energy circulation has been used in input–output tables by connecting the pathways of degraded energy (heat sinks) back as inputs of available energy (Figure 22.6b, which is an arrangement for mathematical purpose contrary to the second thermodynamic law).

Money paid for commodities flows in the opposite direction from the materials (Figure 22.6c). If the flows of money were numerically the same, the table in Figure 22.6a would suffice by swapping the words "from" and "to." In an economic system units in the network are called *sectors* (industrial, agricultural, etc.). The human consumers are the *final demand sector* and are appropriately placed on the right, representing a superior hierarchical position (C in Figure 22.6c).

A Leontief coefficient (input per unit output) can be calculated for each input–output box as the ratio of the input flow to the total output (the production). For example, in Figure 22.6d the input flow from A to B is J_{ab} and the productive output of B is P_b. This Leontief coefficient is $L_{ab} = J_{ab}/P_b = 10/14 = 0.71$. These coefficients measure the structural linking of one unit to another by the flow. If it is believed that the system as a whole is going to grow 10%, without much change in the relationships of the various units, then matrix inversion methods can be used with the coefficient matrix to backcalculate what each intersector flow will become consistent with the increased total growth. Such calculations have been made for national economies with more than 600 boxes (sectors).

◆

22.5   DISCRETE PROCESSES AND MODELS

Many processes and models have numerical steps. For example, stairs in a house have steps. The height of the levels where you place your foot are in a discrete series. With continuous functions the values for a state can have any value. With discrete functions the values are in steps. Most of the models simulated in this book involved continuous functions, although some involved on–off switching actions (example: the FIRE program of Table 15.3).

LOGIC NETWORKS

Logic networks are arrangements of on–off units. For example an AND gate unit is on (has an output of 1) when its inputs are both on (have values of 1). An OR gate is on if either of its inputs is on. A NOT unit is on when the input is off, but off when the input is on. A threshold switch is on when the input of a continuous function is greater than a threshold value (example: water overflowing a river bank). Flip-flops are set by an input pulse of 1 and stay on until turned off by a reset pulse. Some systems have networks of these logic units. They are discrete systems since the values have only two alternative steps, on and off. Models of discrete processes can be simulated by writing a program with a sequence of logic actions. Simulation programs, including BASIC, have the common logic actions AND, OR, NOT, IF-THEN, etc., as part of the language.

Systems of continuous functions can be simulated with discrete networks with results that exaggerate responses. For example, a flip-flop unit can simulate the competition between two species that exclude each other in competition. Depending on input thresholds, there is one species or the other, but not both. Robertson and Muetzelfeldt (1991) use logic programming in ecological systems.

Hardware simulators are available in which you can connect various logic units with jumper wires so that the complex consequence of many units is simulated. The digital computer is such a discrete system built of enormous numbers of electrical logic circuits with values either on (logic 1) or off (logic 0). By having a huge number of such units in the extremely miniaturized network on the computer chip, the digital computer simulates continuous functions, carrying numbers to 40 significant figures.

CUMULATIVE ERROR IN DIGITAL SIMULATION

There is a tiny error that can accumulate in digital simulation of continuous functions, especially in growth processes. Where it is important to have a conservation of something (example: keeping total materials constant), we avoid the error by having the value for material in one unit set equal to the constant total minus that calculated for the other units (INTLIMT, Table 12.3).

NEURAL NETWORKS

The neurons of brains are something like logic circuits in that their outputs are dependent on the combined action of inputs. Network diagrams and mathematics originally made to describe nerve interactions have evolved into a systems language for dealing with other systems, particularly those that have discrete thresholds and actions. Systems are simulated with neural network programs and software. Complex networks are studied to understand how organisms think.

QUEUES OF DISCRETE UNITS

Some systems have separate units passing from one process to another. Examples are cars moving down a manufacturing assembly line and people waiting in line at the bank. These are queuing systems. In industry, where many parts are being processed and combined, the network of flows and junctions may be complex. Programs such as EXTEND have ready-made components and procedures to simulate systems that process discrete units. Simulating the flow of discrete units is important to manufacturing.

22.6   PARALLEL PROCESSING

The first electrical simulations used networks of wires and electrical storage units (capacitors) in which the flows and storage of electric charge were analogous to the flows and storage of water (not unlike the network in Figure 22.1). All parts of the network were operating simultaneously with electrical recorders tapped in to record water levels and flows at points of interest. Later "operational analog computers" had electrical units (operational amplifiers) for each unit in which the voltages were manipulated according to the equations for that unit. The parts of these earlier electrical simulators operated simultaneously in parallel. Except in some industrial uses, electrical analog computers were replaced by the digital computer because of its speed and accuracy. As used in this book, numerical simulation uses the equation for each unit to calculate the values for each unit successively, one by one. It plots points, and then does it again for the next time step. There is one sequence of calculations. New initiatives are now cropping up to develop parallel processing again, but this time using simultaneous digital computations in parallel.

◆

22.7 FRONTIERS AND PERSPECTIVES

Simulation is becoming a mainstream of scientific investigation in old and new fields (Casti, 1998). New areas of application are the study of artificial intelligence (systems that learn and control themselves) and fuzzy logic (systems with equations that allow a range of values; Klir and Folger, 1988). Although simulations are rarely trusted as reliable predictors in public affairs, public and business policies are increasingly seeking the insights of models—for example, energy policy (Bunn and Larsen, 1997), risk analysis (Evans and Olson, 1998), and global futures.

Simulation modeling is a varied and creative field, as observed in the issues of the journals *Simulation, Ecological Modeling and Systems Ecology,* and *Environmental Modeling and Software.* Application is best done by each person modeling and programming his or her own models to learn the behavior and attributes of a particular system of interest. Languages often used include BASIC, DYNAMO, FORTRAN, PASCAL, and C++. For more complex models and simulations involving many repetitive operations, commercial software packages are available such as EXTEND, STELLA, VISUALBASIC, CSMP, GASP, GPSS, and many others, although considerable time is required to learn each. Because principles of materials, energy, and information apply to all scales, apparently similar models and configurations are found in most fields. Understanding, applying, and simulating general systems models is emerging as a unifying branch of knowledge.

OTHER APPROACHES TO SYSTEM SIMULATION

The reference list includes the books and papers cited and some other recent introductions to simulation. These introductions use different approaches, different choices of models believed to be fundamental, and different subject areas of application. Simulation of systems is in a creative stage.

Appendix A

◆

PROGRAMMING ENERGY SYSTEMS BLOCKS FOR SIMULATING WITH EXTEND

Whereas the instruction manual that comes with the program EXTEND fully explains how to make and program your own blocks for use with their standard procedures and libraries (example: Figure 6.9), there are special features in the unique blocks which we developed to simulate energy systems diagrams *without* the user having to write equations. The equations are preprogrammed into the blocks. The following procedure makes a block for EXTEND that is compatible with our energy systems libraries of programmed blocks (Chapters 4 and 5). You may need the main instruction manual to cover items not given in these instructions. Also see Odum and Peterson (1996).

◆

A.1   PROGRAMMING SCREENS

To see the program of an ICON-BLOCK and/or make changes, double click on the block's icon while holding down the option key (Macintosh) or the Alt key (Windows PC). Two screens appear, one for the dialog box of that block and one for the block program. For example, Figures A.1a and A.1b show the screens for the block TANK. Each of the items in the dialog box (Figure A.1b) can be changed. The larger screen (Figure

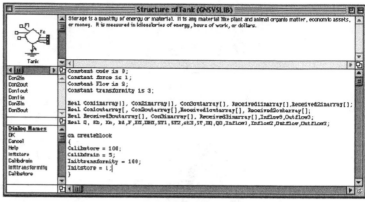

a

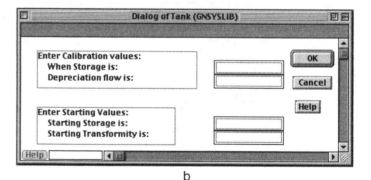

b

FIGURE A.1.
Screen views of the program for
the block named TANK.

A.1a) has a panel for making a picture icon in the upper left, a
panel for help statements in the upper right, and a panel for
writing the main program script on the lower right. Following
instructions in the EXTEND manual, you can enter words and
icon graphics as needed to write a program or change an existing
program. EXTEND program script is written in a language similar to C++.

A.2   ARRAYS SHARING INFORMATION AMONG BLOCKS

The energy systems language symbol blocks for EXTEND are
based on four position arrays for sending information back and

forth between blocks. Figure A.2 shows how the information is passed between blocks in four channels. When one block on screen is connected by means of a line to another block with the mouse, each block program is arranged to share information that is on a four-position array. Position 1 is a code that each block's program can use to take special actions. Position 2 is a force. Position 3 is a flow. Position 4 is a transformity. As the program iterates, the values in the array are changed, some by the upstream block and some by the downstream block.

For example, a store block (see Figure A.2) uses the flow sent by the upstream unit and received in the input connector. It also receives the transformity of that flow so it can calculate the emergy being stored. The storage block determines the force for the output connector, which the next unit uses to determine its use of the outflow. Using information shared through the connectors, the downstream block sends back to the storage block the downstream part of the information on what was used. In the upstream block this information is combined with the information on the stored quantity to determine the amount to be added or subtracted from the storage tally. The outflow array from the storage contains the transformity of the stored quantity that is sent to the downstream unit.

Each block takes array information that was written in connecting blocks, and this information is first placed in receiving arrays, such as "Received1inarray" from upstream and "Received1outarray" from downstream. Then, within the program, these received values of code, force, flow, and transformity are transferred to corresponding positions in arrays that operate within the block. For example, the program statement "Con1in-

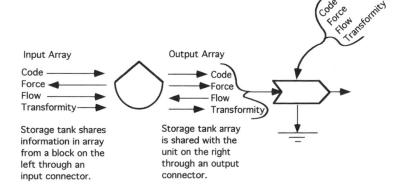

FIGURE A.2.
Interblock information transfers with arrays for a storage block.

Input Array

Code
Force
Flow
Transformity

Storage tank shares information in array from a block on the left through an input connector.

Output Array

Code
Force
Flow
Transformity

Storage tank array is shared with the unit on the right through an output connector.

Code
Force
Flow
Transformity

array[flow]" = "Received1inarray[flow]" says that the values of the "Con1inarray" array become equal to the values in the receiving array. Those array positions that don't receive information from other blocks are left at zero. After the block program has made its computations, its values are passed to an output array and then shared upstream and downstream through the connector arrays.

$$\text{Con1in} = \text{Passarray(Con1inarray)};$$
$$\text{Con1out} = \text{Passarray(Con1outarray)};$$

◆

A.3   STEPS FOR PROGRAMMING THE SCRIPT

1. *Draw what you want on paper first.* Write the equation or equations that you want the block to perform. Decide what input connections are required, what information the inputs are supposed to receive from other blocks, and what the output connections are to transmit to the next block.

2. *Make an icon.* Use any drawing program to make a small picture appropriate to the block you are making. Or use a scanner to put a small picture on disk. Scale the icon to be about 1 cm by 1 cm or smaller. Save the icon to disk to be safe and then to a clipboard. Or use the small drawing program that is part of EXTEND to make the icon after you load EXTEND (see step 3).

3. *Load EXTEND.* After turning on the computer, double click on the hard disk icon to see the various folder and program icons; double click on the EXTEND folder to open it and then double click on EXTEND (triangular icon).

4. *Modify an existing block as a shortcut to making a new one.* Use the FILE menu of EXTEND and open the Gensymb folder. Double click on the worksheet GENSYMB. The screen will be filled with the icons of the present general systems blocks. You can select and reprogram one of these that is somewhat like the one you want to make. This saves many steps and errors. Hold down the option key (Macintosh) or Alt key (Windows PC) as you double click on the icon. The two program screens appear (example: Figure A.1), one behind the other. One is the screen for arranging dialog boxes and instructions; the other has the icon, the help statement, and the program script. You can

make the needed changes and then save the revised program under another name. Using Save As from FILE menu, save the block to a new library. You can decide what changes to make by reading through the procedure for making a new block from the beginning (steps 5–9, next subsection).

MAKING A NEW BLOCK

5. *From the DEFINE menu, select Build a New Block.* A screen appears for you to type in the name of the block and select a library to put it in. After that a pair of screens appears ready for programming. Click on the second screen (behind) to bring it to the front. Either draw an icon in the upper left space or paste an icon that you drew earlier (step 2) from the clipboard (by clicking on the space and then releasing on Paste in the EDIT menu).

6. *Add input connectors.* Using the upper drawing bar, click on the connector box and then click within the space where you want a connector to appear. For each connector a name automatically appears in the connector list on the left side of the screen below the icon. The connector that appears is an input connector (square white connector) and its name ends with "in." Use a mouse to move input connectors to the left of the icon.

7. *Add output connectors.* To make output connectors from an input connector, retype the name of the connector so that it ends with "out." The connector turns black. Move one or more of the output connectors to the right of the icon.

8. *Label the icon.* First click on the "A" box in the upper bar; then click on the place you want a label to go. Type the name of the block and move it underneath the icon.

9. Even before you finish it may be a good idea *to save* what you have done by clicking in the put-away box in the upper left corner of the screen. A screen appears and you can check "Save without compiling." Save with the block name you plan to use and the destination in a new library. To avoid confusion, people working in a class need to designate their libraries with their names.

DESIGNING THE DIALOG PAGE

10. *Make dialog boxes.* In these boxes, users can type in numbers for calibration or for setting initial conditions. Pull the

dialog screen to the front by clicking on it. Using the DEFINE menu, select New Dialog Item. A screen appears with 12 choices. Select Parameter. A screen appears in which you type the word that is to be the program label for the number that is to be entered. For example, you could type WATERSTORE (one word) if the box is supposed to receive a number representing storage of water. Move each parameter box that you put on the screen into the correct position with the mouse. Each parameter label that you enter appears in a list of variables on the icon screen (lower left).

11. *Put text labels next to these dialog boxes.* This is done by selecting New Dialog Item as in step 10, but then selecting the choice Static Text (Label). A box appears on the screen for you to type in the label. Then when this is completed, the label box appears on the dialog screen. Move it into position. You will probably have to expand the box to fit the words by using the mouse pointer to pull at the lower right corner.

WRITING THE SCRIPT

Table A.1 is the script for the tank block of the general systems library GNSYSLIB.lix. Use this script and the screens in Figure A.1 as an example as you go through programming steps 12–15 that follow:

12. *Designate program labels.* Pull the script icon screen into view again (by clicking on the edge of it that is visible). At the top of the script space, you first designate the names of the array positions:

> Constant code is 0;
> Constant force is 1;
> Constant flow is 2;
> Constant transformity is 3;

Next type all the address names that are used in the program on a line starting with a word to indicate the type of variables. Names already in the two lists on the lower left side do not have to be listed. For example, the following would be listed for one input connector and one output connector:

TABLE A.1
Script of Icon-Block TANK in Library GNSYSLIB.lix

```
Constant code is 0;
Constant force is 1;
Constant Flow is 2;
Constant transformity is 3;

Real  con1inarray[], Con2inarray[], received1inarray[],Received2inarray[];
Real  con1outarray[],Con2outarray[], received1outarray[],Received2outarray[];
Real  Con3inarray[], received3inarray[],inflow3;
Real Q, Kh, Km, kd,F,EM,DEM,ST1,ST2,st3,ST, DQ, Q0,Inflow1, Inflow2, Outflow1,
outflow2;

on createblock
{
Calibstore = 100;
Calibdrain = 5;
Inittransformity = 100;
Initstore = 1;
}

on simulate
{
Em = St*Q;

If (Not Getpassedarray(Con1in, Received1inarray))
{
Con1inarray[code] = -1;
Con1inarray[Force] = 0.0;
Con1inarray[Flow] = 0.0;
Con1inarray[transformity] = 0.0;
}
Else
{
Con1inarray[Flow] = Received1inarray[Flow];
Con1inarray[code] = received1inarray[code];
Con1inarray[transformity] = received1inarray[transformity];
}
If (Not Getpassedarray(Con2in, Received2inarray))
{
Con2inarray[code] = -1;
Con2inarray[Force] = 0.0;
Con2inarray[Flow] = 0.0;
Con2inarray[transformity] = 0.0;
}
Else
{
Con2inarray[Flow] = Received2inarray[Flow];
Con2inarray[code] = received2inarray[Code];
Con2inarray[transformity] = received2inarray[transformity];
}

If (Not Getpassedarray(Con3in, Received3inarray))
{
Con3inarray[code] = -1;
Con3inarray[Force] = 0.0;
Con3inarray[Flow] = 0.0;
Con3inarray[transformity] = 0.0;
}
Else
{
Con3inarray[Flow] = Received3inarray[Flow];
Con3inarray[code] = received3inarray[code];
Con3inarray[transformity] = received3inarray[transformity];
}
If (Not Getpassedarray(Con1out, Received1outarray))
```

(continues)

TABLE A.1 (*Continued*)

```
{
Con1outarray[code] = -1;
Con1outarray[Force] = 0.0;
Con1outarray[Flow] = 0.0;
Con1outarray[transformity] = 0.0;
}
Else
{
Con1outarray[Flow] = Received1outarray[Flow];
}
If (Not Getpassedarray(Con2out, Received2outarray))
{
Con2outarray[code] = -1;
Con2outarray[Force] = 0.0;
Con2outarray[Flow] = 0.0;
Con2outarray[transformity] = 0.0;
}
Else
{
Con2outarray[Flow] = Received2outarray[Flow];
}

Inflow1 = Con1inarray[Flow];
Inflow2 = Con2inarray[Flow];
Inflow3 = Con3inarray[Flow];
Outflow1 = Q* Con1outarray[Flow];
Outflow2 = Q* Con2outarray[Flow];

DQ = inflow1  + Inflow2+inflow3 -Kd*Q - outflow1 -outflow2;
Q = Q + DQ * DeltaTime;
If (Q==0)
Q=0.000000001;
If (Q<0)
Q=0.000000001;

St1 = con1inarray[transformity];
St2 = con2inarray[transformity];
St3 = Con3inarray[transformity];

If (DQ/Q> 0.05)
DEM = St1*inflow1 +ST2*inflow2 +st3*inflow3- ST*outflow1 +

If (Dq/Q <0.05 and Dq> 0.0)
DEM = 0;

If (Dq ==0.0)
dEM = 0.0;

If (DQ <0.0)
DEM = ST*DQ;
EM = EM + DEM*Deltatime;
If (EM ==0.0)
EM = 1;

If (EM <0.0)
EM = 1;

ST = EM/Q;
Con1inarray[force] = Q;
Con2inarray[force] = Q;
Con3inarray[force] = Q;

Con1outarray[transformity] = ST;
```

(*continues*)

TABLE A.1 (*Continued*)

```
Con2outarray[transformity] = ST;
Con1outarray[code] = 1;**code 1 is force out
Con2outarray[code] = 1;**code 1 is force out
Con1outarray[Force] = Q;
Con2outarray[Force] = Q;

Con1in = Passarray(Con1inarray);
Con2in = Passarray(Con2inarray);
Con3in = Passarray(Con3inarray);

Con1out = passarray(Con1outarray);
Con2out = passarray(Con2outarray);
}

** Initialize any simulation variables.
on initsim
{
Q = Initstore;
St = Inittransformity;
Kd = Calibdrain/Calibstore;
Em = Q*St;

Makearray(con1inarray,4);
Con1inarray[code] = -1;
Con1inarray[Force] = 0.0;
Con1inarray[Flow] = 0.0;
Con1inarray[transformity] = 0.0;

Makearray(con2inarray,4);
Con2inarray[code] = -1;
Con2inarray[Force] = 0.0;
Con2inarray[Flow] = 0.0;
Con2inarray[transformity] = 0.0;

Makearray(con3inarray,4);
Con3inarray[code] = -1;
Con3inarray[Force] = 0.0;
Con3inarray[Flow] = 0.0;
Con3inarray[transformity] = 0.0;

Makearray(con1outarray,4);
Con1outarray[code] = -1;
Con1outarray[Force] =0.0;
Con1outarray[Flow] = 0.0;
Con1outarray[transformity] = 0.0;

Makearray(con2outarray,4);
Con2outarray[code] = -1;
Con2outarray[Force] =0.0;
Con2outarray[Flow] = 0.0;
Con2outarray[transformity] = 0.0;
}

** User clicked the dialog HELP button.
on help
{
showHelp();
}
```

Real Con1inarray[ ], Con1outarray[ ];
Real Received1inarray[ ], Received1outarray[ ];

On another line type any variable or parameter names used within the block such as:

Real, Q, Water, K1, K2, DQ, DE, Inflow1, outflow1;

13. *Type program lines between the brackets.* Brackets are already provided for the several categories of message handling. Program statements must end with a semicolon. Capitals or lowercase letters are interchangeable. If you want to leave a remark in the program that tells you later what you were doing or why, type a ** or // followed by the remark. The program ignores anything on the same line that follows ** or //. You will need to program three message handling groups which can be in the script in any order:

On Createblock, which gives values to dialog boxes when they open.
On Initsim, which calculates the initial conditions before the run.
On Simulate, which has the program steps during the run.

14. *Scroll the script program space until On CreateBlock appears.* (Use the mouse pointer on the arrow along the right margin.) Type here whatever numbers you want the dialog boxes to have when they are first called to the screen. Type a parameter name (that will be used by the program) showing it equal to the number you want to appear. For example, type

```
{
WATERSTORE = 1000;
}
```

Whenever someone pulls the block from the library onto a worksheet screen, it will have the value 1000. Thereafter the user can change it if needed. If you put in nothing, the dialog box will be empty when it is clicked on.

15. *Next scroll the script space until On Initsim appears.* Here you should set initial conditions and create the input and output arrays that will be used:

```
{
Makearray(con1inarray,4)
Con1inarray[code] = −1;
Con1inarray[Force] = 0.0;
Con1inarray[Flow] = 0.0;
Con1inarray[Transformity] − 0.0;

Makearray(con1outarray,4)
Con1outarray[code] = −1;
Con1outarray[Force] = 0.0;
Con1outarray[Flow] = 0.0;
Con1outarray[transformity] = 0,0;

Eq = 1000; **initial solar emergy storage
}
```

16. *Scroll the script space to find "On Simulate."* Type the simulation program between the brackets. Start with an IF statement as to what values the arrays should receive from connecting blocks and which should be set from the block being programmed. For example:

```
If (not getpassedarray(con1in, Received1inarray))
{
Con1inarray[code] = −1;
Con1inarray[force] = 0.0;
con1inarray[flow] = 0.0;
Con1inarray[transformity] = 0.0;
}
Else
{
(Con1inarray[flow] = Received1inarray[flow];
(con1inarray[transformity] = Recieved1inarray[transformity];
}
```

This statement sets input array positions to zero if nothing is received, but if received, the flow is given the value received from the upstream input.

A similar statement is used to receive downstream values shared by the output array:

```
If (not getpassedarray(con1out, Received1outarray))
{
Con1outarray[code] = −1;
Con1outarray[force] = 0.0;
con1outarray[flow] = 0.0;
Con1outarray[transformity] = 0.0;
}
Else
Con1inarray[flow] = Received1inarray[flow];
```

17. *To write short equations for action, transfer information received to shorter internal names or letters.* For example:

```
Jin = Con1inarray[flow];
Stin = Con1inarray[transformity];
Jout = Con1outarray[flow];
```

18. *Write the equations using the short labels.* For example, if the change in storage is the sum of inflows minus outflows:

```
DQ = Jin -Jout- K1*Q;
```

and the new value of storage is:

```
Q = Q + DQ*deltatime;
```

Deltatime is the iteration interval DT.

For emergy storage (Eq), storage Q, and solar transformity received in (Stin), the transformity of storage St is:

```
St = Eq/Q;
```

and the change in emergy storage sums the products of flows and transformities:

```
DEq = Stin*Jin − St*Jout;
```

and the new value of stored emergy is:

```
Eq = Eq + DEq*deltatime;
```

19. *Next, assign the new values to input and output arrays:*

Con1inarray[force] = Q;

Force is put in an upstream array in case a block is connected that is affected by backforce:

Con1outarray[force] = Q;
Con1outarray[transformity] = ST;

20. By identifying the set of values in arrays with those in connectors, you pass values to connecting blocks:

Con1in = Passarray(con1inarray);
Con1out = Passarray(con1outarray);

Don't forget the closing bracket of On Simulate: }.

21. *Click on the small square to close the program.* A screen appears. Check "To compile if necessary." It may give you error messages for things you have omitted such as semicolons and variables declared in the first lines of the program. Make corrections and repeat.

22. *After it compiles, try to use your block by putting it on a model (worksheet screen).* Use a source and a plotter and see if it generates the output that you planned. During the programming, it may help to print out what you have. If not, return to the program and revise, etc.

PROGRAMMING BLOCKS THAT PUMP
FROM UPSTREAM

Interaction blocks control the inflows from upstream. Their function is a product of a coefficient and the "forces" received. For example, the equation for production $P = K1*N*Q$ is based on the values of forces N and Q received from upstream connectors. For the systems to work, the interaction block is programmed to send back upstream to be subtracted the amounts of N and Q that were used. These flows are included in the array shared with the upstream connector, except the force is omitted and by convention the upstream block adds that force as it receives the incomplete term. Thus, the connector that supplied the force N is sent the following with N omitted:

$$\text{Con1inarray[Flow]} = \text{K2*Q};$$
$$\text{Con1in} = \text{Passarray(Con1inarray)};$$

and the connector that supplied the force Q is sent the following with Q omitted:

$$\text{Con2inarray[Flow]} = \text{K3*N};$$
$$\text{Con2in} = \text{Passarray(Con2inarray)};$$

To be compatible, downstream blocks that control uses must be programmed to send incomplete terms upstream and the upstream blocks must be programmed to add the missing force.

◆

A.4 CONVERTING MACINTOSH BLOCKS FOR WINDOWS PCS

If you develop the block on a Macintosh, it can be converted for use with EXTEND on Windows PCs, by clicking to bring up the conversion program MACWIN (from EXTEND files). The transformation gives the library an extension .lix. Following its instructions, save the transformed block to a "new library" on a disk formatted for the PC. Transfer the library file to a PC and using EXTEND for PC move the new block to the desired library of blocks. After blocks or libraries are put into a new computer, the library has to be recompiled. Open EXTEND, load the library, open the library window, and select Compile from the DEFINE menu. MACWIN is not included with all versions of EXTEND.

Appendix B

◆

NOTES ON COMPUTER USE AND BASIC PROGRAMMING

T o simulate the BASIC models in this book, some form of the program BASIC must be loaded on your computer and operated according to its manual. Computer operating systems are changing so fast that some of these remarks may be irrelevant by the time this book is printed. Software companies encourage the purchase of fancy forms of BASIC like VISUALBASIC and FUTUREBASIC, but much of such programs is distracting for simple simulation, forcing users to learn unnecessary procedures. Sections follow with notes on the use of QUICKBASIC the older BASICA in PC and Chipmunk BASIC for Macintosh. The authors can supply equivalent notes for APPLE II computers.

◆

B.1   QUICKBASIC FOR PCs AND WINDOWS PCs

QUICKBASIC in brief form is included as part of the Microsoft PC DOS and WINDOWS files (In WINDOWS 98: Tools\oldms-dos\). When you load the program by clicking on QBASIC.exe, a screen appears with menus. QUICKBASIC only uses and saves programs in ASCII letter format. Putting numbers at the start of each program line is optional. We use the line numbers so that the lines can be easily referred to and so that the program can be run by older BASICA or transferred to Chipmunk BASIC for Macintosh.

F6 toggles the cursor to or from the program window to one below where you can type in instructions. In the lower window you can determine values such as Q at the end of a simulation run by typing:

PRINT Q or ? Q.

A Screen statement, 10   SCREEN 1,0, is needed in each program. The 1 is for medium resolution graphics with color, and 0 turns color on. The Color statement is 20   COLOR 0,0. After the word COLOR the first number gives the background color from a selection of 15 (0 is black). Numbers 7 and 15 are white backgrounds, which are useful when black on white is desired when printing the screen. The second number designates one of two set of color choices (palettes).

Colors in PCs depend on the version of graphics in the operating system. In simplest form the second digit in the color statement selects one of two palettes each with three colors. 0 selects a palette where 1 in a printing line is green, 2 is red, and 3 is brown. 1 selects a palette were 1 is cyan, 2 is magenta, and 3 is white.

A graph can be printed (screen dumped) from BASIC or QUICKBASIC if the computer operating system and its connected printer have been arranged to print graphics (loading an appropriate printer driver). Some IBM models require the file GRAPHICS to be loaded. Other PCs and printers have printer files that can be designated when setting up DOS supplied on utility disks or can be obtained from the computer companies involved. After the run, print the graph by pressing Shift and Print Screen or just Print Screen. This will copy the graph on the clipboard. Use Alt-Tab to go back to the desktop. Activate WORD or a drawing program and paste the graph from the clipboard. Then label and edit it to get it ready to print.

◆

B.2   PC BASICA

The old standard program BASICA may be needed to access older BASIC programs that were saved in binary form not accessible by QUICKBASIC. It can be operated from PC DOS. After loading and/or running a program on BASICA, it can be saved in ASCII by typing: SAVE "Name", A. Then that program is available to QUICKBASIC.

BASICA.exe was supplied with old versions of DOS. After BASICA is loaded, a simulation program can be loaded by typing LOAD "Name" and then run by typing RUN. To interrupt a program that is running press Ctrl and Num Lock for a temporary interruption. For other computer models press the Pause key. Press Ctrl and Scroll Lock for a permanent interruption. BASICA requires lines to be numbered (i.e., 10, 20, 30, etc.). When a new line is added, it is put in its numerical order automatically. To list the program type LIST. To print a program listing, type LLIST. Delete a program on screen by typing NEW. Names of programs are limited to eight characters and must have the extension.bas (example: NAME.bas).

To list the files while in BASICA, type FILES. To return from BASICA to the PC DOS operating system, type SYSTEM. To read the file list while in DOS, type DIR. DIR/W arranges the file list in a compact horizontal listing. DIR/P lists a full screen of files and stops the scroll until a key is pressed to continue with another screen full of files, etc.

CLS clears the screen and may be the first line in a program (example: 10   CLS). To display graphs from two successive runs on the screen at the same time, run the first. Then make a change in the program and run it again, but start the run after the initial CLS statement (example: RUN 20).

PC-MAC FILE TRANSFERS

The BASIC programs for Macintosh can be transferred to a PC and vice versa with an appropriate file transfer program. Transfers to a PC require file names to have a maximum of eight characters and they must end in .bas. Many newer computers can read ASCII programs on PC-formatted disks and Mac-formatted disks interchangeably. To run a Mac-transferred program on a PC, you have to add two lines: 10   SCREEN 1,0 and 20 COLOR 0,1, which are necessary for PC graphics as described earlier. In transferring a PC program to a Macintosh, these two lines have to be changed and the plotting statements (PSET) replaced as explained in sections B.3 and B.4 that follow. In this book the programs are in form for PC QBASIC (QUICKBASIC).

◆

B.3   MACINTOSH QUICKBASIC

QUICKBASIC (not on the CD) will work on many Macintosh operating systems. It is black and white, with fine grain texture. To use QUICKBASIC in some computers and operating system 7.1, the cache must be turned off using the cache icon in the System folder, and 32-bit addressing turned off using the memory icon in the Control folder. If it is available, load QUICKBASIC. Then use its menu to open a simulation program from the book CD.

To run the program, open the RUN menu and activate RUN. After the program is running, you can stop it with Apple key and a period. You can also use the menu to stop a run. To list the program, open the WINDOWS menu and activate LIST. If the list window covers the graph, put the arrow on the title bar (lines at the top of the window) and hold the mouse down to move the window around. To enlarge the list window, double click the title bar of the list window or the small square box at the upper right corner.

To change numbers or letters in a program listing, highlight the character to change. Then type in the new character. After the first run, you can change a value and rerun with the second graph on the same screen for comparison. Open the WINDOWS menu and activate COMMAND, which opens another window at the bottom of the screen. In the COMMAND window type RUN 40 (if 40 is the number of the first statement below the SCREEN and CLS (clear) statements).

To save a program, open the FILE menu and activate Save As, after which you type in its name and where it is to be saved. The name of a program can have a maximum of 30 characters. It is optional to include the .bas required for PC names. If the name is restricted to eight characters it can be transferred to a PC later without changing the title. Using the SAVE command keeps the same name and destination. To write a new program, open the FILE menu and activate NEW. This clears any previous program and gives you an active list window on which to type.

To print a listing of the program, open the FILE menu and activate PRINT. To print the whole screen (screen dump) on older operating systems, put on the caps lock, and press the Shift key, the Command (Apple) key, and the 4 key. In later operating systems hold down the Shift key, the Command key, and the 3 key. This will copy the screen on the hard disk window in a file named PICTURE1 (PICTURE2, PICTURE3, etc.). Double click the picture. This brings it to the screen with the Macintosh program SIMPLE TEXT. Outline the desired part with the mouse and copy it to the clipboard. From there paste it into WORD or a drawing program for editing and printing.

B.4   MACINTOSH CHIPMUNK BASIC

Because QUICKBASIC for Macintosh is no longer on the market, CHIPMUNK BASIC has become the appropriate program for no-nonsense introduction of programming and minimodel simulation to beginners using Macintosh. The freeware program CHIPMUNK BASIC was developed by Ronald H. Nicholson, Jr., Santa Clara, California and is on the web as CHIPMUNK BASIC Home Page (⟨http://www.nicholson.com/rhn/basic/⟩) CHIPMUNK BASIC has the essence of the original BASIC programs and has been kept current with powermacs and Macintosh operating systems while also being compatible with older Macintosh computers. Some tutorials and manuals are included with the downloaded folder. Although not used in this book, the program has capabilities of displaying moving animations (sprites), sound, and matrix operations. CHIPMUNK BASIC 6.3 is included on this book's disk.

CONVERTING PC QUICKBASIC PROGRAMS TO CHIPMUNK BASIC FOR MACINTOSH

The following are the changes which need to be made to the PC programs listed in this book to run them on CHIPMUNK BASIC. Program lines have to be numbered. To delete a line, simply type the line number without anything else. As with QUICKBASIC the programs are text files and can be read by word processors. The program accepts upper or lower case characters, but converts them to lower case.

Delete the SCREEN statement and replace it with the following that sets up the graphics:

20 graphics 0

Delete the COLOR statement and replace it with graphics color followed by three numbers that indicate foreground line color. 100,0,0 is red, 0,100,0 is green, and 0,0,100 is blue. When all three are zeros the line color is black.

30 color 0,0,0

Instead of the LINE statement at the start of the program, indicate the square frame of the graph as follows:

40 moveto 0,0: lineto 0,180: lineto 320,180: lineto 320,0: lineto 0,0

To plot the graph in color, place the statement on the same line with the plot statement. For example, the following statement plots the line blue in the TANK model simulation in Table B.1:

TABLE B.1
BASIC Program TANK.bas from Tables 1.1 and 9.2 Modified for CHIPMUNK BASIC

```
5 rem CHIPMUNK BASIC for Macintosh
10 rem CHIPTANK.bas (Storage tank with source of steady inflow)
15 cls
20 graphics 0
40 moveto 0,0 : lineto 0,180 : lineto 320,180 : lineto 320,0 : 0,0
50 j = 2
60 q = 1
70 k = 0.02
80 dt = 1
90 t0 = 1
95 q0 = 1
100 rem start of iteration loop
110 graphics color 0,0,100: moveto t,180-q : pset t,180-q
120 dq = j-k*q
130 q = q+dq*dt
140 t = t+dt
150 if t/t0 < 320 then goto 100
```

110 graphics color 0,0,100: moveto t,180-q: pset t,180-q

For plotting of points for Q as a function of time t during the simulation, substitute the following, where t0 and q0 are scaling factors.

110 moveto t/t0, 180-Q/Q0: pset t/t0,180-q/q0

For other details open the file Quick Reference in the Macintosh section of the book CD.

For example consider the tank model in Figures 9.1. The CHIPMUNK BASIC version is given in Table B.1. Load CHIPMUNK BASIC, the program file, CHIPTANK.bas, and RUN the simulation as explained in Chapter 9.

USE OF CHIPMUNK BASIC FOR MINIMODELS

Double click on its icon to Open CHIPMUNK BASIC. Then use its FILE menu to open a BASIC program. To open the file needs the extension .bas. Use the MENU to LIST the program. If the program already has the graphics statements for CHIPMUNK BASIC, type RUN or select RUN from the menu. Otherwise, make the line changes outlined in the previous section.

To edit a line, use the mouse to position the cursor within the line where changes can be made. Then move the cursor to the end of the line and type return. The revised line appears at the bottom of the program, which also means the program has been changed similarly. Type LIST to see the revised program with numbered lines in order.

In the FILE menu use SAVE AS to save the program in the usual way. To print out the program, highlight and copy to the clipboard. Then use SIMPLETEXT or a word processor to bring up the text and print it.

To print out the simulation graphs, find print in the file menu. Also see Section B.3 for directions for screen dumping everything on the screen in a file *Picture 1, Picture 2,* etc. on the hard disk. Then the image can be trimmed and labeled with a drawing program.

With the exception of the different graphics statements, the old classic book by J. C. Kemeny, the inventor of BASIC and T. E. Kurz (1971) is still a good introduction: It is found in many libraries. TRUEBASIC is a later form of BASIC developed by Kemeny.

♦

B.5   HELPFUL HINTS FOR BASIC PROGRAMMING

The zero (0, often printed with a slash) is different from the letter O. It is important to use the right one. Do not put commas in your numbers. 10000 is right; 10,000 is wrong. The scientific notation is like that for many calculators. The number 1234 may be input as 1.234 E3. The number 0.01234 may be input as 1.234 E-2. The E indicates exponent and can be loosely translated as "times ten to the."

In computer equations you indicate multiplication by an asterisk (*) and division by a slash (/). Exponents are indicated with ($\wedge$). For example, Q^2 is either Q*Q or Q$\wedge$2.

To remember what special lines are supposed to do, add a remark. Any remarks typed after REM are ignored by the computer. The remark can be put on the same line as an active statement by putting in a colon first. For example, 50   Q = 10:REM Q is the initial biomass of plants.

To plot a rectangular border around the simulation graph, add the line: 30   LINE (0,0)-(300,180),,B for Macintosh or 30 LINE (0,0)-(300,180),3,B for PC. B indicates a box with the opposite corners designated by the coordinates 0,0 and 300,180. In a PC the number after the coordinates selects a color from the palette that was selected by the COLOR statement.

To plot points, use PSET followed by coordinates in parentheses; for example, PSET (T,180-Q) for the Macintosh. The first letter T is for time on the x axis. Because the origin is in the upper left corner, the variable Q is subtracted from the length of the box 180 so it will plot on the y axis with zero at the bottom. For the PC, add a number for the color; for example, PSET (T,180-Q),2. See section B.4 for Chipmunk BASIC graphics.

To send the program around again and again (iteration), type a program line as follows: IF T*T0<320 GOTO 210, where 210 is the first line with equations that change during iteration.

HOW TO SCALE THE GRAPHS FOR QUANTITY AND TIME

When you are plotting various quantities, you want them to fit within the box drawn on the screen. To do this, multiply each quantity, like Q, by a scaling factor, like Q0. For example, to calculate Q0, you first figure out what the maximum quantity Q will be. Then divide the height of the screen by the maximum Q. The answer will be Q0. For example, if the screen height is 180 and Q increases up to 200: Q0 = 180/200 = 0.9. Sometimes scaling factors are put in the denominator, like Q/Q0 in Table 13.1.

To make a line across the screen to subdivide the graphics 40 lines from the top, add a line to the program. If the width of the screen is 360, type in:

LINE (0,40)-(360,40)

T0 is the scaling factor for time (T) and is used to make the desired simulation time fit within the graphics box on the screen.

When you are writing a program for a certain time span (such as number of days) and want the graph to cover that time, put in two statements: TX = number of days and T0 = number of plotting points across the screen divided by TX. For example, if you want the graph to be in days for a year and the screen width is 320, put in the statements:

TX = 365
T0 = 320/TX

DT is the length of time in each calculation (iteration). For example, if data are in days:

DT = 4 means a 4-day calculation in each iteration.
DT = 0.5 means a calculation is done every half day.
DT and T0 are put into programs as follows:
    PSET (T*T0, 180-Q)
    Q = Q + DQ*DT
    T = T + DT
    IF T*T0 <360 GOTO

For examples of use of these scaling factors, refer to the WETLAND model of Table 16.1.

◆

B.6   USE OF SINE WAVES FOR VARYING INPUTS

Some inflows to a system that vary regularly can be programmed using sine waves. This is especially useful for the daily
and seasonal variations in sunlight that follow sine waves on
clear days (Figure B.1). Seasonal rainfall, evaporation, temperature, etc., can also be programmed this way. Giving a regular
up and down input is a way of studying a system's response and
comparing with others.

As a wheel with a radius of one rotates, a nail generates a
sine wave. The height of a nail on the rim goes from zero to +1
to zero to −1 and back to zero, where measurements are made
from the level of the axle (see Figure B.1a). A right triangle is
formed between the nail, the axle, and the horizontal line of
the axle.

The sine of an angle is the opposite side over the hypotenuse
of a right triangle. As the wheel makes one full revolution, the
sine of the angle A (the angle between the nail spoke and the
horizontal) goes from 0 to +1 to 0 to −1 to 0 again. When the

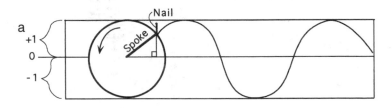

FIGURE B.1.
Sine waves for supplying
diurnal and seasonal solar
inputs.
(a) Sine wave viewed as a line
generated by a nail on a
rotating wheel, with amplitude
plus and minus 1, as generated
by BASIC; (b) graph for
representing annual inputs
using an equation: Y = 3200 +
800*Sin(T*0.017); (c) graph
representing diurnal inputs
using the equation: Y =
4000*Sin(T*0.26).

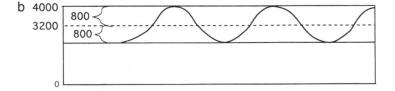

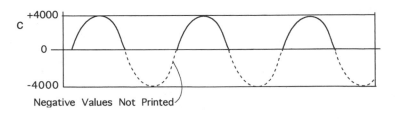

value of the sine A is graphed with time (or the path of the nail up and down graphed with time), a sine wave results.

The angle between the nail spoke and the horizontal as the wheel turns a whole revolution goes from 0 to 90 to 180 to 270 to 360 degrees (0 degrees again). One can also represent the angle in radians. A radian is the arc of the circle the length of the radius. When the wheel makes a full revolution, it passes over a circumference of 2 pi radii. A full revolution of 360 degrees is an angle of 2 pi radians. Since pi is 3.14, 2 pi is 6.28. A full circle that is a full sine wave cycle is 6.28 radians.

Sine waves may be used as a source-driving functions. The ups and downs of solar insolation during the year can be represented with a sine wave. The sunlight during the day can be represented with half a sine wave. The BASIC language has the sine function Y = SIN(X), where X is the angle in radians. As X goes from 0 to 6.28, Y goes through a full cycle (0 to +1 to 0 to −1 to 0 again). To fit a sine wave to a program, substitute time (T) for X. Then multiply or divide T by whatever number is necessary to scale the sine wave to the time needed. For example, if a simulation is being run in days, using the sine function without scaling would generate one cycle every 6.28 days.

EQUATION FOR AN ANNUAL SINE WAVE

To make the sine wave cycle in 1 year (365 days), multiply the time by (6.28/365 = .017). The equation becomes Y = SIN(T*.017). This sine wave will go from +1 to −1 and back in 1 year. To scale this up or down, multiply by the range of sunlight (or whatever is being represented by the sine wave). If this range is R, where R is the amount that the quantity (of sunlight) varies from its average V, then the equation becomes Y = R*SIN(T*.017).

To move the graph completely into the positive range, add an amplitude equal to the average value of sunlight or whatever is being represented. If this average value is V (like the average per day for a year), then the equation for a seasonal source results: Y = V + R*SIN(T*.017). See the example shown in Figure B.1b.

EQUATION FOR A DAILY SINE WAVE SOURCE

If the time unit is in hours and there are 24 hours a day, then scale the sine function so that one cycle takes 24 time units.

TABLE B.2
Example of an Array

BASIC program	Explanation
DIM A(12)	Spaces for 12 items of data are set for array A.
DATA 2300, 2000, 2800, etc.	Sets 12 values of sunlight for 12 months.
FOR M = 0 TO 11	M is the number of the month.
READ A(M)	Tells computer to read the values in the DATA line.
NEXT	Takes next number next time around.
I = A(M)	Input of sunlight from array A for the month (M).
Y = INT(T/365)	Calculates the number of years since the start. (INT is the command to round to a whole number.)
W = 12*Y	Calculates total number of months in previous years.
M = INT(T/30.4) − W	The number of the month in the current year is the total months since the start minus number of months in the previous years.

Therefore, multiply the time by (6.28/24 = .26). The equation becomes $Y = R*SIN(T*.26)$.

To represent sunlight during daytime, put in an IF statement that cuts out Y (the sunlight) when the sine goes negative (night period): IF $Y < 0$ THEN $Y = 0$. See the example shown in Figure B.1c.

◆

B.7   USE OF AN ARRAY TO ENTER A SERIES OF INPUTS

An array is a way to input a set of numerical data, like a different value for each period of time. For example, in the WET-LANDS model of Figure 16.1, a different value of sunlight is entered for each month. Statements needed to input sunlight figures for 12 months are given in Table B.2, with explanations at the right. T is time in days.

Appendix C

◆

ANSWERS TO "WHAT IF" EXPERIMENTAL PROBLEMS

CHAPTER 4   SIMULATION WITHOUT EQUATIONS

4.2   Pond Water Model (Program PONDWATR in EXTEND)

4.2–1 The dialog box of the source icon has a value of 100. If the source is made 10, then the stored quantity levels off at a low value, less than 200.

4.2–2 If the outflow is set to zero, growth is on a straight line and does not level. This is called a ramp.

4.2–3 If the starting value in the dialog box of pond storage is changed to 2000, the graph of storage descends and levels off at the same lower value reached when the starting quantity was small.

4.2   Model of Exponential Growth (Program PICEXPO in EXTEND)

4.2–3 The city block that starts with a greater value grows faster.

4.2–4 If the fossil fuel source is reduced, the city grows more slowly.

4.2   Model of the Earth System (Program EARTHPIC in EXTEND)

4.2–5 If the ocean carbonate sink is disconnected, with fossil fuel in use the CO_2 in the atmosphere increases to much higher values, since it is no longer absorbed by the ocean.

4.2–6 With twice the fossil fuel the CO_2 and consumer assets grow much faster and reach higher levels.

4.2–7 Without fossil fuel and with half the sunlight, plants are reduced, causing a reduction of organic matter and assets. CO_2 increases for several years and then decreases to a balance again.

CHAPTER 5   SIMULATION WITH GENERAL SYSTEMS BLOCKS FOR EXTEND

5.2   Model of Storage (Program TANK in EXTEND)

5.2–1 With a high starting value of 100, storage declines reaching the same steady state as simulation started with low value.

5.2–2 With no inflow, the storage declines toward zero but at a decreasing rate.

5.2–3 Increasing the outflow from storage causes the steady state to be less.

5.2–4 In the model EXPO, storage drains if the input force is very low, or the starting quantity very small.

5.5   Ecological Model (Program PC&CYCLE in EXTEND)

5.5–1 With half the nutrient input, steady state takes 300 years, twice as long. Level of organic storage reached is less.

CHAPTER 9   PROGRAMMING IN BASIC

9.5   Model of Growth of Storage with Inflow and Outflow (Program TANK2.bas, Figure 9.1)

9.5–1 The graph starts high and then levels at the same place because that is where the inflow of water equals the outflow through the drain.

9.5–2 The graph of Q will grow up as before, but when T =

96, Q will start to decrease, quickly at first, and then more slowly.

9.5–3 With a faster outflow, the quantity of leaves never gets as high as with the original outflow.

9.7   Model of Exponential Growth (Program EXPO.bas, Figure 9.4)

9.7–1 If you make the food more concentrated, the growth of the population Q will be faster and the graph of Q will go up more steeply. If you reduce the concentration, the graph of Q will increase more gradually—at the end of the screen (279 weeks) Q will be less than when E was 1.

9.7–2 When K = 0.08, the graph of Q will be steeper, reaching the top of the screen sooner. When K is reduced <0.07, the graph will grow more gradually.

9.7–3 To increase the death rate, increase K4. You can type 80   K4 = .06. The graph of Q will grow more slowly. If you decrease the death rate (80   K4 = .04), then Q will grow faster.

9.7–4 When the growth rate is equal to the death rate, the population will neither decrease nor increase; it is at equilibrium.

9.8   Model of Dispersal of Storage (Program DRAIN.bas, Figure 9.5)

9.8–1 The leaves would last longer.

9.8–2 Your bank account would only last half as long.

9.9   Model of Production, Consumption, and Recycle (Program PC&CYCLE.bas, Figure 9.6)

9.9–1 The increase in consumers makes a big change in the graph of organics and nutrients. There are fewer nutrients in the soil because there are fewer plants since they are being consumed, and there is less growth of plants because there are fewer nutrients.

9.9–2 Half the sunlight produces half the plant growth. More nutrients are left in the soil at first, but they level off at the same place as in the original.

9.9–3 Growth of organics and nutrients in the soil increase. However, they come to the same steady state as in the

original graph. They level at the same place because the
inflows and outflows have not changed.

9.9–4 Changing the inflow and outflow of nutrients does not
visibly affect the graph. Therefore, the limiting factor in
this model is the energy source (sun and rain).

CHAPTER 10   SIMULATING WITH STELLA

10.1   TANK (Figure 10.3)

10.l–1 The tank value starts at 100, but levels off to the same
value as it had before the change.

10.1–2 The level of the tank is lower than before

10.2   Sensitivity Analysis (EXPO, Figure 10.9)

10.2–1 Q Quantity reaches a higher value when E = 2 than
when E = 1.

10.3   Production–Consumption–Recycle Model (PC& CYCLE, Figure 10.10)

10.3–1 Nutrients N and organics Q reach much higher steady-
state levels when K4 = 0.05.

10.3–2 The level of organics goes up and the level of nutrients
goes down.

CHAPTER 11   SIMULATING EMERGY AND TRANSFORMITY

11.3   Emergy and Transformity in Storage (Program EMFISH.bas, Figure 11.1)

11.3–1 With a higher percent of fish removed from stock because
of the larger K2, the stock is at a lower level with less
accumulation of input. The emergy level is less.

11.3–2 Transformity measures the quality of the stock. With
fewer inputs the stock represents less accumulated work.
Transformity is less.

11.4   Emergy in Simple Storage (Program EMGTANK.bas, Figure 11.2)

11.4–1 With faster depreciation, stock accumulates less value. The emergy and transformity of the stock are lower.

11.4–2 Increasing the transformity of the initial condition and inputs moves the model to a higher position in the energy hierarchy. Emdollars are obtained by dividing the emergy by the emergy per unit money of the economy. Higher position yields more emdollars.

11.5   Emergy in Autocatalytic Process (Program RENEMGY.bas, Figure 11.3)

11.5–1 The output transformity is higher because of the energy transformation work within the model. Emergy is transmitted, whereas the energy transmitted is much less.

11.5–2 Without an input, the stock drains away. Its quantity decreases but n⁴ its transformity.

CHAPTER 12   MODELS OF PRODUCTION AND RECYCLE

12.1   Net Production (Program NETPROD.bas, Figure 12.1)

12.1–1 Increasing the consumption decreases the net production.

12.1–2 The efficiency of gross production (fraction of sunlight transformed) is the ratio of production K1*S to input energy S if both are given in energy units. In other words the efficiency is K1. Sometimes the fraction transformed is multiplied by 100 to express the efficiency in percent. Increasing K1 increases gross production and net production.

12.1–3 With large initial storage, consumption exceeds gross production so that the net production at first is negative. When net production is negative it is below the zero line in the middle graph.

12.2   Limiting Factors (Program FACTORS.bas, Figure 12.2)

12.2–1 The result is a set of curves that goes a little higher because the limiting inflow J is larger. Thus increased production is possible with an increase in the limiting factor. Limits of production due to J occur at a higher level when the nonlimiting factor E is increased.

12.2–2 With higher K1, more of the input energy goes into higher output production P without affecting the rest of the factors or flows involved.

12.2–3 Efficiency P/E is the slope of the curves in Figure 12.2b. The slope is steepest (and efficiency greatest) when M is least limiting because value of material inflow J is high and value of E is small.

12.3   Internal Limiting Factors (Program INTLIMIT.bas, Figure 12.3)

12.3–1 Production is increased by adding materials when the internal cycle is limiting.

12.3–2 When recycle coefficient K2 is increased, the storage is less and the recycle and turnover rates are increased.

12.3–3 The organic storage is less but the flows are greater. Since the turnover time is less, the organisms are smaller.

12.4   Production and Consumption (Program DAYP&C.bas, Figure 12.4)

12.4–1 With a smaller pool of available nutrient materials M is less and the storage is less. Although the energy source can be more limiting in this model compared to INT-LIMIT, the response to lowered materials for cycling was similar.

12.4–2 During the energy pulses, the surge of growth of Q increases and materials are reduced more. When the energy inflow is off, there is a larger surge of consumption and material regeneration.

12.4–3 With reduced consumption, organic storage stays high, the material pool stays small and the production rate is less.

12.4–4 Stored quantity is high and steady while the nutrient materials available are held to a low level. Production rate is less, but there is more production because of the longer lighted period.

12.5   Diurnal Oxygen in Aquatic Ecosystems (Program OPENAQ.bas, Figure 12.5)

12.5–1 Respiration rate exceeds photosynthetic production and oxygen goes down.

12.5–2 With continuous light, oxygen stops fluctuating and rises to a higher level after which a new balance develops between continuous production and continuous respiration without net production.

12.5–3 With fewer nutrients for photosynthesis, production decreases, temporarily less than respiration. Later a new balance begins to develop. When metabolic processes are regulated by ecosystem feedbacks so as to restore a balance of processes, the process is called *homeostasis*.

12.5–4 With the diffusion coefficient increased the range of dissolved oxygen is reduced. The bubbling devices in an aquarium are analogous to wind and waves in a pond. With a smaller diffusion coefficient and organic matter added, very low oxygen values result that can kill fish.

12.6   Logistic Autocatalytic Production and Recycle (Program AUTOCYCL.bas, Figure 12.6).

12.6–1 Adding TN increases the growth of assets.

12.6–2 Increasing energy has some positive effect. With more nutrients added first so that their limitation is less, the energy has a greater effect.

12.6–3 The autocatalytic logistic production function (Figure 12.6b) generates S-shaped curves because it accelerates growth at first before leveling, whereas the simpler interaction of energy and nutrients has a decreasing rate of growth from the start, causing the curve to be convex (Figure 12.3b).

CHAPTER 13   MODELS OF GROWTH

13.1   Growth on a Renewable Source (Program RENEW.bas, Figure 13.1)

13.1–1 A system that has a source that stays available no matter how much is used can grow faster and faster; this produces exponential growth. In contrast, a system that has a source with a steady flow that cannot be changed by the system will grow up only to the level where it uses most of the available source. The kind of source determines the kind of system. The limited source energy $R = J/(1 + K0*Q)$, which is a limiting expression;

whereas, with exponential growth the energy E = 1 is a constant not affected by growth, however large.

13.1–2 When J is increased, the graph of Q increases faster and levels off higher. When J is decreased, the graph increases more slowly and levels off lower.

13.1–3 The graph will start partway up the vertical axis and grow up to the same level as the original graph. You have changed the starting point but the leveling point is dependent on J, the energy source.

13.1–4 Increase K4 in line 80. The graph of the changed forest model grows slower and levels off at a lower level than the original graph. With the same production rate and a higher decomposition rate, it cannot develop as large a storage.

13.2   Slowly Renewable Growth (Program SLOWREN.bas, Figure 13.2)

13.2–1 Until the reserve E can build up, the unit Q cannot grow. After the reserve builds up, the assets unit Q grows, pulling the reserve down again. In this scenario the assets never get as large as the growth that started with large E, but the later steady state is the same.

13.2–2 Because higher inputs maintain higher levels of the reserve, the dependent Q unit can grow to a higher sustainable level also.

13.2–3 Without an inflow the reserve would be drained away and the dependent unit would have only a brief period of growth based on the initial storages, as will be considered in model NONRENEW (Figure 13.3a).

13.3   Growth on a Nonrenewable Source (Program NONRENEW.bas, Figure 13.3)

13.3–1 Q (beetles) grow faster and develop larger stock, but they don't last as long since the resource (log) E is used up faster. If this model applies to the civilization's use of fossil fuels, finding new deposits may accelerate the time of running out.

13.3–2 Starting with more beetles causes the log to be used up faster and more beetles to be produced more quickly. If this model were the mining town, it would mean that the town and some people were already there when the mine was opened. The ore would be mined faster and the town developed more since it already had some people and structure.

13.3–3 When beetles are able to get more growth from the wood of the log, the beetles grow faster to a larger population, which uses up the log faster. All of the log is consumed. If the growth rate is less efficient, the beetles grow more slowly to a smaller population; the log is not completely consumed.

13.3–4 Change K4 to .05. With a higher death rate, Q is less and uses up E more slowly.

13.4 Growth on Two Sources (Program 2SOURCE.bas, Figure 13.4)

13.4–1 By increasing the nonrenewable reserve E, the growth peak occurs sooner, is higher, and uses up the source faster. This property of autocatalytic designs helps maximize power.

13.4–2 Assets Q develop more slowly and use up the fuels (E) more slowly. There is a lower steady state after nonrenewables are gone.

13.4–3 With an increased depreciation rate assets decrease more rapidly, use fewer resources, and level off with fewer assets. There are fewer, older assets for use.

13.4–4 With more assets at the start, Q grows faster and higher and uses up nonrenewable energy E quickly. For example, societies that had initial developments on renewable energies had the technology to implement the industrial revolution first.

13.5 Logistic Growth (Program LOGISTIC.bas, Figure 13.5)

13.5–1 The population grows faster and reaches a higher level steady state, in proportion to the energy.

13.5–2 The population quickly decreases to the steady-state level based on its energy E. The deaths from crowding (K4*Q*Q) are high because the number of mice Q is high.

13.5–3 K4 is the coefficient that represents the crowding effect. Decrease K4 and more mice are sustained.

13.5–4 With the term for linear losses added, the growth is a little less but the simulation graphs are similar.

CHAPTER 14   MODELS OF COMPETITION AND COOPERATION

14.1   Competition for Limited Source (Program EXCLUS.bas, Figure 14.1)

14.1–1 Both decrease some, but one prevails at the same level as the original graph.

14.1–2 The cooperative population prevails, driving the other population quickly.

14.1–3 Set K6 equal to K5.

14.2   Two Populations in Exponential Growth (Program TWOPOP.bas, Figure 14.2)

14.2–1 The run in Figure 14.2b results when everything is similar except efficiency of production. Changing the death rates can cause the same graph, with one population outdistancing the other.

14.2–2 With K1 and K2 equal, the only difference is in the starting condition. The one with the larger start rapidly outdistances the other.

14.2–3 The population with quadratic feedback overgrows the other quickly.

14.3   Two Populations with Competitive Interactions (Program INTERACT.bas, Figure 14.3)

14.3–1 Each population reaches a level population because of its own quadratic drain without influence on the other. The source is not depleted by the growth. The populations coexist, unlike Figure 14.3b.

14.3–2 With a lower energy availability, both populations have lower levels but without influencing each other. They coexist. Population Q2 is most numerous.

14.3–3 With a lower interaction coefficient, population Q1 is somewhat reduced but both populations coexist.

14.3–4 With K6 at 0.002 the species population are about the same. With K6 at 0.003, Q2 is displaced.

14.4   Model of Two Populations That Cooperate (Program CO-OP.bas, Figure 14.4)

14.4–1 Populations die out. The growth equations contain both Q1 and Q2. Each is dependent on the other.

14.4–2 Although starting with different quantities, the populations reach the same steady state as before, since they are using the same energy source and growth rates.

14.4–3 Set K1 = K2, and K5 = K6. With all the coefficients the same the populations grow identically.

CHAPTER 15   MODELS OF SERIES AND OSCILLATION

15.1   Model of Prey–Predator Oscillation (Program PREYPRED.bas, Figure 15.1)

15.1–1 If you double the growth rate of the carnivores, they eat more herbivores, reducing the quantity of herbivores to about half of the original. The pulse is more frequent because the herbivores are eaten faster.

15.1–2 When you reduce the death rate of the carnivores from 0.3 to 0.1 the pulses are much longer, with fewer quantities of both carnivores and hervbivores. Because they live longer, they eat more herbivores.

15.1–3 Cutting the energy source in half halves the quantities of carnivores and herbivores.

15.2   Model of Oscillation at Three Levels (Program OSCILLAT.bas, Figure 15.2).

15.2–1 Oscillations become more frequent. With more energy, more of everything can grow and eat faster. To see the plants P on the screen, change P0 to 50. With less energy the oscillations dampen out. There is not enough energy to keep them going. The small amount of life remains almost constant year after year.

15.2–2 Like the model PREYPRED in the previous section, animal oscillations continue with constant energy availability.

15.2–3 With no carnivores, the herbivores increase and level off, and their steady grazing keeps plants at a low level. With faster turnover time, oscillations are more frequent. The consumers C are small in quantity, and the plants are steadier.

15.3   Model of Switching Oscillation (Program FIRE.bas, Figure 15.3)

15.3–1 With more energy, growth is faster and fires more frequent. In this model, an increase in nutrients has little effect on the growth of grass because nutrients are not limiting.

15.3–2 With reduced nutrients, growth is less and the threshold for fire is not reached.

15.3–3 The grass will grow until all the nutrients are bound up in the grass. Then the graph levels off, as does the growth of grass.

15.3–4 The system never grows enough biomass to reach the threshold to support fire.

15.4   Model of Pulse and Recycle (Program PULSE.bas, Figure 15.4)

15.4–1 With inadequate materials, the production decreases and oscillation stops. With excess materials, the oscillations continue.

15.4–2 With reduced energy, production is less and oscillation stops. With doubled energy, oscillations are smaller and more frequent. With larger energy, the production is large and storage of products is large without oscillation. The energy loading that sustains pulsing is intermediate and may use energy better. By controlling their chlorophyll, green leaves adapt to intermediate energy, allowing excess energy to pass through for other leaves to use.

15.5   Model of Destruction and Restoration (Program DESTRUCT.bas, Figure 15.5)

15.5–1 D represents the strong winds, A represents forest tree structures or buildings, and M the soil nutrients or building materials that can be used.

15.5–2 More source energy increases the assets A while decreasing unbound materials M. In the forest example, if more nutrients are part of the trees, there are fewer in the soil.

15.5–3 A large impulse D destroys the storage A quickly; it takes longer to regrow.

15.6   Model of One Pulsing Pair Coupled to Another (Program CHAOS.bas, Figure 15.6)

15.6–1 With much less energy available, the oscillation at the lower level becomes small. The upper pair of units oscillate so the chaos is eliminated. Chaos generally increases with energy flow.

15.6–2 Without the connection between the two pulsing pairs, steady states develop without chaos; the phase plane curves become simple loops, not varying.

15.6–3 With the larger DT, points appear jumping all over the screen because of artificial chaos. Because of very high values off scale, the program may stop with an "overflow" message.

CHAPTER 16   MINIMODELS OF SUCCESSION AND EVOLUTION

16.1   Model of Wet Grassland (Program WETLAND.bas, Figure 16.1)

16.1–1 When J is increased, the vegetation thickens, and more peat is produced. Nutrient levels in the soil increase and remain high.

16.1–2 Double K6 to 3.2 E-3. There is less peat buildup. Nutrients pass through more rapidly with lower concentrations and less vegetation.

16.1–3 Under the simulated arctic conditions, production is less but consumption is much less, recycling fewer nutrients. Peat accumulates rapidly.

16.2   Model of Succession (Program CLIMAX.bas, Figure 16.2)

16.2–1 Biomass increases, but diversity remains small.

16.2–2 Diverting the biomass energy causes standing biomass, diversity production, and diversity to be less.

16.2–3 Production and diversity are less and growth slower.

16.3   Model of Nutrients and Diversity (Program NUTRSPEC.bas, Figure 16.3)

16.3–1 With intermediate nutrient inflow, an initial surge of weeds occurs, but some diversity develops later.

16.3–2 With nutrients low, the more seeding (increased S) the more species develop in the mature complex.

16.3–3 With high nutrients, a very large seeding effort (S = 10) is required to develop any diversity. This condition may be like a plant nursery where seeding and nutrient additions are intense.

16.4   Model of Species and Area of Resources (Program SPECAREA.bas, Figure 16.4)

16.4–1 Greater resources support higher diversities for the same areas.

16.4–2 Higher rates of depreciation or species loss cause a lower diversity curve, less information per unit energy used.

16.4–3 The graph with a logarithmic horizontal scale curves upward, a property observed in field data on species distributions. There is no tendency to level off.

16.5   Models of Island Speciation and Diversity (Program SPECIES.bas, Figure 16.5)

16.5–1 Diversity that can be sustained is larger.

16.5–2 Diversity decreases to a level that can be sustained.

CHAPTER 17   MODELS OF MICROECONOMICS

17.1   Model of Sales (Program SALES.bas, Figure 17.1)

17.1–1 Change statement 80 to K1 = 0.05. The quantity of goods will be less so the price will be higher. The money coming in will not change.

17.1–2 Increase J to 2. Changing production does not change the money income. However, it does increase the products the consumer can buy for the dollar.

   In this case the value of production to the business is not the same as the value to the economy. The more product made, the lower the price and the more the consumers get for their money.

17.2   Model of Economic Use of Mined Resources (Program TANKSALE.bas, Figure 17.2)

17.2–1 With higher sale price there is more money accumulating faster so the reserves are mined faster and more is processed before money runs out. With price set as 20, only negligible reserve remains after the mining pulse.

17.2–2 Some initial capital is required to start. With more initial capital, mining develops sooner, but the final result is the same.

17.2–3 With more reserves available, more money is obtained and ultimately the reserves are mined to a lower level before money runs out.

17.2–4 With small availability, other things being equal, there is no net gain in money and mining does not get started.

17.3   Model of Loans, Interest, and Banking (Program BANK.bas, Figure 17.3)

17.3–1 If the higher interest is paid to depositors, enough money is diverted from the bank's capital so that the accounts spiral down to bankruptcy.

17.3–2 With half the depositor income there are also some reduced costs and interest paid out; net profit is maintained and the growth spiral develops sooner than before. With no depositors JD is set to zero but the spiral of growth continues because of favorable interest from loans.

17.3–3 With its main income source from loans reduced, bankruptcy results. Adding the lines to make a ceiling to the money out in loans produced a steady state.

17.3–4 The run with less interest from loans causes the bank to crash unless the bank also decreases the interest it pays depositors.

17.4   Model of Money Countercurrents in Production and Consumption (Program ECONP&C.bas, Figure 17.4)

17.4–1 When the environmental energy source is half, the production and environmental products are less, but the circulation of money and city assets are the same.

17.4–2 Increasing the money supply in this particular condition increased the circulation of the economy, causing more of the rural production to go into city assets.

17.4–3 Increasing the wages (price of goods and services) with the same amount of circulating money in this condition increased the city assets because outflow to the rural sector was less.

17.5   Model of Economic Use of Environmental Resource (Program ECONUSE.bas, Figure 17.5)

17.5–1   Type 150   PG = 3.0. Money is down, assets are down, but the quantity of natural product stays up. The business will not do as well.

17.5–2   Products, assets, and money are less. It might be better to use the start-up money in another venture.

17.5–3   Type 90   M = 100, and to show both graphs on some computers type in the command panel RUN 60. The money does not increase as fast, because it is used to buy assets quickly. These assets process the product faster and keep it from growing up as much as it did when money started low. After a time the business gets to the same steady state.

17.5–4   When you halve the price of your product (100   PE = .5), less of the product gets sold and there is not enough money to exploit the resource very much. When you double the price of your product, money goes up, assets go up, and the resource is used up faster.

17.5–5   With price rising as the storage is used up, the money going into more effort increases and the environmental structure (storage Q) is pulled below its ability to sustain itself. There is a very high short-range profit but ultimately the business will collapse without sustainable environmental resources.

17.6   Model of the Economic Role of Biodiversity Reserve Areas (Program RESERVE.bas, Figure 17.6)

17.6–1   The coefficient for land moving out of the economic production area is K1. Using an increased rate of soil deterioration reduces the economic yield and increases the area of land that is bare. More reserve area is needed to get maximum yield.

17.6–2   Maintaining the economic yield with a small reserve area and small area of bare land implies very rapid seeding: large K2. Seeding rates are increased by having large populations of birds, mammals, and bats and/or favorable winds and climatic conditions for germination.

CHAPTER 18   MODELS OF
MACROECONOMIC OVERVIEW

18.1   *Model of Money Driven Growth (Program MONEYGRO, Figure 18.1)*

18.1–1 Growth of both money and assets is slower, but it is still exponential, with no limit.

18.1–2 Growth of money and assets is faster.

18.1–3 With a higher rate of adding money, money and assets grow faster; with a lower rate, they grow more slowly.

18.1–4 Assets and money increase more rapidly. If the company is in competition, it is an advantage to start with more assets (for example, by borrowing).

18.2   *Model of Price in Relation to Unsustainable Growth (Program BUYPOWER.bas, Figure 18.2)*

18.2–1 Assets go higher and prices lower, but they both come to the same level as before when the fuels are used up.

18.2–2 At first assets will go up faster and prices down, but since fuel will be used up faster, assets go higher and reach the steady state sooner. Faster conversion makes use more rapid.

18.2–3 Assets will go higher and prices will be lower, but these differences will be permanent since they are based on the renewable sources.

18.3   *Model of Textbook Macroeconomics (Program MACROEC.bas, Figure 18.3)*

18.3–1 A higher depreciation rate decreases the rate of growth. If the depreciation rate is made high enough, the assets decay without growth.

18.3–2 More savings and reinvestment accelerate conversion of available energies into stored assets. With unlimited energy (E constant), growth is faster. With flow-limited resources, more capital assets develop.

18.3–3 More available energy generates higher growth rates and higher levels of developed assets.

18.3–4 More available energy increases Q and Y, but they still level off because the energy is limited.

18.4   Model of an Economy in Relation to Land Use (Program ROTATION.bas, Figure 18.4)

18.4–1 With no seedlings and environmental restoration, urban development goes to the maximum as all the area becomes urban (A2 = 1.00). As soils and other environmental storages are used up, urban development Q declines.

18.4–2 With large starting storage of S, its levels are higher, causing more assets Q to develop.

18.4–3 With less area, all of the variables are reduced in quantity.

CHAPTER 19   MODELS OF INTERNATIONAL RELATIONS AND TRADE

19.1   Model of Cooperation through Trade (Program ENTRADE.bas, Figure 19.1)

19.1–1 The assets of A go up, and those of B down, because there is only a limited inflow of energy. The total empower is increased, since A is being more efficient.

19.1–2 If you change K3 to 0.2, the assets of A are much less than B, and the total power with trade is slightly less.

19.2   Model of Impact of Global Economy on a Nation. Programs FREEMARK.bas and EMEXCHANG.bas, Figure 19.2

19.2–1 With the investment sequence debt (top panel of plot) increases and decreases again. The borrowed money allows an acceleration of growth of assets.

19.2–2 Lowering the sale price of environmental products or increasing the price of purchased goods–services–fuels reduces the nation's growth of assets, because fewer assets can be imported. Little general benefit comes when developed countries selling raw products use money to buy expensive goods. With these unfavorable prices, running the borrowing sequence reduces growth.

19.2–3 Even though the prices are favorable enough to cause some growth of assets and emergy (EmA), the net benefit in emergy terms is negative. The prices must be 40 times more favorable to get an emergy exchange ratio greater than one.

19.2–4 Selling assets (finished products) and using the money to buy other finished products from outside is a benefit only if the prices for emergy sold are greater than the prices for emergy purchased. For example, selling manufactured goods and using the money to buy fossil fuels is a net benefit because of the high emergy/money ratio in fuels.

19.3   Model of Warring Competition (Program, WAR.bas, Figure 19.3)

19.3–1 If you make all K the same for each country, they coexist. Change K6 to 0.01 and K7 to 0.05.

19.3–2 If you increase I to 1.2, A increases faster and more, the total production is greater, and B is destroyed sooner. Since A increases more and faster, its power of destruction is greater.

19.3–3 If you increase K10 to 0.06, nation B wins the war. You have increased the proportion of B's assets going into defense enough to outcompete A.

19.4   Model of Sharing Information (Program INFOBEN.bas, Figure 19.4)

19.4–1 With less energy the whole system is reduced. When fossil fuels become scarce, medical information will be reduced in both countries and the total power and shared information will be correspondingly reduced.

19.4–2 Without the sharing of information, competition drives out one unit, and total power and information are reduced.

19.4–3 With only one nation contributing information, the productivity and assets of both nations are reduced, but coexistence is still supported.

CHAPTER 20   MODELS OF THE GLOBAL GEOBIOSPHERE

20.1   Model of World Carbon Dioxide (Program WORLDCO2.bas, Figure 20.1)

20.1–1 When you double the rate of forest production, CO_2 goes down, forest organic matter increases, the bicarbonates increase, and the limestone increases slightly. The refor-estation uses the CO_2 in its photosynthesis.

20.1–2 When continuous forest cutting is increased, CO_2 in the air and in the water bicarbonates increases. The forest biomass does not change much since the cut area grows back.

20.1–3 Change statement 505 to FC = 2*(FCS − .00035*YR). At first the CO_2 goes higher from the burning of fossil fuel; later, as fossil fuels decrease, the CO_2 decreases. The CO_2 from the fuel burning goes into the bicarbonates and not much into the plants, because CO_2 is not the limiting factor for the plant growth in this model.

20.2   Model of World Population (Program PEOPLE.bas, Figure 20.2)

20.2–1 Assets and population grow up slowly to level off at the same place as the original, but there is no spurt of growth because there is no fossil fuel to stimulate the production of assets or population.

20.2–2 If you increase K9, the medical care, to .01, the population increases a lot and the assets increase a little during the time of fossil fuel use. Later the population levels off a fraction higher and the assets a fraction lower than in the original graph: Any medical care uses assets.

20.2–3 With the lower birthrate the population goes to almost nothing. The population is so low the fossil fuels are not completely used.

20.2–4 If you increase the death rate 10 times, the population and assets do not get as high as in the original graph; they level off at about the same place. If you then cut medical care to 0, the population and assets grow very little before the population decreases to a level about 20% lower than in the original graph. Much of the fossil fuel was not used.

20.3   Overview Model of Geological Processes in Earth Evolution (Program EARTHGEO.bas, Figure 20.3)

20.3–1 Increasing the solar energy increases the energy in the hydrologic cycle, ultimately generating more continental mass.

20.3–2 Reducing water inflow reduces the size of the ocean and the land mass.

20.3–3 Reducing earth energy reduces the continental mass that develops.

20.4   Model of a State Driven by a World Minimodel (Program STATECON.bas, Figure 20.4)

20.4–1 Pattern is similar but the assets are fewer and less water is used.

20.4–2 More sunlight produces more world assets, using up fuels faster. The state's assets grow faster and to a slightly higher level, using up the water more quickly.

20.4–3 At T = 400, both world and state assets have a short pulse of growth but decline later as this new fuel is used up.

CHAPTER 21   MODELING PROJECTS AND COMPLEXITY

21.1   Model of Main Components in a Shallow Marine Lagoon (Program LAGOON.bas, Figure 21.1)

21.1–1 With more nutrient inflow, production, detritus, stock of shrimp, and frequency of oscillations increase. Small consumers are reduced.

21.1–2 Without mangroves, there is less filtration of inflowing nutrients. There are more oscillations, detritus, and consumers.

21.1–3 Without the larger consumers, the systems loses its oscillation. Small consumers become more abundant and levels of detritus are reduced.

Appendix D

◆

USE OF ENERGY SYSTEMS SYMBOLS*

System Frame A rectangular box is drawn to represent the boundaries that are selected. Boundaries selected must include three dimensions. For example, the analysis of a city would probably include its political boundaries, a plane below the ground surface (example: 10 m), and a plane above the city (example: 100 m).

Source Any input that crosses the boundary is a source, including pure energy flows, materials, information, genes, services, and inputs that are destructive. All of these inputs are given a circular symbol. Sources are arranged around the outside border from left to right in order of their solar transformity, starting with sunlight on the left and information and human services on the right. No source inflows are drawn in to the bottom.

*Modified from Odum, 1996, with permission of John Wiley, New York.

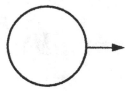

Pathway Lines   Any flow is represented by a line, including pure energy, materials, and information. Money is shown with dashed lines. Where material flows of one kind are to be emphasized, dotted lines are suggested (or color). Barbs (arrowheads) on the pathways mean that the flow is driven from behind the flow (donor driven) without appreciable backforce from the next entity. Lines without barbs flow in proportion to the difference between two forces; they may flow in either direction.

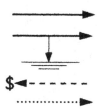

Heat Sink   The heat sink symbol represents the dispersal of available energy (potential energy) into a degraded, used state, not capable of further work. Representing the second energy law, heat sink pathways are required from every transformation symbol and every tank. At the start, one heat sink may be placed at the center bottom of the system frame. Then two lines at about 45 degrees to the bottom frame border are drawn to collect heat sink pathways. Using finer lines or yellow lines for heat sinks keeps these from dominating the diagram. No material, available energy, or usable information ever goes through heat sinks, only degraded energy.

Outflows   Any outflow that still has available potential, materials more concentrated than the environment, or usable information is shown as a pathway from either of the three upper system borders, but not out the bottom.

Storage Tank   Any quantity stored within the system is given a tank symbol, including materials, pure energy (energy without accompanying material), money, assets, information, image, and quantities that are harmful to others. Every flow in or out of a tank must be the same type of flow and measured in the same units. Sometimes a tank is shown overlapped by a symbol of which it is part.

Adding Pathways   Pathways add their flows when they join or when they go into the same tank. No pathways should join or enter a common tank if they are of different type or transformity or measured in different units. A pathway that branches represents a split of flow into two of the same type (Figure 2.2).

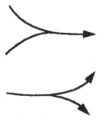

Interaction   Two or more flows that are different and both required for a process are connected to an interaction symbol. The flows to an interaction are drawn to the symbol from left to right in order of their transformity, the lowest quality one connecting to the notched left margin. The output of an interaction is an output of a production process, a flow of product. These should usually go to the right, since production is a quality-increasing transformation.

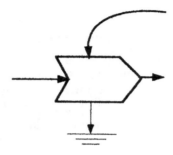

Constant Gain Amplifier   A special interaction symbol is used if the output is controlled by one input (entering symbol from left), but most of the energy is drawn from the other (entering from the top).

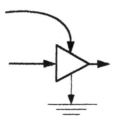

Producers   Producer symbols are used for units on the left side of the systems diagram that receive commodities and other inputs of different types interacting to generate products. The producer symbol implies that there are interactions and storages within. Sometimes it may be desirable to diagram the details of interactions and processes inside Figure 2.9. Producers include biological producers such as plants and industrial production.

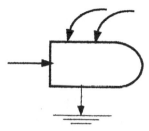

Miscellaneous Box   The rectangular box is used for any subsystem structure and/or function. Often these are appropriate for representing economic sectors such as mining, power plants, and commerce. The box can include interactions and storages with products emerging to the right. Details of what goes on within the consumer is not specified unless more details are described or diagrammed inside.

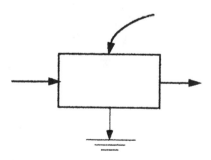

Small Box   A very small box on a pathway or on the side of a storage tank is used to initiate another circuit that is driven by "force" in proportion to the pathway or storage. This is sometimes called a *sensor* when it delivers its action without draining much energy from the original pathway or tank.

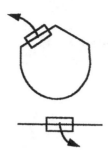

Consumers   Consumer symbols are used for units on the right side of the systems diagram that receive products and feedback services and materials. Consumers may be animal populations or sectors of society, such as the urban consumers. A consumer symbol usually implies autocatalytic interactions and storages within (Figure 2.10). However, the consumer symbol is a class symbol (refers to many similar but different units), and details of what goes on within the consumer are not specified exactly unless more details are diagrammed inside.

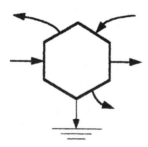

Counterclockwise Feedbacks   High quality outputs from consumers, such as information, controls, and scarce materials, are fed back from right to left in the diagram. Feedbacks from right to left represent a diverging loss of concentration, the service usually being spread out to a larger area. These flows should be drawn with a counterclockwise pathway (up, around, and above the originating symbol—not under the symbol). These drawing procedures are not only conventions that prevent excess line crossing and make one person's diagrams the same as another's, but they allow the diagrams to be used as a way of representing energy hierarchies.

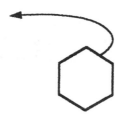

Switch   The concave sided box represents switching processes, those that turn on and off. The flows that are controlled enter and leave from the sides. The pathways that control the switches are drawn to the top. This includes thresholds and other information. Switching occurs in natural processes as well as with human controls. Examples are earthquakes, reproductive actions, and water overflows of a river bank.

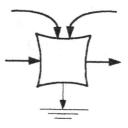

Exchange Transaction   Where quantities in one flow are exchanged for those of another, the transaction symbol is used. Most often the exchange is a flow of commodities, goods, or services exchanged for money (drawn with dashed lines). Often the price that relates one flow to the other is an outside source of action representing world markets; it is shown with a pathway to the top of the symbol.

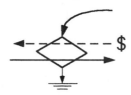

Material Balances   Because all inflowing materials either accumulate in system storages or flow out, each inflowing material, such as water or money, needs to have outflows drawn.

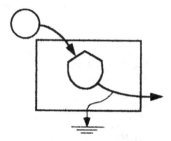

Loop Limited Conversion   This short symbol represents a cycle of material that interacts with, captures, and transforms an input flow. The internal cycle becomes a limiting factor as the amount of input increases (Explanation in Chapter 12). For example, light energy is captured by the green chloroplasts in plant cells transferring the energy to a biochemical web beyond the symbol. The chlorophyll cycles in a closed loop in the process. Increasing the light intensity has a diminishing effect because of the lag in materials circulating within the unit.

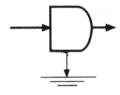

Appendix E

◆

CONTENTS OF THE CD SUPPLIED WITH THIS BOOK

TABLE E.1
Contents of the CD Supplied with This Book

Folders	Description	PC files	Macintosh files
Extend Folder			
	EXTEND4—Program, Imagine That, Inc.	EXTEND4.exe	EXTEND4
	Mathsim—Folder using generic blocks from Imagine That, Inc., Chapter 6:		
	Generic library of blocks	GENERIC.lix	GENERIC LIB
	TANK simulation, math blocks	MATHTANK.mox	MATHTANK
	IntrModl—Folder for introductory files and pictorial icon blocks, Chapter 4:		
	Icon blocks for Chapter 4	PICTLIB.lix	PICTLIB
	Storage model, gen. symbols*	TANK.mox	TANK
	Pictorial exponential model	PEXPO.mox	PEXPO
	Storage model in pictures	PONDWATR.mox	PONDWATR
	View of EDM icons	BIOICON.mox	BIOICON
	EarthSys—Folder for model of earth system, Chapter 4:		
	Instruction manual in WORD	ESMANUAL.doc	ESMANUAL
	Library of picture blocks	ESLIB.lix	ESLIB
	View of pictorial icons	ESICONS.mox	ESICONS
	View of GENSYS blocks used*	ESGSYMB.mox	ESGSYMB
	EARTHSYS model in pictures	EARTHPIC.mox	EARTHPIC
	EARTHSYS in gen. symbols*	EARTHGS.mox	EARTHGS

(continues)

TABLE E.1 (*Continued*)

Folders	Description	PC files	Macintosh files
Biosphr2—Folder for model of Biosphere 2, Chapter 4:			
	Instruction manual, WORD	SBMANUAL.doc	SBMANUAL
	Library of Biosphr2 blocks*	SB2LIB.lix	SB2LIB
	View of pictorial icons	SB2PICONS.mox	SB2PICONS
	View of gen. system blocks used*	SB2GSYMB.mox	SB2GSYMB
	SimBio2 model in pictures	SIMBIO2P.mox	SIMBIO2P
	SimBio2 model in gen. symbols*	SIMBIO2G.mox	SIMBIO2G
EDM—Folder for BioQUEST "Environmental Decision Making," Chapter 4:			
	Instruction manual in WORD	EDMANUAL	EDMANUAL
	View of pictorial icons	See BIOICON in IntrModl folder	
	Library of EDM blocks*	BIOQULIB.lix	BIOQULIB
	Fishing model in pictures	FISHPIC.mox	FISHPIC
	Fishing in gen. symbols*	FISHGSYM.mox	FISHGSYM
	Grass-fire model in pictures	GRASSPIC.mox	GRASSPIC
	Grass-fire, gen. symbols*	GRASSGS.mox	GRASSGS
	Logging model in pictures	LOGGINGP.mox	LOGGINGP
	Logging in gen. symbols*	LOGGINGG.mox	LOGGINGG
	Fishsale model in pictures	FISHSALE.mox	FISHSALE
Gensymb—Folder for general systems models (GENSYS), Chapter 5:			
	Library of GENSYS blocks*	GNSYSLIB.lix	GNSYSLIB
	Model of exponential growth*	EXPO.mox	EXPO
	Model of prod., cons., recycle*	PC&CYCLE.mox	PC&CYCLE
	Model of Earth, gen. symb.*	EARTHGS.mox	EARTHGS
	View of GENSYS blocks*	GENSYMB.mox	GENSYMB
Spreadtab—Folder for spreadsheet programs, Chapters 7 and 8:			
	Spreadsheet calibrating FISH	FISHNUMB.wk1	FISHNUMB
	Spreadsheet simulating FISH	FISH.wk1	FISH
Basic—Folder with models and programs to simulate in BASIC			
Chipmunk Basic 3.4.6—MacIntosh Freeware Program			
	Program, Ronald H. Nicholson, Jr.	—	Chipmunk Basic
	Notice about updates	—	Readme
	Tutorial from T. S. Ferrel	—	Tutorial
	Manual-text file	—	basic.man
	Quick reference, Graphics	—	quick-ref
Basicprg—Model simulation programs Chapters 1, 9, 11, Part III, and Chapter 21:			
	Autocatalytic growth—two sources	2SOURCE.bas	2SOURCE
	Material limited autocat. growth	AUTOCYCL.bas	AUTOCYCL
	Processes in a bank	BANK.bas	BANK

(continues)

TABLE E.1 (Continued)

Folders	Description	PC files	Macintosh files
	Production and prices in growth	BUYPOWER.bas	BUYPOWER
	Simple storage of money	CASH.bas	CASH
	Chaos of oscillators on two scales	CHAOS.bas	CHAOS
	Model of succession	CLIMAX.bas	CLIMAX
	Two cooperating units	CO-OP.bas	CO-OP
	Diurnal metabolism	DAYP&C.bas	DAYP&C
	Destruction-assisted recycle	DESTRUCT.bas	DESTRUCT
	Exponential outflow of storage	DRAIN.bas	DRAIN
	Minimodel of earth evolution	EARTHGEO.bas	EARTHGEO
	Money circulation in economy	ECONP&C.bas	ECONP&C
	Resource use interface	ECONUSE.bas	ECONUSE
	Emergy in internatl. exchange	EMEXCHNG.bas	EMEXCHNG
	Emergy in a storage with yield	EMFISH.bas	EMFISH
	Storage emergy without outflow	EMGTANK.bas	EMGTANK
	Cooperation through trade	ENTRADE.bas	ENTRADE
	Competition for limited source	EXCLUS.bas	EXCLUS
	Exponential growth	EXPO.bas	EXPO
	External limiting factors	FACTORS.bas	FACTORS
	Switching control of consumption	FIRE.bas	FIRE
	Global influence by free market	FREEMARK.bas	FREEMARK
	Emergy in market influence	INFOBEN.bas	INFOBEN
	Interactive competition	INTERACT.bas	INTERACT
	Internal limiting factors	INTLIMIT.bas	INTLIMIT
	Model of tropical lagoon	LAGOON.bas	LAGOON
	Crowding limited logistic growth	LOGISTIC.bas	LOGISTIC
	Classical macroeconomics model	MACROEC.bas	MACROEC
	Exponential economy	MONEYGRO.bas	MONEYGRO
	Net production, seasonal changes	NETPROD.bas	NETPROD
	Growth on a stored resource	NONRENEW.bas	NONRENEW
	Species diversity and resource	NUTRSPEC.bas	NUTRSPEC
	Metabolism in open aquarium	OPENAQ.bas	OPENAQ
	Three-level prey–predator model	OSCILLAT.bas	OSCILLAT
	Production, consump., and recycle	PC&CYCLE.bas	PC&CYCLE
	Global population model	PEOPLE.bas	PEOPLE
	Classical prey–predator model	PREYPRED.bas	PREYPRED
	Pulsing use of accumulations	PULSE.bas	PULSE
	Emergy in autocatalytic growth	RENEMGY.bas	RENEMGY
	Source-limited growth	RENEW.bas	RENEW
	Optimum diversity reserves	RESERVE.bas	RESERVE
	Economy and land rotation	ROTATION.bas	ROTATION
	Production and sales	SALES.bas	SALES
	Slowly renewable growth source	SLOWREN.bas	SLOWREN
	Model of species with area	SPECAREA.bas	SPECAREA
	Species development on islands	SPECIES.bas	SPECIES
	Global-driven state economy	STATECON.bas	TATECON

(continues)

TABLE E.1 (*Continued*)

Folders	Description	PC files	Macintosh files
	Simple storage model	TANK.bas	TANK
	Storage with scaling	TANK2.bas	TANK2
	Sales from a resource reserve	TANKSALE.bas	TANKSALE
	Two units with unlimited energy	TWOPOP.bas	TWOPOP
	Essence of war	WAR.bas	WAR
	Wetland processes	WETLAND.bas	WETLAND
	Global carbon dioxide	WORLDCO2.bas	WORLDCO2

* Programs and libraries with general systems symbols are marked with an asterisk.

Also included in the disk are the book's BASIC programs for Macintosh converted to run on Chipmunk Basic. The converted files have the prefix "Chip" added in front of the model name. For example, the program TANK for Chipmunk BASIC is labeled Chiptank. Load Chipmunk Basic and use its menu to open and run these programs.

REFERENCES AND SUGGESTED READING

Alexander, J. F. 1987. Energy basis of disasters and cycles of order and disorder. Ph.D. dissertation, Environmental Engineering Sciences, University of Florida, Gainesville.

Bala, B. K. 1998. "Energy and Environment, Modeling and Simulation." Nova Science Publishing, Commack, NY.

Banks, J., Carson II, J. S., and Nelson, B. 1996. "Discrete Event System Simulation." Prentice Hall, NJ.

Barnsley, M. F., and Hurd, L. P. 1993. "Fractal image compression." A. K. Peters, Wellesley, MA.

Beltrami, E. J. 1997. "Mathematics for Dynamic Modeling," 2nd Ed. Academic Press, San Diego, CA.

Bennett, D. J. 1998. "Randomness." Harvard University Press, Cambridge, MA.

Bossel, H. 1994. Modeling and Simulation. A. K. Peters, Natick, MA.

Bunn, D., and Larsen, E. R. 1997. "Systems Modeling for Energy Policy." John Wiley & Sons, New York.

Campbell, D. E., and Newell, C. R. 1998. "MUSMOD, a production model for bottom culture of the blue mussel, *Mytilus edulis L. J. Exper. Marine Biol. Ecol.* **219,** 171–203.

Cartwright, J. J. 1993. "Modeling the World in a Spread Sheet: Environmental Simulation on a Microcomputer." Johns Hopins Univ. Press, Baltimore, MD.

Casti, K. 1998. "Would-be Worlds: How Simulation Is Changing the Frontiers of Science." John Wiley & Sons, New York.

DeAngelis, D. L. 1992. "Dynamics of Nutrient Cycling and Food Webs." Chapman and Hall, New York.

De Wit, C. T., and Goudriaan, J. 1978. "Simulation of Ecological Processes." John Wiley & Sons, New York.

Evans, J. R., and Olson, D. L. 1998. "Introduction to Simulation and Risk Analysis." Prentice Hall, Englewood Cliffs, NJ.

Forrester, J. W. 1961. "Industrial Dynamics." The MIT Press, Cambridge, MA.

Gilman, N., and Hails, R. 1997. "An Introduction to Ecological Modeling, Putting Practice into Theory." Blackwell Science, London, UK.

Grant, W. E. 1986. "Systems Analysis and Simulation in Wildlife and Fisheries Sciences." John Wiley & Sons, New York.

Grant, W. E., Pedersen, E. K., and Main, S. L. 1997. "Ecology and Natural Resource Management, Systems Analysis, and Simulation." John Wiley & Sons, New York.

Hall, C. A. S., and Day, J., ed. 1977. "Ecosystem Modeling in Theory & Practice, An Introduction with Case Histories." J. Wiley, NY.

Hall, C. A. S., Taylor, M. R., and Everham, E. 1992. A geographically-based ecosystem model and its application to the carbon balance of the Luquillo Forest, Puerto Rico. *Water Air Soil Pollut.* **64,** 385–404.

Hannon, B., and Ruth, M. 1994. "Dynamic Modeling." Springer-Verlag, New York.

Haefner, J. W. 1996. "Modeling Biological Systems: Principles and Applications." Chapman and Hallm, NY.

Hurst, C. J., ed. 1996. "Modeling Disease Transmission and Its Prevention by Disinfection." Cambridge Univ. Press, London.

Jacobsen, M. Z. 1998. "Fundamentals of Atmospheric Modeling." Cambridge University Press, Cambridge, MA.

Jorgensen, S. E. 1986. "Fundamentals of ecological modeling." Developments in Environmental Modeling, Vol. 9. Elsevier, Amsterdam.

Jorgensen, W. E., Halling-Sorensen, B., and Nielsen, S. N., ed. 1995. "Handbook of Environmental and Ecological Modeling." Lewis Publ, Boca Raton, FL.

Keen, R. E., and Spain, J. D. 1992. "Computer Simulation in Biology, a Basic Introduction." John Wiley & Sons, New York.

Kheir, N. A., ed. 1995. "Systems Modeling and Computer Simulation." M. Dekker, NY.

Kemeny, J. S., and Kurtz, T. W. 1971. "Basic Programming." 2d ed. John Wiley and Sons, New York.

Kirkby, M. J., Naden, P. S., Burt, T. P., and Butcher, D. P. 1993. "Computer Simulation in Physical Geography." John Wiley & Sons, New York.

Klir, G. J., and Folger, T. A. 1988. "Fuzzy Sets, Uncertainty, and Information." Prentice Hall, Englewood Cliffs, NJ.

Koch, A. L., J. A. Robinson, and G. A. Milliken, ed. 1997. Mathematical Modeling in Microbial Ecology. Kluwer, Academic, Boston, MA.

Kulkarni, V. G. 1995. "Modeling and Analysis of Stochastic Systems." Chapman and Hall, NY.

Law, A. M., and Kelton, W. D. 1990. "Simulation Modeling and Analysis." McGraw Hill, NY.

Lendrum, D. 1986. "Modelling in Behavioral Ecology." Timber Press, Portland, OR.

Logan, M. O. 1998. A simulation model of health care in the United States (STELLA). Ph.D. dissertation, College Mursing and Health Science, George Mason University, Fairfax, Virginia.

Logofet, D. O. 1993. "Matrices and Graphs." CRC Press, Boca Raton, FL.

Mandelbrot, B. B. 1983. "The Fractal Geometry of Nature." W. H. Freeman, San Franciso, CA.

Mangel, M., and Clark, C. W. 1988. "Dynamic Modeling in Behavioral Ecology." Princeton University Press, Princeton, NJ.

Meadows, D. H., Meadows, D. L., and Randers, J. 1992. "Beyond the Limits." Chelsea Green Publishing Co., Post Mills, VT.

Meadows, D. H., Meadows, D. L., Randers, J., and Behrens, W. W. 1972. "Limits to Growth: A Report for the Club of Rome's Project on the Predicament of Mankind." A Potomac Associates Book. Universe Books, New York.

Meinhardt, H. 1982. "Models of Biological Pattern Formation." Academic Press, NY.

Nelson, B. L. 1995. "Stochastic Modeling: Analysis and Simulation." McGraw Hill, NY.

Odum, H. T. 1971. "Environment Power and Society." J. Wiley and Sons, New York.

Odum, H. T. 1996. "Environmental Accounting, Emergy, and Decision Making." John Wiley & Sons, NY.

Odum, H. T. 1983, 1993. "Systems Ecology" (reprinted as "Ecological and General Systems"). University Press of Colorado, Niwot, CO.

Odum, H. T. and Munroe, M. 1993. Simulation of the Laguna Joyuda ecosystem in Western Puerto Rico. *Acta Científica* 7(1–3), 189–227.

Odum, H. T. and Peterson, N. 1996. Simulation and evaluation with energy systems blocks. *Ecol. Model.* **93**, 155–173.

Oster, G. F. and Auslander, D. 1971. Topological representation of thermodynamic systems. I. Basic concepts. *J. Franklin Inst.* **292**, 1–16.

Richardson, G. P., and Pugh III, A. L. 1981. "Introduction to Systems Dynamics Modeling with DYNAMO." MIT Press, Cambridge, MA.

Richardson, J. 1987. Spatial patterns from pulsing simulation. Ph.D. dissertation. Environmental Engineering Sciences, University of Florida, Gainesville.

Ripley, B. D. 1987. "Stochastic Simulation." John Wiley & Sons, New York.

Robertson, D., Muetzelfeldt, P., and Bundy, A. 1991. "Ecologic: Logic-Based Approaches to Ecological Modeling. Logic Programming." The MIT Press, Cambridge, MA.

Ross, S. M. 1992. "Applied Probability Models with Optimization Applications." Dover, New York.

Roughgarden, J. 1998. "Primer of Ecological Theory." Prentice Hall, Upper Saddle River, NJ.

Ruth, M. and Hannon, B. M. 1997. "Modelling Dynamic Economic Systems." Springer-Verlag, New York.

Schroeder, M. 1991. "Fractals, Chaos, Power Laws." W. H. Freeman, New York.

Sheldon, R. M. 1996. "Simulation: Statistical Modeling and Decision Science." Academic Press, San Diego.

Smith, R. F. 1970. The vegetation structure of a Puerto Rican rain forest before and after short-term gamma irradiation. In "A Tropical Rain Forest" (H. T. Odum and R. F. Pigeon, eds.) Div. Technical Information, Atomic Energy Commission, Oak Ridge, TN. TID-24270 (PRNC-138), Clearinghouse for Federal Scientific and Technical Information, National Bureau of Standards, U.S. Dept. of Commerce, Springfield, VA 22151.

Swartzman, G. L. and Kaluzny, S. P. 1987. "Ecological Simulation Primer." Macmillan, NY.

Taylor, H. M., and Karlin, S. 1984. "An Introduction to Stochastic Modeling." Academic Press, San Diego.

Vose, D. 1996. "Quantitative Risk Analysis, a Guide to Monte Carlo Simulation Modeling." John Wiley & Sons, New York.

Weber, T. C. 1994. Spatial and temporal simulation of forest succession with implications for management of bioreserves. M.S. thesis. Environmental Engineering Sciences, University of Florida, Gainesville.

Yount, J. L. 1956. Factors that control species numbers in Silver Springs, Florida. *Limnol. Oceanogr.* **1**, 286–295.

Index

◆